Library of Congress Cataloging-in-Publication Data

Names: Rowland, Thomas W., editor. | American College of Sports Medicine,
  editor. | North American Society for Pediatric Exercise Medicine, editor.
Title: Cardiopulmonary exercise testing in children and adolescents / Thomas
  W. Rowland, American College of Sports Medicine, North American Society
  for Pediatric Exercise Medicine, editors.
Description: Champaign, IL : Human Kinetics, [2018] | Includes
  bibliographical references and index.
Identifiers: LCCN 2016050870 (print) | LCCN 2016051582 (ebook) | ISBN
  9781492544470 (print) | ISBN 9781492544487 (e-book)
Subjects: | MESH: Exercise Test | Child | Adolescent
Classification: LCC RC669 (ebook) | LCC RC669 (print) | NLM WG 141.5.F9 | DDC
  616.1/062--dc23
LC record available at https://lccn.loc.gov/2016050870

ISBN: 978-1-4925-4447-0 (print)

The web addresses cited in this text were current as of April 2017, unless otherwise noted.

Senior Acquisitions Editor: Amy N. Tocco; **Developmental Editor:** Judy Park; **Managing Editor:** Anna Lan Seaman; **Copyeditor:** Kevin Campbell; **Indexer:** May Hasso; **Permissions Manager:** Dalene Reeder; **Graphic Designer:** Whitney Milburn; **Cover Designer:** Keri Evans; **Photo Asset Manager:** Laura Fitch; **Photo Production Manager:** Jason Allen; **Senior Art Manager:** Kelly Hendren; **Illustrations:** © Human Kinetics, unless otherwise noted; **Printer:** Sheridan Books

Printed in the United States of America   10 9 8 7 6 5 4 3 2 1

The paper in this book is certified under a sustainable forestry program.

Human Kinetics
Website: www.HumanKinetics.com

United States: Human Kinetics
P.O. Box 5076
Champaign, IL 61825-5076
800-747-4457
e-mail: info@hkusa.com

Canada: Human Kinetics
475 Devonshire Road Unit 100
Windsor, ON N8Y 2L5
800-465-7301 (in Canada only)
e-mail: info@hkcanada.com

Europe: Human Kinetics
107 Bradford Road
Stanningley
Leeds LS28 6AT, United Kingdom
+44 (0) 113 255 5665
e-mail: hk@hkeurope.com

For information about Human Kinetics' coverage in other areas of the world,
please visit our website: www.HumanKinetics.com   E6943

Contents

Preface

The pediatric exercise testing laboratory is playing an increasingly important role in the diagnosis and assessment of children and adolescents with—or suspected of having—heart or lung disease. This text represents an effort to consolidate information about our growing experience with exercise testing in young persons so that those who conduct these tests can have a guide and a reference book. Recent scientific statements (1) and review articles (2, 4) about clinical exercise testing of children have given us new material since the publication of my earlier book, *Pediatric Laboratory Exercise Testing: Clinical Guidelines* (3). It is the goal of this text to extend coverage of the topic in order to

- provide up-to-date guidance for the performance of exercise stress testing in youth, and
- document our current knowledge about interpreting the physiological variables measured during these tests.

The extent of the knowledge and experience shared in this book, compared to its much thinner predecessor published over 20 years ago, is a testament to the growing importance of pediatric exercise testing.

The central theme of this text is that clinical exercise testing in children differs from that conducted in the adult stress testing laboratory. The clinical questions being addressed in these two populations reflect obvious differences in forms of cardiopulmonary disease, and the protocols for testing adults must be modified to satisfy the requirements of a wide range of subject sizes and degrees of physical development as well as the intellectual and emotional immaturity of the child.

The first part of this book outlines the testing procedure and its measurement variables, followed by chapters that offer practical approaches to patient complaints that are commonly encountered in the exercise testing laboratory. Throughout these discussions, we emphasize the value of determining gas exchange variables to supplement the traditional measurement of blood pressure and electrocardiogram.

This book is directed toward those who conduct exercise testing of children and adolescents in the clinical laboratory setting. Appropriately, then, the consideration of testing methods will take into account the typical wide variety of ages, sizes, levels of physical fitness, and body composition of patients as well as states of cardiac and pulmonary health. Still, those involved in exercise testing of youth for research purposes will undoubtedly find useful material in these pages.

A practical guidebook such as this one is obligated to present normative data for exercise testing variables. While such published norms will be included in these chapters, one must—as will be repeatedly emphasized—accept such information with a high level of caution. Normative data for one laboratory often differ from those of another because of variability in types of measuring equipment, subject population, testing protocols, and staff characteristics. There are few data that can be confidently considered normal for a general pediatric population. Consequently, normal values for variables measured during exercise testing are best established for one's own laboratory.

As much as possible, the discussions in these chapters will be based on published, evidence-based observations of youth during exercise testing. In some cases, however, the authors of these chapters will draw observations and recommendations from their own professional experience. These authors all have a high level of expertise and extensive experience in exercise testing of children and adolescents; they each represent established pediatric clinical testing laboratories in major medical centers.

eBook available at HumanKinetics.com

The authors expect that this book will prove useful to physicians and exercise scientists as a source of current testing information and practical guidelines for performing exercise testing in young patients. Beyond this, we hope that this book will serve as a means of focusing and unifying approaches to such testing and that it will serve as a foundation for the future development of innovative approaches to exercise testing in the health care of children and adolescents.

Thomas W. Rowland, MD
Editor

Notice and Disclaimer

Care has been taken to confirm the accuracy of the information presented and to describe generally accepted practices. However, the authors, editors, and publisher are not responsible for errors or omissions or for any consequences from application of the information in this book and make no warranty, expressed or implied, with respect to the currency, completeness, or accuracy of the contents of the publication.

Application of this information in a particular situation remains the professional responsibility of the practitioner; the clinical treatments described and recommended may not be considered absolute and universal recommendations.

The authors, editors, and publisher have exerted every effort to ensure that drug selection and dosage set forth in this text are in accordance with the current recommendations and practice at the time of publication. However, in view of ongoing research, changes in government regulations, and the constant flow of information relating to drug therapy and drug reactions, the reader is urged to check the package insert for each drug for any change in indications and dosage and for added warnings and precautions. This is particularly important when the recommended agent is a new or infrequently employed drug.

Some drugs and medical devices presented in this publication have Food and Drug Administration (FDA) clearance for limited use in restricted research settings. It is the responsibility of the health care provider to ascertain the FDA status of each drug or device planned for use in their clinical practice.

PART

I

INTRODUCTION

There is no single, standardized approach to clinical exercise testing of children and adolescents. Each study must be designed to accommodate the age and fitness of the subject while producing the information needed. Still, all such tests require proper equipment, experienced staff, and safe methods. Following an introductory overview of age-appropriate testing, the chapters in this section provide guidelines for conducting exercise tests in children and adolescents and describe the available options for testing protocols.

Clinical Applicability
of the Pediatric Exercise Test

Thomas W. Rowland, MD

You can gain an appreciation for the diagnostic relevance of exercise stress testing by performing a simple experiment. Take an old 1993 Volvo station wagon and park it next to a Formula One racing car. Now start the engines of both. Question: How can you tell which vehicle will perform better? Answer: You will be hard put to tell. Certainly, the racing car has a bigger engine, complex gearbox, smooth tires, and so on. But, the point is, you can't tell anything about how each of these two cars will perform—how much power they can produce, how fast and how long they can go—until you take them onto the road.

Think, too, as you contemplate these two vehicles sitting side by side, their motors humming: How can you tell which one has its tires badly out of balance? Which one has a faulty fuel injector or a slow leak in the cooling system? Once again, you must put these two vehicles on the highway to detect malfunctions that will only become apparent when their systems are stressed.

This is the rationale for clinical exercise testing. Lying at rest, with inert leg muscles and the heart and lungs "idling," there is little to say about the functional capacity of the human organism. Only by putting this machine to work, forcing the need for increased coronary blood flow, oxygen delivery, heat dissipation, and so on, will one be able to detect individual differences in functional capacity. Only by revving up its "motor" can one detect weaknesses in any of the mechanisms that make it function normally. We need to take the body "out on the road"—put it to work on a cycle ergometer or on a treadmill in the testing laboratory—to uncover functional limits that define the level of fitness or the liabilities of cardiopulmonary disease. Most of us do not spend our lives immobile on an examining table. Clinicians need to know how the heart and lungs function in real life, surrounded by constant physical demands. And so we have clinical exercise testing.

Starting with the basics, why should monitoring a bout of motor activity, essentially a repetitive set of skeletal muscle contractions, provide us with any information about cardiopulmonary health? The answer, of course, lies in the dependence of the musculature on an adequate set of functional responses from the heart and lungs and a host of supportive systems for the performance of endurance exercise.

The performance of endurance exercise relies on an appropriate rise in minute ventilation, cardiac output, and circulatory blood flow to exercising muscle. Depressed cardiopulmonary functional reserve results in limitations on muscle endurance during exercise. The former must provide adequate oxygen supply to satisfy the energy requirements of the latter. But that's only part of the story. The end product of cellular aerobic metabolism—carbon dioxide—must be eliminated. Accumulating lactic acid must be buffered to prevent unacceptable metabolic acidosis. Blood flow must be directed to the cutaneous circulation for thermoregulation. Oxidative substrate in the form of glycogen and fatty acids needs to be supplied to contracting muscle, and hormonal stimulation must occur via circulating sympathomimetic amines.

In short, satisfactory cardiopulmonary responses are essential to the performance of endurance exercise. Without such increases in pulmonary and circulatory function, one would have difficulty running to the end of the block. Assessing a patient's ability to perform endurance exercise therefore serves as an accurate marker of the effectiveness of cardiac and pulmonary reserve function.

This information may have significant clinical importance, whether expressed as minutes of treadmill endurance time, maximum work performed (watts [W]) on a cycle ergometer, or physiological fitness (maximal oxygen uptake, or $\dot{V}O_{2max}$). Measurement of $\dot{V}O_{2max}$, for instance, serves as a prognostic indicator in young patients with cystic fibrosis, and a pattern of change over time of $\dot{V}O_{2max}$, a surrogate of maximal cardiac output, has been used in the timing of cardiac transplantation in children.

One can look at this exercise performance–physiological function relationship from the other direction, too. The metabolic demands of endurance exercise require that the myriad components of cardiopulmonary function be operating at full efficiency. Deficiencies in such a parts-of-the-whole system may only become apparent under the stresses of exercise. Considering the number of individual contributors, the list of potential weaknesses in the response of heart and lungs to exercise is a long one. Coronary blood flow must increase unimpeded to serve the rising metabolic requirements of the cardiac musculature. Sinus node function must accelerate appropriately to generate cardiac output. Increases in myocardial systolic and diastolic function are required to sustain stroke volume. There can be no significant obstruction or leakage of cardiac valves. Airflow though the lungs must be unimpeded, and gas exchange at the alveolar–capillary interface must be unimpeded. All of these critical issues—and many others—are critical to normal cardiopulmonary responses to endurance exercise. Stress testing provides the opportunity to detect such weaknesses, the magnitude—or even existence—of which may not be apparent in the resting state.

Depressed exercise performance and the detection of abnormalities in cardiac and pulmonary functional responses, then, serve as the basis for the utility of exercise stress testing. The findings from exercise testing are often important to clinical decisions such as the timing of surgical interventions, the dosage of medications, and the evaluation of anomalies that may present a risk for exercise. These will be outlined in the chapters that follow. A normal exercise test can also serve the important purpose of reassuring patient, parent, and physician that certain symptoms (chest pain, syncope) during physical activity do not reflect cardiopulmonary disease.

The theme throughout this book is that while the preceding basic tenets hold true for both children and adults, there are unique considerations for successful exercise testing in young people. To start with, the clinical questions that bring children to the exercise testing laboratory differ from those of adult patients. The enormous variability in age, size, and physical fitness during the growing years pose challenges to formulating optimal approaches to the exercise test, and the immaturity of the pediatric subject demands a great deal of sensitivity and special encouragement from an experienced testing staff. "Normative" values of physiologic variables are often different in children and adults, and they progressively change during the course of the pediatric years.

Development of Pediatric Exercise Testing

Clinical exercise testing first developed in the early part of the 20th century after it was recognized that the appearance of electrocardiographic changes during exercise, specifically ST-T wave depression, could effectively identify adults with angina pectoris and underlying coronary artery disease (CAD). Not surprisingly, then, the history of stress testing closely parallels the evolution of electrocardiography.

In 1903, Willem Einthoven, a Dutch physician and physiologist, invented the first practical electrocardiograph (for which he received the Nobel Prize in 1924), and within the next three decades several investigators found that ST changes on the electrocardiogram (ECG) were observed when patients experienced episodes of angina pectoris triggered by exercise.

Acceptance of exercise as a means of clinically evaluating patients with chest pain, however, was slow to develop. Indeed, exercise testing of an adult suspected of having coronary artery disease in 1930 was not altogether a simple matter. Patients were asked to exercise by running up and down flights of stairs, or performing sit-ups (with varying resistance applied to the chest), or lifting barbells, immediately after which two hands and one foot were plunged into separate buckets of electrolyte solution connected to electrode wires. The electrocardiogram was then recorded (using a string galvanometer) with an apparatus that weighed 600 pounds and required five workers to operate.

Early exercise testing also focused on defining aerobic fitness, usually by recovery heart rate after step exercise, as a marker of cardiac functional capacity. In 1942, Arthur Master published data indicating the utility of electrocardiography immediately after two-step exercise to recognize ischemic changes in adults with coronary artery disease. However, the safety of performing such tests, which were often poorly tolerated by patients, remained in question, and clinical acceptance was limited.

The subsequent development of modern-day electrocardiography and motor-driven treadmills opened the door for greater utilization of exercise stress testing for the diagnosis of coronary artery disease in adults. Robert Bruce and his colleagues at the University of Washington brought such testing into the modern era with the development of a progressive, multistage treadmill protocol in the early 1960s. Now testing could identify symptoms and electrocardiographic changes *during* exercise. Moreover, with the progressive increments of treadmill speed and slope of the Bruce protocol, exercise stress testing could be more easily performed by patients of varying levels of fitness.

Exercise stress testing in adults, eventually supplemented by radionuclide angiography and postexercise echocardiography, rapidly became accepted as a standard component of the diagnostic armamentarium, not only for coronary artery disease but also for an assortment of other clinical issues surrounding dysrhythmias, hypertension, and cardiac function. Karlman Wasserman and coworkers at UCLA demonstrated, too, how the acquisition of gas exchange variables measured during exercise could further delineate and differentiate abnormal cardiac and pulmonary responses to exercise.

The use of exercise testing in pediatric populations, whose members do not normally suffer from coronary artery disease, initially developed in the shadow of this story about adult patients. Early exercise studies in youth were performed in the research setting. They were designed to examine physiological differences that separate children from adults. Sid Robinson provided the first such treadmill-derived data in the Harvard Fatigue Laboratory in Boston in the 1930s, demonstrating the progressive changes in metabolic and physiological responses that normally occurred between the ages of 6 and 91. Similar exercise data in healthy children were subsequently provided by other investigators in the middle of the 20th century, including Per-Olof Åstrand in Sweden, Simon Godfrey in Great Britain, and Gordon Cumming in Canada. When Oded Bar-Or published

his landmark book *Pediatric Sports Medicine for the Practitioner* in 1983, he was able to accumulate a large base of normative data from these earlier studies to outline aspects of physiological responses to exercise in youth and how these developed during the growing years.

While such research was designed to reveal the normal development of physiological responses to exercise in children, this information also served as normative data for those who developed exercise testing for the clinical assessment of heart and lung disease in pediatric patients. As studies dealt, for the most part, with those with congenital heart disease, early exercise stress testing in young patients involved a more diverse approach than that of the traditional adult laboratory focused on the detection of coronary artery disease. The assessment of ischemic changes on the ECG was still an issue, particularly in assessing the severity of aortic outflow obstruction, but exercise testing in young patients also involved a wider range of information, such as blood pressure responses (in coarctation of the aorta, systemic hypertension), endurance capacity (postoperative cyanotic heart disease), and rhythm responses (complete heart block).

Useful clinical testing methodologies and clinical findings were described by a number of key early pioneers, such as Fred James at Cincinnati Children's Hospital, David Driscoll at the Mayo Clinic, Bruce Alpert and William Strong at the Medical College of Georgia, Rolf Mocellin in Germany, and Tony Reybrouck and Dirk Matthys in Belgium. The importance of gas exchange measures, including $\dot{V}O_{2max}$ and $\dot{V}O_2$ kinetics, was highlighted by the early reports of exercise testing in patients with congenital heart disease by Hans Wessel at Children's Memorial Hospital in Chicago. During this time, too, exercise testing became established in both children and adults as a useful means of assessing bronchospasm and lung function in patients with asthma and other respiratory diseases (particularly via the early experience reported by Hans Stoboy, Gerd Cropp, and Svein Oseid).

In many cases, clinical exercise testing in children was performed in adult laboratories, using protocols, exercise equipment, and monitoring systems (ECG, blood pressure) traditionally used to test adults with suspected coronary artery disease. A number of developments have now expanded the role of exercise testing in youth and have identified the need for more specific approaches to exercise testing for this population of patients.

- Perhaps most importantly, the past several decades have witnessed a dramatic expansion in the scope and nature of patients cared for by pediatric cardiologists. Children with complex forms of congenital heart disease, particularly those characterized by marked unilateral ventricular hypoplasia (hypoplastic left heart syndrome, tricuspid valve atresia), once had little hope for long-term survival. Now, thanks to remarkable progress in surgical techniques, these young patients not only often survive but also live productive and fulfilling lives. The physicians caring for these survivors as they grow toward the adult years are confronted with new issues, such as myocardial dysfunction, stubborn tachyarrhythmia, hypoxemia, and pulmonary hypertension. These problems have required new diagnostic and therapeutic approaches, often using information available in the exercise stress-testing laboratory. Similarly, in patients with lung diseases such as cystic fibrosis, improvements in patient care have successfully extended survival and have at the same time introduced new clinical questions that can be assessed through exercise testing. It is likely that the future will continue to bring steady improvements in the survival of young patients with both cardiac and pulmonary disease that will be paralleled by expanded indications for clinical exercise testing.

- A growing understanding of the pathophysiology of cardiac malformations and factors influencing risk stratification have created a need to expand the information obtained during exercise stress testing in young patients. It is true that many issues can be adequately examined by a limited study involving a traditional bout of progressive exercise accompanied by electrocardiographic monitoring and measurement of blood pressure. Assessment of possible ischemic changes in a child after Kawasaki disease, for example, or determination of blood pressure responses after medical treatment of a hypertensive young wrestler could be adequately performed using this approach.

However, the clinical insights gained from exercise testing can be improved by the measurement of gas exchange variables, which are now readily obtained with user-friendly commercial metabolic systems. The changes in the oxygen and carbon dioxide content of expired air during exercise reflects similar gas exchange dynamics at the cellular level. With this approach, for example, the measurement of $\dot{V}O_{2max}$ provides an objective physiological assessment of aerobic fitness, and the determination of ventilatory variables (minute ventilation, $\dot{V}CO_2$) offers insights into pulmonary responses as well. As will be outlined in the chapters that follow, it is often the calculation of relationships between these variables that provides clues into the relative importance of cardiac and pulmonary etiologies of exercise limitation.

Termed an "integrative cardiopulmonary test" by Wasserman et al. (20), this approach expands the utility of exercise testing by providing data that can help to answer questions about cardiac and pulmonary issues in youth. In a teenage boy with moderate aortic valve insufficiency, does his cardiac disease explain the shortness of breath that limits his ability to exercise? What mechanisms lie behind a star athlete's inability to perform well after an extended viral illness? Is syncope of an anxious child during running related to hyperventilation? Is breathlessness during exercise in a markedly obese child caused by excess body fat, exercise-induced bronchoconstriction, or cardiac dysfunction? These types of issues are best addressed by a full examination of gas exchange variables during clinical exercise testing.

- A growing recognition of the effects of exercise on electrophysiological function has created new roles for exercise testing in youth. Assessment of changes in ventricular ectopy during exercise is a traditional indication for exercise testing. Newer indications include the use of responses of rate of ventricular repolarization (QT interval) and conduction down accessory pathways (WPW syndrome) during exercise as means of patient risk stratification.

- The increased use of pediatric exercise testing has also been stimulated by the concerns of parents, coaches, and physical education instructors over the occurrence of symptoms of chest pain, dizziness, syncope, or palpitations in young people during exercise. Such concerns have been fueled by the tragic occurrences of sudden unexpected death of young, presumably healthy athletes during sports training or competition. While such symptoms are highly unlikely to reflect the rare diseases that pose a risk of sudden death, the youngster with occult hypertrophic cardiomyopathy, coronary artery anomalies, or repolarization abnormalities that can predispose to fatal dysrhythmias can present with such complaints. Findings on exercise testing have thus become part of the assessment of symptomatic children and athletes to rule out these anomalies.

- A normal exercise test can provide clearance for sports play in young patients with heart disease or in those who have suffered illnesses such as viral myocarditis. Exercise testing also plays a role in assessing risk and exercise capacities in young patients who are enrolled in cardiac and pulmonary rehabilitation programs.

We can expect that clinical exercise testing in children and adolescents will continue to expand as the value of exercise is recognized in the assessment of not only heart and lung disease but also metabolic and musculoskeletal disorders. Such trends will undoubtedly follow improvements in medical and surgical treatment of these patients. We can also expect to see new techniques for performing exercise tests (miniaturization of metabolic systems permitting field testing, for example) and assessing their results (three-dimensional echocardiography, myocardial strain Doppler studies).

Unique Features of Exercise Testing in Children

The approach to clinical exercise testing of children and adolescents differs from that of the laboratory dedicated to testing adults. One need only consider the various approaches needed to perform a satisfactory exercise test first in a 15-year-old cross country runner who experienced precordial chest pain in her last race, followed by a test looking for heart rate response in a 5-year-old youngster with complete heart block, and then a 12-year-old obese boy with a dilated cardiomyopathy and progressively worsening shortness of breath with exercise.

The pediatric exercise testing laboratory must accommodate wide variations in patient age, size, and fitness, and that means that testing protocols and equipment must be similarly adjusted. The indications for testing are much broader than those in the adult lab, so the questions to be answered must be carefully considered before the exercise begins.

Most children can be easily motivated to give exhaustive efforts during exercise testing, but it requires charismatic skill from staff members experienced with the emotional and physical responses of children during treadmill or cycle exercise. It has been said that perhaps the single most important factor in a successful exercise test in the pediatric laboratory is the staff administering the test.

Pediatric exercise testing, then, is distinguished by the need for a creative approach to each patient. The staff must know what information is needed to address the clinical question being asked, the proper modality—cycle or treadmill—to obtain that answer, and the optimal protocol for the subject's age and fitness level.

The physiological mechanisms underlying the cardiac and pulmonary responses to a bout of progressive exercise are no different in children (at least those over age 6) and adults. Nonetheless, certain quantitative measurements (heart rate, blood pressure, endurance time) are different in children, and these must be recognized in the testing of immature subjects.

For example, resting and maximal heart rates in an exhaustive exercise test are greater in children than in adults. As will be discussed in chapter 5, peak heart rate depends on testing modality and protocol, and there is considerable variability between individuals. It is important to recognize that there is generally no specific "target heart rate" for an exercise test. Importantly, too, the maximal heart rate during exercise testing in a given subject does not change over the course of childhood. Only at about age 16 does this value begin to decline. Thus, age-related formulae for predicting a maximal heart rate, such as "220 minus age," do not apply to youths.

The concept of metabolic equivalents, or METs, as a measure of energy expenditure during exercise is commonly used with adult subjects but is fraught with difficulty in children and adolescents, and it is best avoided in the pediatric exercise laboratory. METs is a means of expressing the oxygen requirement of a physical activity relative to an assumed resting value. One MET, or resting $\dot{V}O_2$, in an adult is considered to be $3.5 \text{ ml} \cdot \text{kg}^{-1} \cdot \text{min}^{-1}$; thus, when walking on a treadmill at a certain speed and slope that is expected to demand $17.5 \text{ ml} \cdot \text{kg}^{-1} \cdot \text{min}^{-1}$, a patient is exercising at a level of 5 METs.

The difficulty with this concept in youth is that resting energy expenditure is not constant but evolves throughout childhood during physical growth. As would be expected, absolute values of resting $\dot{V}O_2$ rise with the accrual of body mass. When adjusted for body mass, or body surface area, basal or resting values of energy expenditure decline progressively during the pediatric years. When expressed as calories per meter of square body surface area per hour, the basal metabolic rate declines by about 20% between the ages of six and the mid-teen years.

In considering a mass-relative definition of a MET in children, the story is more exaggerated. Harrell et al. reported that the resting $\dot{V}O_2$ per kg in a group of 8- to 12-year-old children was almost 50% higher than that of 16- to 18-year-old subjects and 70% higher than that expected in adults (8). The use of the MET as defined in adults as a "currency" or multiplier of energy expenditure in children, then, clearly would introduce large errors in defining $\dot{V}O_2$ levels during exercise—and the extent of the error would be different depending on the age and size of the child.

Normative Values

As has been emphasized before—and will be again in future chapters—the use of published "normative exercise data for children and adolescents" should be approached with a good deal of caution. Variation in such results may be strongly influenced by important differences in equipment, protocols, subject population, laboratory environment, and, especially, testing staff. In some cases data may be selective—defining exercise response, for instance, in recruited rather than random populations in which the results may be restricted to those of youth willing to participate in an exercise study. For these reasons, each testing laboratory should create its own normative data for healthy young males and females.

This caveat notwithstanding, a number of authors have published results of maximal exercise tests in nonselective large numbers of healthy children and adolescents that can be assumed to reflect particular populations at large. These are outlined in table 1.1.

Adjusting Values for Body Size

A number of important variables recorded during exercise stress testing are measures of *volume*—oxygen uptake, minute ventilation, cardiac output, stroke volume. These, in turn, are manifestations of body size. Consequently, during the childhood years the progressive growth of lungs, heart, blood volume, and muscle mass are reflected in a steady increase in the absolute values of these measures. When a boy reaches the age of 15 years, his maximal oxygen uptake has almost doubled from when he was 5.

To permit comparison of such physiological variables over time in the same patient, or to assess values obtained relative to established "norms," it is necessary to adjust these absolute values for body size. Just how this is best accomplished, however, serves as a challenge to both pediatric exercise scientists and clinicians in the exercise testing laboratory. There exist a number

Table 1.1 **Key Studies of Maximal Exercise Test Results in Healthy Children and Adolescents**

Study	Location	Age	Number	Modality	Protocol
van der Cammen-van Zijp et al. (18)	Netherlands	4–5	80	Treadmill	Bruce
Cumming et al. (6)	Canada	4–18	327	Treadmill	Bruce
Ahmad et al. (1)	USA	5–18	347	Treadmill	Bruce
Lenk et al. (9)	Turkey	10–15	80	Treadmill	Bruce
Riopel et al. (14)	USA	4–21	288	Treadmill	Balke
Armstrong et al. (3)	Great Britain	11–16	220	Treadmill	Intermittent
Washington et al. (19)	USA	7–12	151	Cycle	James
Ten Harkel et al. (17)	Netherlands	8–18	175	Cycle	Ramp
Armstrong et al. (3)	Great Britain	11–16	200	Cycle	Intermittent

of candidate measures by which absolute physiological or anatomic variables might be "normalized": body mass, body mass raised to a particular allometrically derived exponent, lean body mass, height (raised to the 1.0 or 2.0 or 3.0 power), and body surface area.

In the clinical exercise laboratory, certain variables have historically been adjusted by particular size-adjusting measures. Specifically, oxygen uptake has traditionally been expressed relative to body mass in kilograms, and cardiac output and stroke volume have been normalized by body surface area (as cardiac index and stroke index, respectively). There are appropriate reasons for continuing this practice, first from a practical standpoint—both are easily measured— and second, values expressed as mass-relative $\dot{V}O_{2max}$ or cardiac index facilitate comparisons to those obtained in other laboratory or published normative data. Still, exercise physiologists have concerns about the validity of such measures to truly and accurately normalize values for body size (21). Often these are problematic more from a scientific than a practical perspective, but certain weaknesses, particularly of body mass in normalizing values of $\dot{V}O_{2max}$, need to be recognized by clinicians in the testing laboratory.

Perhaps the most pertinent issue is the influence of body composition—particularly body fat content—on values of $\dot{V}O_{2max}$. Body fat resides in the "per kg" denominator, but, being physiologically inert during exercise, it does not influence absolute $\dot{V}O_{2max}$. Thus, $\dot{V}O_{2max}$ per kg will be reduced in the obese subject yet inflated in the lean subject simply due to variation in body composition rather than any differences in maximal cardiopulmonary function. That is, $\dot{V}O_{2max}$ per kg provides information regarding *both* body composition and cardiovascular fitness.

Consequently, this effect of body fat needs to be taken into account when interpreting $\dot{V}O_{2max}$ per kg in an individual patient compared to "norms" for aerobic fitness. Serial measurement of $\dot{V}O_{2max}$ per kg in a patient over repeated tests as a measure of changes in cardiopulmonary function are safely made only if the patient's body composition remains relatively stable. A sudden loss of weight due to obesity treatment, however, will likely cause a rise in $\dot{V}O_{2max}$ per kg, which may have no bearing on improvement of cardiac or pulmonary status but will simply reflect a decrease in body fat content.

Tyranny of "Maximal" Testing

Much of the information obtained from clinical exercise testing (particularly $\dot{V}O_{2max}$) rests on the assumption that the pedaling, running, or walking subject has provided an exhaustive effort—has pushed to the limits, or at least nearly to the limits, of his or her muscular endurance. This defines the functional reserve, or cardiac or pulmonary function, and it serves to establish whether a given patient achieves "normal" values of each. Certain criteria have been established to document such a maximal effort by heart rate, respiratory exchange ratio (RER), and plateau of oxygen uptake, as well as by subjective signs of exhaustion (hyperpnea, sweating, effort strain). Studies involving "supramaximal" tests have confirmed that a ceiling of $\dot{V}O_2$ is achieved when such criteria are met.

With proper encouragement, most healthy children and adolescents can achieve an exhaustive peak effort when using standard progressive exercise testing protocols. Aiming to have a patient perform to a work rate that satisfies criteria for a maximal test has certain advantages. Obviously, the cardiopulmonary systems are stressed to a peak level of work, which may have particular importance depending on the clinical question being addressed. Also, physiological values obtained during a maximal test are now "standardized," allowing (a) valid comparisons with outcomes on repeat studies in the future and (b) comparison to certain normal values that are based on maximal tests.

It is not always necessary to push a subject to achieve maximal exercise criteria to address certain clinical questions. A high-intensity but not maximal test may be sufficient, for instance, to examine electrocardiographic changes in a teenage athlete with a past history of Kawasaki disease, or ventricular rate response in complete heart block, or QT interval duration in a patient with syncope.

The difficulty arises, however, in that patients with heart or lung disease—as well as otherwise healthy youths with low cardiovascular fitness—are often incapable or unwilling to perform exhaustive exercise to achieve standard criteria for a maximal test. It is difficult to interpret the results when patients claim to be unable to continue at a point far short of an exhaustive effort. For example, consider a treadmill exercise test in which one wishes to assess possible cardiac-based limitations in a sedentary, significantly obese patient using $\dot{V}O_2$ as a surrogate for cardiac output. At a heart rate of 140 bpm and RER of 0.92, values far below maximal criteria, the patient is uncomfortable, breathless, and complains that he cannot continue. The test is terminated. At this point his $\dot{V}O_2$ is only 22 ml $\cdot$ kg^{-1} $\cdot$ min^{-1}. By all standard criteria, he did not even approach an exercise intensity that would have maximally taxed his cardiovascular system. Here one would have to resort to an "operational" definition that his "peak $\dot{V}O_2$" was the value obtained at a work level to which the subject was willing to exercise. But what does this mean? Did he have to stop simply because he is less able to cope with the discomfort of high levels of exercise intensity compared to a nonobese, active child? Or does he have myocardial dysfunction and a limited cardiac functional reserve? Or did something else limit his exercise independent of cardiac function, such as myopathy of skeletal muscle, or biomechanical abnormalities, or a painful knee, or diminished ventilatory reserve? Did the problem lie in the difficulty a nonathletic, physically awkward child has in adjusting to treadmill exercise? It may be difficult to distinguish among these possibilities.

Given the frequency of this dilemma, considerable attention has focused on identifying ways to provide information about cardiopulmonary fitness from submaximal measures that do not require an exhaustive exercise effort. These will be examined in detail throughout this text. For example, the ventilatory anaerobic threshold (VAT), the point where the rate of change in minute ventilation exceeds that of oxygen uptake due to buffering of lactic acid (usually at about 50%-60% of maximal effort) may reflect level of aerobic fitness. The slopes of change in oxygen uptake plotted against either work rate or heart rate provide insights into cardiac function. The oxygen efficiency slope ($\dot{V}O_2$ versus the logarithm of minute ventilation) during submaximal exercise has been linked to both $\dot{V}O_{2max}$ and VAT (4). Similar relationships between minute ventilation and both $\dot{V}O_2$ and CO_2 can reflect possible ventilation:perfusion imbalance.

As Cooper has contended, "the most current [maximal] protocols . . . which are profoundly effort dependent . . . are inadequate for children and limit the valuable clinical and developmental information that can be gained from in-laboratory testing" (5, p. 1156). Alternative, submaximal measures, he noted, would be of particular utility in expanding clinical exercise testing to very young children as well as patients with physical disabilities.

The future application of exercise testing in the evaluation and management of patients with heart and lung disease may depend on the development of such submaximal markers that bear both diagnostic and predictive value.

Safety of Clinical Exercise Testing

Assuring the safety of the exercising subject during cycle or treadmill testing is paramount, and all measures need to be in place to

- prevent physical injury and
- recognize and manage cardiac and pulmonary complications that can arise.

This is done by close attention to the testing milieu—normal function of monitoring and testing equipment, use of appropriate exercise protocol, availability of resuscitation materials, presence of trained personnel, and proper instructions to the patient.

Means of dealing with potential complications of cardiac rhythm and hypotension need to be defined in advance of testing. Resuscitation protocols should be established, and practice drills in responding to such events should be conducted on a regular basis.

Contraindications for exercise testing in certain high-risk patients need to be considered. Similarly, the appearance of certain findings should signal a need to discontinue the exercise. We will discuss a number of additional issues surrounding exercise testing that have a bearing on subject safety. Should parents be allowed in the room during the test (might their vocal interjections distract the subject)? How many staff persons should be present? Should holding onto handrails be permitted?

With such measures in place, experience indicates that clinical exercise testing of children and adolescents is extraordinarily safe. Rhodes et al. indicated that over an eight-year period at Children's Hospital in Boston, "almost 15,000 exercise tests have been undertaken at our institution without encountering a serious testing-related complication" (13, p. 1963).

Alpert et al. reported an overall incidence of complications during 1,730 cycle exercise tests in children of 1.79% (2). The most common were chest pain (0.69%), dizziness or syncope (0.29%), fall in blood pressure (0.35%), and dangerous arrhythmias (0.35%). These findings mimic those of an informal survey by Freed of the experience of 87 pediatric cardiologists during over 6,000 exercise tests (7). Significant complications were reported in 1.7% (with no deaths), and only 0.3% required treatment.

This strong safety record appears to extend even to testing subjects who are expected to be at higher risk for complications triggered by exercise. Smith et al. reported their experience in maximal exercise testing of 27 pediatric patients (mean age 12.5 yr) with pulmonary hypertension who had depressed aerobic fitness (average $\dot{V}O_{2max}$ of 23.3 ± 5.4 ml $\cdot$ kg^{-1} $\cdot$ min^{-1}) (16). A dysrhythmia was observed in a third of these patients during exercise, and one in five demonstrated some degree of ST segment changes on an electrocardiogram. Oxygen saturation fell, on the average, to $85 \pm 16\%$ at peak exercise. Despite these changes, there were no significant adverse events (such as syncope, dizziness, or chest pain).

Hypoxemia often develops during exercise in patients with cystic fibrosis, who may also exhibit associated elevation in pulmonary artery pressure and right heart strain. Despite reductions observed in arterial oxygen saturation, Ruf et al. described no significant adverse reactions in 713 exercise tests of patients with cystic fibrosis (15).

Young patients with hypertrophic cardiomyopathy (HCM) are susceptible to dysrhythmias triggered by exercise that can cause sudden death. Olivotto et al. reported their experience in 243 symptom-limited cycle ergometer exercise tests in adults with HCM (11). Early termination of exercise was necessitated because of light-headedness in eight patients, and some dysrhythmia was evident in a third. But no syncope, cardiac arrest, or malignant dysrhythmia occurred in any case.

Conclusion

Many gaps exist in our present knowledge of the clinical applicability of cardiopulmonary exercise testing in children and adolescents. Most particularly, we need a better understanding of the clinical "meaning" of findings during cycle and treadmill exercise, especially regarding the strength of their ability to predict clinical outcomes. How does the value of $\dot{V}O_{2max}$ in a child with dilated cardiomyopathy help define risks and benefits for the timing of cardiac transplantation? How do markers of myocardial function with exercise provide insights into the efficacy of cardiac rehabilitation programs? How do rhythm responses to exercise help define the risk of sport participation? With continued progress in medical and surgical management of heart and lung diseases, such issues will become increasingly pertinent.

New technologies such as infrared spectroscopy, assessment of $\dot{V}O_2$ kinetics, and echocardiographic measures of myocardial deformation during exercise will undoubtedly add to the value of the information gained. How such measures can be used to assess the cardiopulmonary status of individual patients remains to be seen.

A better understanding of the physiological factors that dictate performance during exercise testing of youth with cardiac or pulmonary disease will be essential to this progress. For instance, the critical laboratory measure, $\dot{V}O_{2max}$, is considered to be an expression of the combined elements of the oxygen delivery and utilization chain during exercise. According to the traditional concept, any limitation of part of this chain—be it myocardial performance, or heat rate response, or lung function—will be expressed as a depression in $\dot{V}O_{2max}$ values. The clinical use of $\dot{V}O_{2max}$ as a marker of the severity of heart or lung disease is based on this construct.

The central question is what is it that limits endurance exercise in healthy or diseased patients? This applies to defining the physiological limitations of the elite high school cross country runner as well as the young patient with a dilated cardiomyopathy. It has been suggested, in fact, that cardiac factors may not be exercise-limiting, even in those with heart disease. In adult patients with congestive heart failure, for example, exercise tolerance appears often to be related most closely to peripheral factors—especially skeletal muscle myopathy—rather than myocardial contractile function (12). This issue has not yet been addressed in pediatric patients.

Some would argue that the limits of exercise performance are most directly defined by the central nervous system rather than by physiological factors. They posit the presence of an evolutionary, subconscious "governor" within the brain that creates sensations of fatigue—dizziness, breathlessness, discomfort—at "peak" exercise that cause one to stop exercising as a means of protecting himself or herself from the ultimate risks of coronary insufficiency, muscle tetany, hyperthermia, and even bone fractures (10). According to this concept, then, no exercise performance is ever truly maximal.

Much remains to be learned about the physiological, anatomic, and psychological factors that limit endurance exercise performance in healthy individuals as well as those with cardiac and pulmonary disease. As we continue to learn more, we can expect to see an enhanced role for clinical exercise testing.

Conducting the Pediatric Exercise Test

Amy Lynne Taylor, PhD

Exercise testing is a clinically valuable tool that can be used in the diagnosis and management of disease in children. Children can be tested as effectively as adults when certain adaptations unique to pediatrics are made. The successful pediatric exercise test must be carefully tailored with respect to physical environment, testing modality and protocol, staff approach, and the individual needs of the patient. This chapter details some of the distinctive components of the pediatric exercise test.

Pediatric Exercise Laboratory Environment and Equipment

The required physical size of the exercise testing laboratory is related to the types of testing being performed and resultant equipment needs. Normally this should range between 400 and 500 ft^2 (37-47 m^2) (15, 23). In addition to the equipment, the physical space also must be able to accommodate a response to an emergency situation. The laboratory should be kept at a neutral climate with temperature between 20 and 23 °C (68-72 °F) and a relative humidity between 50% and 60% (11, 15, 23).

The laboratory may need to account for the following:

- Permanent equipment—equipment that will stay in the laboratory at all times (treadmill, cycle ergometer, metabolic cart, patient gurney)
- Transient equipment—equipment that may move in and out of the laboratory based on the type of test being performed and laboratory workflows (ultrasound machine for stress echocardiography or pulmonary function cart for provocation testing)
- Flexible staff work space—desk, computer and chair
- Ancillary equipment—sink, storage solutions, linen, scale, stadiometer

A pediatric exercise testing laboratory must have equipment that can be scaled to the size of the child (16, 23). Blood pressure cuffs should be available in multiple sizes. Many treadmills have an option for adjustable handrails that can be moved to accommodate the height of the patient. Cycle ergometers should have seats that can be adjusted for height, adjustable cranks, and handlebars that can be positioned for each child.

The choice between treadmill and cycle testing is dependent on a multitude of factors ranging from patient comfort and safety to the clinical question being addressed. There are benefits and drawbacks to both treadmill and cycle ergometer testing. Whichever is chosen, most protocols should be designed to last approximately 8 to 12 min. More than half of pediatric stress testing labs in the United States have both a treadmill and a cycle ergometer (3).

Modern treadmills and cycle ergometers allow for both direct operator control and preprogrammed exercise protocols based on the type of testing being performed. A warm-up period is typically employed that can range in length from several seconds to several minutes, depending on the patient's exercise testing experience.

The treadmill is the most commonly used modality for delivering exercise physiological stress (15). The stimulus is adjusted by varying the speed or the grade of the belt. Walking and running on a treadmill can be challenging for some. Patients should be assessed for their ability to safely walk and run on a treadmill prior to stress testing. A demonstration by a staff member may be helpful.

Advantages of the treadmill include the potential to reproduce symptoms that have occurred during running. Treadmill testing will yield a maximal oxygen uptake that is about 10% greater than cycle ergometry (10, 13, 21). Disadvantages of the treadmill include the inability to calculate work rate, cost, and movement artifacts that can affect blood pressure and electrocardiogram recordings (15, 23). Generally speaking, treadmill testing is not considered as safe as cycle ergometer testing because of the risk of falling (15, 18).

There are two types of cycle ergometers available for testing. A mechanically braked ergometer uses friction bands to increase the work rate, while an electronically braked ergometer increases the work rate with electromagnetic forces. The former carries the disadvantage of requiring the subject to maintain a particular pedaling cadence. Electronically braked ergometers require a cadence only within a specific range, and they are typically more accurate at reporting power output. Use of a cycle ergometer permits an accurate quantification of workload that is not possible during treadmill exercise because of individual differences in gait, body size, and stride length.

Optimizing Safety

Maintaining a safe environment for the performance of the exercise test is paramount. The very low incidence of serious adverse events during maximal exercise testing is a testament to the combined efforts of careful selection and screening of patients, skilled oversight by staff, and the development of specific training programs for those performing these procedures, among other factors.

Laboratory Staffing

Laboratories should be under the direction of a physician who has received training in exercise testing and exercise physiology (15, 23). Both the American College of Cardiology and the American Heart Association have documented clinical competencies for physicians who direct exercise testing laboratories (16, see also http://circ.ahajournals.org/content/130/12/1014). The physician may perform day-to-day management of the laboratory or delegate this to a non-physician who has also completed specialized training in exercise physiology. Ideally the non-physician manager should have a master's degree in exercise physiology (23). This person is responsible for training staff, maintaining equipment, guiding testing procedures based on the patient's unique needs, and making timely reports of results (15, 16, 23). Quality assurance methodologies must be used to continuously assess data reliability, reproducibility, and comparisons to normative values (4).

Laboratory staff performing the exercise test should understand broad testing indications, normal and abnormal physiological responses to exercise, and the monitoring used during testing (4, 15, 16, 23). A formal list of competencies for those who perform stress testing has been presented by others (16). A minimum of two people should perform the exercise test—one to directly monitor the patient and the other to monitor the data collection, including the electrocardiogram, oxygen saturation, and blood pressure (15, 23).

Staff should be certified in basic life support (BLS) at a minimum (15, 18, 23). Requirements for advanced cardiac life support (ACLS) and American College of Sports Medicine (ACSM) certification may vary given the differences between adult and pediatric practices. Emergency protocols should be prepared, documented, updated, and practiced regularly.

Indications for Pediatric Exercise Testing

Cardiopulmonary exercise testing (CPET) in children is usually done to document physical working capacity, to act as a provocation challenge, or to provide an additional diagnostic tool in the medical management of the child (1). Specific indications for pediatric stress testing are presented in multiple publications (8, 15, 16, 18, 23) and elsewhere in this text. The question to be answered for each patient should be precisely stated so that appropriate protocol, equipment, and measured variables can be used.

Contraindications to Pediatric Exercise Testing

Contraindications to exercise testing in adults are well-documented. Similar guidelines were initially developed for children (11, 15, 20), but experience has indicated when this type of test can be performed safely in both healthy and diseased children (17, 18, 24), including those previously considered high risk by adult standards (7, 15). Therefore, over time, the absolute and relative contraindications for exercise testing in children have been modified (23).

The guiding principle is that testing should not be planned in situations where the risk of maximal exercise outweighs the benefit of the information gained (23). Many pediatric testing laboratories have utilized minimal contraindications for testing, most of which center around the presence of an acute disease process (18). Examples of test contraindications may include those children with acute cardiac disease (such as pericarditis, rheumatic fever with cardiac involvement, or myocarditis), severe cardiac valve stenosis, unstable arrhythmias with concurrent hemodynamic compromise, or severe congestive heart failure or hypertension (15, 23). Careful assessment of the risk-to-benefit ratio for all patients is paramount. Each laboratory should develop criteria related to contraindications to testing based on their staffing, experience, and comfort level (23).

Test Termination

Given that exercise testing is commonly used to assess maximal cardiopulmonary capacity or to trigger clinical symptoms, it is important that the subject exercise to a high—if not exhaustive—level of exercise intensity. There are, however, some situations when an exercise test should be terminated early: when conclusions have been reached and the test will not yield any further additional information, when there is a failure in monitoring equipment, and when continuing the test would compromise the child's safety or well-being (23).

Clinical judgment and experience are important in determining if a test should be stopped due to symptoms reported by the child. Any decision to terminate a test early should be based on an assessment of the entire data picture (15). The decision to terminate a test can be variable and patient dependent (18).

Other Safety Considerations

Exercise testing in children is considered safe, even in children who carry a diagnosis that would stratify them as high risk (7). A debated safety consideration for pediatric exercise testing is the need for the presence of a physician during testing. Communication and pretest planning are essential in this decision process, with the safety of the child being of utmost importance. A physician should be immediately available for testing deemed to be low risk and should normally be physically present when testing the high-risk child (23). Generally, patients who are asymptomatic or clinically stable are on the low end of the risk spectrum. More structured guidelines for low-risk and high-risk patient populations are available (23); however, the final decision and planning for risk should occur before the test is performed.

According to the American Heart Association Council on Cardiovascular Disease in the Young, Committee on Atherosclerosis, Hypertension,

Should Parents Be in the Laboratory During Their Child's Test?

Pediatric exercise physiologists disagree about whether to allow parents to be in the room during testing. In this author's experience, pediatric labs are split 70 to 30 on letting parents directly observe their child's exercise testing, with more letting parents in than keeping parents out.

Parental presence may be distracting for some children and may provide reassurance for others. Shy, very young, or nervous children may be comforted by having their parent present. Children whose parents do not accept less than an Olympic record for exercise testing (or any other athletic endeavor for that matter) will not be comforted by having their parent present. Some parents can help their children to describe their symptoms to the exercise staff before testing, while others will take over describing their child's symptoms to the staff while the child is exercising. Some parents will sit quietly in their designated space during the test, while others will practically try to help conduct the test. Some children can verbalize their symptoms to the exercise staff without their parents, while others only have symptoms in front of their parents.

At the very least, the family should be actively engaged in the discussion of the testing procedures, expectations, risks, and benefits. The entire family should be encouraged to ask questions and discuss any concerns they may have prior to the procedure. If the laboratory uses an orientation visit or a practice testing session for young or new patients, the family should be present for the entire session. They then could be absent for the test itself. Depending on the size of the laboratory, there may not be room for the parents to sit and observe, let alone the grandparents, aunts, uncles, and cousins who may accompany the child.

and Obesity in Youth (23) (see also the AHA position statement at http://circ.ahajournals.org/content/113/15/1905), high-risk patients include those with a history of

- pulmonary hypertension,
- long QTc syndrome,
- dilated or restrictive cardiomyopathy with congestive heart failure or arrhythmia,

- hemodynamically unstable arrhythmia,
- hypertrophic cardiomyopathy symptoms with left ventricular outflow tract obstruction (greater than mild),
- hypertrophic cardiomyopathy symptoms with documented arrhythmia,
- airway obstruction on baseline pulmonary function tests (greater than moderate),
- Marfan syndrome with activity-related chest pain (suspected noncardiac cause),
- myocardial ischemia with exercise (suspected), and
- syncope with exercise (unexplained cause).

Equipment to manage patient emergencies is essential. Laboratories should have a fully stocked resuscitation cart with a defibrillator, oxygen, and suction in the laboratory. Oxygen and suction capabilities may be on the resuscitation cart, wall mounted, or built-in.

Preparing the Child for an Exercise Test

Compared to adult patients, there are some unique pretest considerations for children. Special care should be taken to orient the children and their adult guardians to the testing site, procedures, and protocols.

Pretest Considerations

It is often helpful to discuss the test with the child's guardian(s) several days before the test. This can take the form of a verbal conversation or a mailed letter. See *Sample Pretest Letter* for an example. Written instructions may increase compliance (20). Information provided should include the following:

- The child should be wearing clothing appropriate for exercise. The importance of proper footwear (no street shoes, sandals, or flip flops) should be emphasized. A loose-fitting, short sleeve shirt is preferred.
- It is recommended that children do not eat for 2 to 3 hr prior to an exercise test. If the test is the first thing in the morning, however, the child should eat a light breakfast. (It only takes one experience attempting to test a teenager who has entered the lab immediately after consuming fast food, chips, and soda to understand the impor-

tance of this recommendation.) Also, the child should refrain from highly vigorous activity 24 hr before the test.
- Patients may need to be instructed to either take or hold routine medications depending on the type of test being performed (e.g., holding a routine bronchodilator the morning of a provocation test) or the purpose of the test (e.g., holding a beta blocker to assess heart rate response).
- A brief, high-level description of what to expect during the test.

Laboratory Orientation

There are certain times when performing an orientation to the laboratory, staff, and equipment is beneficial. This is formally required in some research protocols. This allows children and their families to see, touch, and experience the equipment in a non-pressure situation (not right before the actual test). It also allows time for children to ask additional questions that they may have about the procedure. The child then arrives for testing in a more relaxed state. This author has performed multiple successful orientation sessions with children who are younger than the age range normally tested in our laboratory.

Informed Consent

It has been suggested that pediatric exercise testing laboratories use a written consent form for the procedure (see *Sample Consent Form*) (4). It is important to discuss the use of an additional consent form (in addition to a standard consent for treatment form) with institutional leadership because different institutions have different policies regarding consent for medical procedures. It is also important to understand local and state laws on this issue. At this author's institution, an additional consent form for exercise testing is used because the test is considered to present a level of risk above that of a standard clinic visit. Whether or not a formal consent document is used, a thorough discussion of the procedure, the risks and benefits, and the expectations should occur prior to the test. Documentation of this conversation may be warranted.

Testing Protocols

A number of different cycle and treadmill protocols are available for testing children, each with their own benefits and drawbacks. These will be discussed in detail in chapter 3.

Sample Pretest Letter

You/your child has been scheduled for an exercise test. This test allows your doctor to learn more about how your heart and lungs work when you exercise. There is nothing painful about this test, but you will be asked to exercise on a treadmill or bicycle until you are very tired. Most children exercise for about 8 to 12 minutes, but there is no formal time limit for the test. It is important to try your very best during the test. Before, during, and after the test, you will be attached to several monitors that take measurements of you.

- Electrocardiogram (ECG): This measures your heart rate and rhythm.
- Blood pressure: This measures your blood pressure in your arm.
- Oxygen saturation: This measures the amount of oxygen in your blood.

- Breathing: You may be asked to breathe into a mouthpiece or face mask when you exercise.
- How you are feeling: The people performing your test will ask you about how you are feeling.

Other instructions:
- Please wear comfortable exercise clothes and sneakers to the test.
- Please do not eat for 2 hours prior to the test. If your test is first thing in the morning, however, you should eat a light breakfast.
- Please do not participate in vigorous physical activity 24 hours before the test.
- Please take your medicine as you normally would unless your doctor gives you other instructions.

From T.W. Rowland, American College of Sports Medicine, and North American Society for Pediatric Exercise Medicine, 2018, *Cardiopulmonary exercise testing in children and adolescents* (Champaign, IL: Human Kinetics).

Multistage incremental protocols increase in intensity every 2 to 3 min. The most common multistage incremental protocols used in pediatric exercise testing are the Bruce (treadmill) and the James and McMaster (cycle ergometer). The Bruce protocol can be applied to children of almost any age. Highly trained children may become quite bored with the first three to four stages of this protocol, however. The 3 min stages may also cause boredom in some children. On the other hand, the work increments between stages may be too challenging for some patients. Normative data on children ages 4 to 14 are available (6).

The James and the McMaster cycle protocols each use three different protocols based on specific patient characteristics. The James protocol is selected based on gender and body surface area, whereas the McMaster protocol is chosen based on the patient's gender and height. The James protocol uses 3 min stages while the McMaster uses 2 min stages. Normative data are available for the James protocol (12, 22).

Cycle ergometer testing can also use a progressive incremental protocol, where the workload increases every minute, or a continuous ramp protocol, where the workload increases constantly. These two protocols provide similar physiological responses (14, 19, 25). These types of protocols are highly effective at yielding diagnostic data within 10 to 12 min (5). The Godfrey test was the first 1 min stage test used in children. The Godfrey test utilizes three protocols based on the height of the child and then uses a work interval of either 10 or 20 W. Normative data are available (9). Cooper described the first continuous ramp protocol used in children. The slope of the ramp is adjusted to the child's size and physical abilities and is typically assessed on a patient-by-patient basis.

Test Communication

Performing an exercise test on a child is often more art than science. The science aspects of protocol design, safety procedures, and equipment maintenance are covered before the child enters the laboratory and are generally not seen by the patient. The art comes into effect during the actual

Sample Consent Form

Purpose and explanation of the test: _____ is being asked to perform an exercise test to determine the presence or absence of any limitation related to the heart and lungs and/or to evaluate the effectiveness of any past surgical or medical intervention. The exercise intensity will begin at a low level and will be increased over time. While _____ is exercising, blood pressure, electrocardiogram (EKG) readings, and oxygen levels will be constantly monitored and recorded. _____ will also be monitored for a short recovery period after exercise. _____ may be asked to breathe through a mouthpiece so that oxygen intake can be sampled and studied.

Patient risks and discomforts: By reviewing preliminary information about _____ health and by careful observation during and after the exercise test, every effort will be made to conduct the test in such a way to minimize discomfort and risk. However, as with other types of procedures, there are potential risks involved. Risks during exercise testing include but are not limited to feeling lightheaded, fainting, chest discomfort, and muscle cramps. Sudden death and problems that would require hospitalization are extremely rare but possible. Emergency equipment and trained personal are available to handle unusual situations should they occur.

Participant responsibilities: Each patient will be asked by the supervising staff if there are any symptoms or discomfort that they are experiencing during the exercise test. Patients will tell the supervisory staff what medication(s) they are currently taking.

Benefits to be expected: The results obtained from the exercise test may help in diagnosing illness, evaluating the effect of medication or surgical intervention, or evaluating what kinds of physical activities the patient can perform.

Opportunity for questions: I have read the preceding information and have been informed about the exercise testing procedure. All of my questions have been asked and answered. I know I can ask further questions at any time about the exercise test.

Freedom of consent: I hereby consent to perform an exercise test. My permission is given voluntarily.

I have read this form and I understand the proposed test procedure and the risks and discomforts of the test procedure. Knowing the risks and discomforts and having had my questions answered to my satisfaction, I consent for my/my child's participation in this test.

Signature of Parent/Legal Guardian or Patient _____   Date _____

I have explained the contents of this document to the patient and have answered all of the patient's questions, and to the best of my knowledge I feel the patient has been adequately informed and has consented to the procedure(s) detailed above.

Signature of Physician or Trained Delegate _____   Date _____

From T.W. Rowland, American College of Sports Medicine, and North American Society for Pediatric Exercise Medicine, 2018, *Cardiopulmonary exercise testing in children and adolescents* (Champaign, IL: Human Kinetics).

test. Figuratively speaking, getting a child to perform a maximal effort during testing is an exercise in successful sales and marketing. The atmosphere of the laboratory must be relaxed, warm, and inviting. Pediatric exercise physiologists must be able to successfully motivate and praise a child during the procedure. This is particularly important in children who are not active and not used to pushing themselves physically to exhaustion. These children may feel they are at their maximal capacity while the objective data do not indicate a maximal effort. Verbal encouragement becomes even more important during this portion of the test in these children. In essence, the staff must achieve buy-in from the child that the test is fun and that the child's performance is special.

Communication with the child during the test should be a balance between confirming the presence or absence of symptoms and distracting the child from feelings of discomfort. Frequent interaction with the child during the test is important and positive. This interaction should focus on how wonderful the child is performing (phrases such as "you're doing a great job" or "I'll bet you can do one more minute") instead of how tired the child is becoming. The key is to refocus the child away from feeling uncomfortable to finishing as much as possible.

Communication plans should be discussed before the test begins, particularly when a mouthpiece is used for metabolic measurements, because the child will not be able to speak normally during the test. Options may include using a thumbs-up or thumbs-down or using a symptom rating scale (0 is not at all, 10 is the worst ever) during the test to allow for nonverbal communication about symptoms.

Children should be alerted that at the end of test they should not stop exercising or jump off the treadmill because there will be a cool-down period. Applause from the staff at the end is supportive. Providing the child with an achievement

Music During Pediatric CPET: Motivational or Distracting?

Several pediatric exercise labs play popular music in the background during CPET. There are reports in the literature of decreased ratings of perceived exertion when music was played during standardized exercise compared to a no-music control situation. Music during testing may be distracting in some children while others may find it motivational.

Communication between the exercise staff and the patient throughout the test is paramount in maintaining a safe environment. If a patient is listening to music through earphones during the test, then this communication could be hampered, especially in situations where communication must be rapid. If background music is played, the volume should not obstruct ongoing communication between the child and the exercise staff.

Another consideration is the effect of playing music on longitudinal data obtained on the same patient. If the patient was listening to a favorite artist for one test (and hypothetically was more enthusiastic) and did not do this for a follow-up test, there is potential for the data to be skewed.

certificate or other type of award after the test may be useful.

Conclusion

Maximal exercise testing in pediatrics can be performed safely in both healthy children and children with disease. Proper training of staff is essential. There are multiple testing protocols that are well suited for children. It is important to recognize that there is an art to pediatric exercise testing that is needed to achieve quality results.

EXERCISE TESTING METHODOLOGY

An exercise test can only be expected to yield useful clinical information when measurement of appropriate case-specific physiological variables are incorporated into the study design. Testing experience over the past several decades has greatly expanded the number of useful variables that can be measured, including aerobic fitness, cardiac output, myocardial function, and ventilatory capacity. Part II provides an overview of current cardiopulmonary variables that are available during exercise testing.

Exercise Testing Protocols

Richard J. Sabath III, EdD, David A. White, PhD, and Kelli M. Teson, PhD

Over the past three decades there has been a significant increase in the use of pediatric exercise testing both in the clinical setting and in the research arena. Clinicians and researchers have benefited from the development of testing equipment that has greatly enhanced our ability to monitor the patient's electrocardiogram (ECG), oxygen saturation, blood pressure, and echocardiogram during exercise. The development and steady improvement in breath-by-breath oxygen consumption technology has made the accurate measurement of cardiorespiratory parameters possible without the use of cumbersome Douglas bags and Scholander devices that were once necessary. These technological advances have aided the clinician in diagnosing disease, assessing more precisely a patient's functional capacity, and making decisions about future medical management. Researchers have also benefited from the enhanced assessment techniques and new methodologies that have provided fresh insights into the physiological responses of children during exercise.

The purpose of exercise testing is to stress the central oxygen delivery mechanisms in a controlled and monitored setting to allow identification of abnormalities of the cardiopulmonary systems that may not be evident at rest. Exercise testing provides information about exercise capacity, indications for future surgery, the need for therapy or additional tests, the efficacy of medicines, and the risk potential for future disease or complications from existing disease. It is potentially helpful in instilling confidence in parents and children (8, 73). The basic tenets of exercise testing, however, have remained unchanged despite the advances that have been made. One of the first decisions to be made by the test administrator is the mode of exercise to be used and the specific protocol for that testing mode. The mode of exercise is determined by the objective of the test, the age and health status of the patient, and the available equipment. The protocol must be capable of producing valid, reproducible results.

A significant amount of research has focused on the effect of exercise testing protocols on exercise tolerance, diagnostic accuracy, and gas exchange parameters (57). Myers (57) has stated that a child's performance on an exercise test largely depends on the mode of exercise and the test protocol selected. Stage duration and the work rate increment of each stage are also important considerations. Stage duration has no apparent effect on maximal heart rate or maximal oxygen uptake (89). Traditionally, protocols with 3 min stages or longer have been used to assess submaximal variables. However, protocols with 2 min stages may be sufficient for testing children since they typically adapt to changes in the external work rate more quickly than adults do (42, 77). Several studies have shown that peak oxygen uptake is relatively unaffected by protocol variations within a given mode of exercise (76, 77, 81).

Information about a child's physiological response to exercise is only as good as the methods used to obtain it. Therefore, it is very important that appropriate protocols and testing techniques be used. There are several unique considerations in testing children, who are emotionally and physically immature compared to adults (73). Exercise testing of children demands more personal attention by the laboratory staff and requires more time to prevent fear, misunderstanding, or inadequate understanding of test requirements (73). Ideally, the test protocol should be designed to maximally tax the patient's central delivery mechanisms within 8 to 12 min (63) of exercise. Some researchers and clinicians have suggested that test protocols for young children should be 6 to 10 min in duration (40), but Wasserman et al. (86) stated that testing protocols of less than 10 min in duration may result in poor aerobic performance secondary to increased muscle fatigue caused by the higher work rates required to reduce test duration.

Typically, pediatric exercise testing guidelines recommend a total test duration of 10 ± 2 min (41, 63, 84). Many experienced clinical pediatric exercise physiologists report, with few exceptions, that children 5 to 6 years of age can exercise for 8 to 12 min on various types of treadmill protocols without significant difficulty. Each exercise test protocol should begin with a 2 to 3 min warm-up. The exception to this rule is when the objective of the test is to evaluate a patient for possible exercise-induced bronchospasm. In this case the onset of exercise should be very brisk, elevating the patient's heart rate to 160 to 180 bpm relatively quickly (as described in chapters 11 and 13).

In tests where oxygen consumption is measured using a breath-by-breath metabolic system, the same recommendations for protocol duration apply. Some youth tend to terminate exercise early due to discomfort with the headgear, mouthpiece, and nose clips associated with measurement of $\dot{V}O_2$. The silicone oronasal face mask is generally accepted as an appropriate alternative to the older mouthpiece and nose clip system. When using oronasal face masks, it is vital to ensure there is a proper seal, particularly around the patient's chin and the sides of the nose, to prevent any leaking that would affect the measurements. Although many advanced metabolic systems can measure oxygen consumption breath-by-breath, this frequency of measurement may make the data too variable, potentially altering the accuracy of the data. Adjusting the metabolic system settings to 10 to 30 s averages will allow for ample exercise data while minimizing variability. The patient should avoid verbal communication while oxygen consumption is being measured. Thus, the use of predetermined hand signals and ratings of perceived exertion (RPE) scales are effective modes of communication during an exercise test. Although the traditional Borg scale is the most commonly used and well-studied RPE scale, children often struggle to comprehend the unorthodox 6 to 20 rating system. The Children's OMNI RPE scale may be more appropriate for this population due to the easy-to-understand 0 to 10 scale and the scale's use of numerical, pictorial, and verbal descriptors (38). The OMNI scale is valid and reliable for youth with versions specific for running and cycling.

Exercise Testing Modality

The two most common modes of exercise testing are the mechanically or electronically braked cycle ergometer and the motor-driven treadmill. Debate over the advantages and disadvantages of cycle ergometer and treadmill testing protocols is ongoing. General perceptions are that the treadmill is noisy and may frighten small children. There is also concern about increased risk of injury on the treadmill (63). Experience reveals that very young children (aged 3-5) have not been hesitant to exercise on the treadmill. In fact, many appear eager because they often have been restricted from treadmill use by their parents or guardians. With proper instructions, the risk of injury with the use of treadmill protocols is very low. Many treadmills designed for exercise testing can be modified for a younger pediatric population. These modifications include adjustable side rails and a fixed or adjustable front handlebar that can be placed at the child's level. These modifications not only help children feel more comfortable while exercising (which may lead to longer treadmill test time) but also may help with ECG quality by allowing them to place their hands in a more appropriate position. Furthermore, many younger children may fail to start moving their feet at the initiation of exercise and may need to be reminded to start walking as the belt moves. Appropriate treadmills for pediatric exercise testing should be able to start at speeds less than 1.0 mph (1.6 km/h), which will allow the younger patient to become acclimated to treadmill walking before exercise intensity increases.

Stephens and Paridon have published guidelines regarding the mode of exercise that may be most useful in evaluating different types of patients (78). They recommend cycle ergometer protocols for the evaluation of patients with repaired or unrepaired aortic stenosis or insufficiency, transposition of the great arteries, tetralogy of Fallot, single ventricle physiology, and coronary artery anomalies in which the detection of ischemia and arrhythmias is of primary importance. Treadmill protocols are recommended for the assessment of pacemaker rate responsiveness, exercise-induced bronchospasm, and chest pain. Assessment of peak aerobic exercise capacity may be made using either a cycle ergometer or treadmill protocol. However, the treadmill is preferred over the cycle ergometer for testing young children (≤7 yr) because of concerns that relatively underdeveloped knee extensors lead to leg fatigue and early test termination during cycle testing (9, 73). The cycle ergometer is preferred by many pediatric laboratories due to more stable ECG tracings, ease of blood pressure measurement, and increased patient safety (63). Numerous studies have provided reference values for children and adolescents performing cycle ergometry (1, 28, 32, 85). Reference values for treadmill testing have also been published (11, 22, 83, 87).

As with treadmills, cycle ergometers for pediatric exercise testing may need to be modified to fit a variety of children. The most common pediatric modification for an adult-sized ergometer is adjusting the pedal crank arms. These crank arms allow the pedals to be moved to different positions in order to shorten the diameter of the pedal revolution for smaller children. Many cycle ergometers also have rotatable handlebars. The handlebars should be placed in a position that is comfortable for the child's or adolescent's seat height. Many children will report significant seat discomfort during cycle ergometer exercise. Cycle saddles are produced in many different shapes and sizes; it is best to find one that is wide in the back to provide support and stability without being too large. If the seat is too large, the child's range of motion may be disturbed, and he or she may tend to slide off the front of the seat. An adult-sized wide seat is recommended for adolescent populations. Some manufacturers produce smaller cycle ergometers designed specifically for the pediatric population. These ergometers typically are brightly colored and adjustable to fit younger children comfortably.

A third mode of exercise not commonly used in most pediatric exercise testing laboratories is arm ergometry. In the past, arm ergometry has been used to evaluate nonambulatory patients, but current clinical practice seems to favor pharmacologic testing of nonambulatory patients or those who are too deconditioned or too ill to exercise safely on a treadmill or cycle ergometer (63). Arm ergometry is used in the evaluation of patients with spinal cord injuries, various congenital defects, or leg amputations due to disease or injury.

Protocol Design

Protocols for all modes of exercise may be divided into three categories: multistage, intermittent, or ramp. Multistage protocols (also known as incremental protocols) consist of continuous, progressive changes in slope or speed at specific time intervals typically 2 to 3 min in duration. Examples of multistage protocols include the Bruce and Balke treadmill protocols and the McMaster and James cycle ergometer protocols.

Intermittent protocols have been used to determine maximal oxygen uptake and to measure multiple submaximal variables. In intermittent protocols the subject exercises at a given intensity for 3 to 5 min and is then given a short rest period. The protocol continues with additional intermittent bouts of exercise until the subject can no longer continue. While this type of protocol is very good for assessment of submaximal parameters and the determination of $\dot{V}O_{2peak}$, it is very time consuming and is therefore not recommended for daily clinical use. Intermittent protocols have found some utility in the measurement of stress echocardiography using an upright cycle or an ergometer specifically designed for stress echo. Immediately after exercise, youths will typically have a very short heart rate recovery period, which drastically limits the time available to acquire images. Intermittent protocols provide multiple bouts of exercise followed by bouts of rest, which will provide the sonographer with many chances to acquire the necessary images.

The third type of testing protocols is ramping protocols. This category includes both standard and individualized continuous ramp protocols as well as those protocols that employ stages 1.5 min or less in length. Although a significant number of studies have presented the advantages of using a ramping-style exercise protocol, its use is not widespread. For patients undergoing serial exercise testing over time, it is recommended that the same protocol and mode of exercise be used at each test session because different types of tests can produce slightly different results. Slightly greater peak heart rates (5%-10%) and higher peak oxygen consumption values of 10% to 15% are typically seen when comparing the results of treadmill tests to those of cycle ergometer tests (12, 52).

Brief History of the Bruce Treadmill Protocol

Dr. Robert A. Bruce is often referred to as the father of exercise cardiology (74). Before he developed the Bruce treadmill protocol, there were no standardized protocols that were safe, valid, and appropriate for the evaluation of cardiopulmonary parameters in exercising patients. The enhancement of electrocardiographs and the development of quality-motorized treadmills in the 1940s helped Dr. Bruce to develop a standardized clinical assessment tool. Dr. Bruce's first study in 1949 used a single-stage protocol of 10 min duration and a constant walking speed (18). In 1950 Dr. Bruce became the first head of cardiology at the University of Washington, where he continued the single-stage studies primarily to examine the predictive value of treadmill testing in determining the success of surgery for valvular and congenital heart disease (74). In 1962, due in part to his single-stage studies, Dr. Bruce was invited to participate in a symposium on exercise fitness testing organized by the American Academy of Pediatrics that addressed concerns about low physical fitness levels in American children (74). As a result of the symposium, Dr. Bruce and his coworkers published the first report of their research using a multistage treadmill protocol in *Pediatrics* in 1963 (14). This initial multistage protocol had only four stages: stage I: 1.7 mph (2.7 km/h), 10% grade; stage II: 3.4 mph (5.4 km/h), 14% grade; stage III: 5.0 mph (8 km/h), 18% grade; and stage IV: 6.0 mph (9.6 km/h), 22% grade. Over time, Dr. Bruce's original protocol evolved into the current seven-stage protocol in use today. Beginning in 1971, Dr. Bruce and another University of Washington cardiologist Dr. Harold T. Dodge studied thousands of individuals participating in the Seattle Heart Watch program (74). These studies established the efficacy, safety, and prognostic value of the Bruce protocol (15-17).

Treadmill Protocols

In contrast to the wide use of the Bruce protocol in the adult population, there is no single standardized testing protocol established for children. This has often made it difficult to compare clinical or research test data from one facility to another. However, many treadmill protocols have been validated in the pediatric population and have been recommended for specific outcomes. The following section will describe in detail two of the most common treadmill protocols used for young people, followed by a general review of other multistage and ramping-style treadmill test protocols. Although intermittent protocols can be beneficial in some cases, that form of exercise testing will not be covered in this section and is not recommended for general pediatric exercise testing.

Bruce Treadmill Protocol

The most commonly used treadmill protocol for the evaluation of adults and children in the clinical setting has been the Bruce protocol. Over several decades of use, the Bruce protocol has produced well-validated functional and prognostic normative values for many different cardiovascular and metabolic conditions (57). The Bruce protocol consists of seven 3 min stages (table 3.1). Despite its widespread use, a number of potential problems have been identified with the use of the Bruce protocol to test small children, patients with marked obesity, and those with low functional capacity. The large and uneven workload changes may cause significant errors in estimating exercise capacity based on total exercise time or workload (9, 63, 72). Highly fit patients may find the protocol boring since they must exercise for 12 min before encountering running speeds. The 3 min stages may cause the test to become too long for young children (recommended exercise test duration of 10 ± 2 min) who may lose interest and motivation to give a full effort. The steep grades on the Bruce protocol may cause patients to stop due to lower leg fatigue and may tempt patients to hold onto the handrails, significantly affecting both the estimated and directly measured oxygen consumption (63).

Various modifications have been made to the original protocol over the years to overcome some of its deficiencies. Changes in initial slope, speed, and stage duration have been the most frequent modifications to the Bruce protocol. Modified Bruce protocols often use 1.5 min (half Bruce) or 2 min stage times. As shown in table 3.1, stages A and B may be added to the standard Bruce protocol to provide a less strenuous initial workload in patients who are unfit, obese, or have advanced cardiac or lung disease.

Should Holding Onto Handrails Be Permitted?

There is considerable disagreement about whether a patient should be allowed to hold onto the handrail during clinical treadmill exercise testing (personal communications). There is a rather general agreement that, except in extenuating circumstances, it would be better if patients did not grasp the handrails. In fact, most published exercise testing guidelines indicate that subjects should not hold onto the handrails during testing because this leads to a reduction in metabolic cost, decreases submaximal heart rate and oxygen uptake responses, and prolongs test duration. However, while test time is prolonged, some data suggest that there are no apparent differences in peak heart rate or peak oxygen uptake in adults or children during treadmill exercise with or without handrail use (66, 75). Green and Foster's study showed a relationship between decreasing metabolic cost and how tightly the handrail was grasped (34). On the other hand, many testers allow holding onto the handrails for subject stability and safety. A recent study of 78 children ages 4 and 5 by van der Cammen-van Zijp et al. concluded that young children should be allowed to hold the handrails to give them "the opportunity to achieve maximal performance successfully" (83).

However, many laboratories allow handrail use. Patients are told they may elect not to use the handrail if they are comfortable in doing so. It is usually readily apparent during the warm-up and early stages of the test if the patient is comfortable exercising on the treadmill. If the patient is going to use the handrail, he or she is instructed to grasp the handrail as lightly as possible. Due to time constraints in many clinical settings, it is not realistic to allow the patient to become fully acclimated to the treadmill in order to exercise hands-free. Our primary concern is patient safety and comfort. In a research setting, the investigators should provide enough time in their protocol to allow the subject to become acclimated to the treadmill and should encourage hands-free exercise. Each laboratory will have to weigh the concerns of patient safety against time constraints in establishing guidelines for handrail use.

Table 3.1   Bruce Treadmill Protocol

	Stage duration (min)	Cumulative time (min)	Speed (mph)	Grade (%)
Stage A (0.0)	3	-	1.7	0
Stage B (0.5)	3	-	1.7	5
Stage 1	3	3	1.7	10
Stage 2	3	6	2.5	12
Stage 3	3	9	3.4	14
Stage 4	3	12	4.2	16
Stage 5	3	15	5.0	18
Stage 6	3	18	5.5	20
Stage 7	3	21	5.5	22

Stage A and stage B are also known as Bruce stage 0.0 and 0.5 respectively. Cumulative times do not include stages A and B.

Balke Treadmill Protocol

The Balke protocol was first published in 1959 (7). Balke protocols are performed at constant walking speed with changes in slope every 1 to 3 min. Most commonly the stages are 2 min in duration. The protocol is useful in evaluating patients who are obese, unfit, very young, or chronically ill (63, 73, 83). Variations of the Balke protocol using walking speeds of 3.0 mph to 3.5 mph (4.8 to 5.6 km/h) and slope increases of 2% or 2.5% per stage have been used in studies of children (68, 77). The primary disadvantage of the Balke protocol is that it may progress too slowly for fit children, leading to unduly long test durations (often 18-20 min or more). This has led to the development of protocols with increased initial slopes of 6% to 10% and faster treadmill speeds. Investigators have used running speeds (5 mph [8 km/h]) in an attempt to keep treadmill slope lower and test duration shorter (4, 71). A running Balke protocol has been shown to be very suitable for the measurement of peak oxygen uptake in pre- and postadolescents (3). If changes to slope and speed are made to fit the patient's age and fitness level, normative data should be developed by the laboratory staff (84). Two common Balke protocols are shown in tables 3.2 and 3.3. Tables 3.4 and 3.5 show common multistage and ramping treadmill protocols, respectively, which have been used to evaluate children.

Cycle Ergometer Protocols

Two types of cycle ergometers are used in exercise testing:

1. mechanically braked, which apply external resistance to the flywheel through frictional bands, and
2. electronically braked, which increase resistance electromagnetically.

Mechanically braked ergometers require the patient to maintain a very precise pedaling cadence, normally 50 or 60 rpm, whereas electroni-

Table 3.2 Balke Treadmill Protocol

	Stage duration (min)	Cumulative time (min)	Speed (mph)	Grade (%)
Stage 1	2	2	3.5	2
Stage 2	2	4	3.5	4
Stage 3	2	6	3.5	6
Stage 4	2	8	3.5	8
Stage 5	2	10	3.5	10
Stage 6	2	12	3.5	12
Stage 7	2	14	3.5	14
Stage 8	2	16	3.5	16
Stage 9	2	18	3.5	18
Stage 10	2	20	3.5	20
Stage 11	2	22	3.5	22
Stage 12	2	24	3.5	24
Stage 13	2	26	3.5	26

Table 3.3 Modified Balke Protocol

	Stage duration (min)	Cumulative time (min)	Speed (mph)	Grade (%)
Stage 1	2	2	3.0	0
Stage 2	2	4	3.0	2.5
Stage 3	2	6	3.0	5
Stage 4	2	8	3.0	7.5
Stage 5	2	10	3.0	10
Stage 6	2	12	3.0	12.5
Stage 7	2	14	3.0	15
Stage 8	2	16	3.0	17.5
Stage 9	2	18	3.0	20
Stage 10	2	20	3.0	22.5

Table 3.4 Other Common Multistage Treadmill Protocols

Protocol TIME	Modified Bruce I 2 MIN	Modified Bruce II 3 MIN	Modified Bruce III 3 MIN	Cornell 2 MIN	Ellestad STAGE 1, 3 MIN; STAGES 2–7, 2 MIN	Naughton (Three speed choices) 2 MIN			Mod Naughton 2 MIN	Kattus 3 MIN
0.0	1.7 mph, 10%	1.7 mph, 0%	1.2 mph, 0%	1.7 mph, 0%	1.7 mph, 10%	2.0 mph, 0%	3.0 mph, 0%	3.4 mph, 2%	1.0 mph, 0%	2.0 mph, 10%
0.5										
1										
1.5										
2	2.5 mph, 12%			1.7 mph, 5%		2.0 mph, 3.5%	3.0 mph, 2.5%	3.4 mph, 4%	1.5 mph, 0%	
2.5										
3		1.7 mph, 5%	1.2 mph, 3%		3.0 mph, 10%					3.0 mph, 10%
3.5										
4	3.4 mph, 14%					2.0 mph, 7%	3.0 mph, 5%	3.4 mph, 6%	2.0 mph, 3.5%	
4.5				1.7 mph, 10%						
5					4.0 mph, 10%					
5.5										
6	4.2 mph, 16%	1.7 mph, 10%	1.7 mph, 6%			2.0 mph, 10.5%	3.0 mph, 7.5%	3.4 mph, 8%	2.0 mph, 7%	4.0 mph, 10%
6.5				2.1 mph, 11%						
7					5.0 mph, 10%					
7.5										
8	5.0 mph, 18%					2.0 mph, 14%	3.0 mph, 10%	3.4 mph, 10%	2.0 mph, 10.5%	
8.5				2.5 mph, 12%						
9		2.5 mph, 12%			5.0 mph, 15%					4.0 mph, 14%
9.5										
10	5.5 mph, 20%					2.0 mph, 17.5%	3.0 mph, 12.5%	3.4 mph, 12%	3.0 mph, 4.3%	
10.5				3.0 mph, 13%						
11					6.0 mph, 15%					
11.5										
12		3.4 mph, 14%				2.0 mph, 21%	3.0 mph, 15%	3.4 mph, 14%	3.0 mph, 7.5%	4.0 mph, 18%
12.5				3.4 mph, 14%						
13					7.0 mph, 15%					
13.5										
14							3.0 mph, 17.5%	3.4 mph, 16%	3.0 mph, 10%	
14.5				3.8 mph, 15%						
15										4.0 mph, 22%
15.5										
16							3.0 mph, 20%	3.4 mph, 18%	3.0 mph, 12.5%	
16.5				4.2 mph, 16%						
17										
17.5										
18							3.0 mph, 22.5%	3.4 mph, 20%	3.0 mph, 15%	
18.5				4.6 mph, 17%						
19										
19.5										
20										

Protocols are presented as speed in miles per hour (mph) and percent treadmill grade per stage. Protocol names presented with length of stages in minutes.

Table 3.5 Common Ramp Treadmill Protocols

Protocol TIME	Bruce ramp 1 MIN	Balke ramp 1 MIN	Fitkids 1.5 MIN	German Society of Paediatric Cardiology protocol 1.5 MIN	CMH max STAGE 1, 3 MIN; STAGES 2–14, 1 MIN	Ramp protocol 30 S
0.0	1.0 mph, 0%	3.3 mph, 1%	2.17 mph, 0%	1.24 mph, 0%	3.0 mph, 0%	0.5 mph, 0%
0.5						1.0 mph, 0%
1	1.3 mph, 5%	3.3 mph, 2%				1.5 mph, 0%
1.5			2.17 mph, 1%	1.55 mph, 0%		2.0 mph, 0%
2	1.7 mph, 10%	3.3 mph, 3%				2.5 mph, 0%
2.5						3.0 mph, 0%
3	2.1 mph, 10%	3.3 mph, 4%	2.48 mph, 3%	1.86 mph, 3%	4.0 mph, 0%	3.0 mph, 1%
3.5						3.0 mph, 2%
4	2.3 mph, 11%	3.3 mph, 5%			4.0 mph, 2.5%	3.0 mph, 3%
4.5			2.79 mph, 5%	2.17 mph, 6%		3.0 mph, 4%
5	2.5 mph, 12%	3.3 mph, 6%			4.0 mph, 5%	3.0 mph, 5%
5.5						3.0 mph, 6%
6	2.8 mph, 12%	3.3 mph, 7%	3.10 mph, 7%	2.48 mph, 9%	5.0 mph, 5%	3.0 mph, 7%
6.5						3.0 mph, 8%
7	3.1 mph, 13%	3.3 mph, 8%			5.5 mph, 5%	3.0 mph, 9%
7.5			3.41 mph, 9%	2.79 mph, 12%		3.0 mph, 10%
8	3.4 mph, 14%	3.3 mph, 9%			6.5 mph, 5%	3.0 mph, 11%
8.5						3.0 mph, 12%
9	3.8 mph, 14%	3.3 mph, 10%	3.72 mph, 11%	3.11 mph, 15%	7.0 mph, 5%	3.0 mph, 13%
9.5						3.0 mph, 14%
10	4.1 mph, 15%	3.3 mph, 11%			7.0 mph, 7.5%	3.0 mph, 15%
10.5			4.04 mph, 13%	3.41 mph, 18%		3.0 mph, 16%
11	4.2 mph, 16%	3.3 mph, 12%			7.5 mph, 7.5%	3.0 mph, 17%
11.5						3.0 mph, 18%
12	4.5 mph, 16%	3.3 mph, 13%	4.35 mph, 15%	3.72 mph, 21%	8.0 mph, 7.5%	3.0 mph, 19%
12.5						3.0 mph, 20%
13	4.8 mph, 17%	3.3 mph, 14%			8.5 mph, 7.5%	3.0 mph, 21%
13.5			4.66 mph, 15%	4.03 mph, 21%		3.0 mph, 22%
14	5.0 mph, 18%	3.3 mph, 15%			8.5 mph, 10%	3.0 mph, 23%
14.5						3.0 mph, 24%
15	5.3 mph, 18%	3.3 mph, 16%			9.0 mph, 10%	3.0 mph, 25%
15.5						
16	5.6 mph, 19%	3.3 mph, 17%				
16.5						
17	5.8 mph, 20%	3.3 mph, 18%				
17.5						
18		3.3 mph, 19%				
18.5						
19		3.3 mph, 20%				
19.5						
20						

Protocols are presented as speed in miles per hour (mph) and percent treadmill grade per stage. Protocol names presented with length of stages in minutes.

cally braked ergometers allow for a much wider range of cadences by the patient while maintaining the correct workload. While treadmills are easily calibrated, electronically braked cycle ergometers require specialized equipment to calibrate. Although mechanically braked cycle ergometers are not difficult to calibrate, it is often required. Cycle ergometry is highly dependent on patient motivation and comfort. Early test termination may occur because of leg fatigue in young and poorly conditioned patients.

Cycle ergometry offers several distinct advantages over treadmill testing. Cycle ergometers are safer, less intimidating to pediatric patients, have a lower initial cost, and require less physical space, which is helpful in clinical laboratories with space limitations. Cycle ergometer protocols may be more appropriate than treadmills in the testing of patients with neuromuscular disorders (8). Cycle ergometer exercise is less dynamic than treadmill exercise, which allows for less motion artifact on the ECG tracings. Blood pressure measurement at peak exercise may be obtained more accurately compared to the treadmill. Electronically braked cycle ergometers make determining physical working capacity straightforward, while differences in walking and running economy make the assessment of physical working capacity quite difficult on the treadmill (72).

As previously stated, children younger than age 6 usually require a pediatric-sized cycle ergometer or modifications to a standard cycle ergometer. These modifications typically include alterations in the pedal arm length, seat size, seat height, and handlebar angle (84). Young children may have problems maintaining the proper pedaling cadence on a mechanically braked ergometer even when the cycle ergometer is appropriately adjusted to their size (63). Paridon et al. (63) have suggested work rate increments during cycle ergometry of 20 to 25 W/min for fit adolescents and 10 W/min for young and unfit patients. Commonly a load increase of 0.25 watts per kilogram per min is employed for ramp protocols. The following section will describe three of the most common cycle ergometer protocols in youth.

McMaster Cycle Ergometer Protocol

The McMaster protocol divides patients into five categories based on gender and height (9). Each test stage is 2 min long. The patient is given strong verbal encouragement to exercise until reaching volitional exhaustion. The pedaling cadence is normally set at 50 rpm, but some laboratories use a pedaling cadence of 60 rpm. The primary outcome measures are peak mechanical power and peak oxygen uptake. If the test is terminated before the final stage is completed, peak power is prorated based on the time completed in the final stage (9). Children with significant heart, lung, or muscular disease may require reductions in the initial workload and stage increments (9). Total test time is 8 to 12 min for most patients. The specifics of the McMaster protocol are shown in table 3.6.

James Cycle Ergometer Protocol

The James protocol divides patients into three categories based on body surface area. Each stage is 3 min in duration with pedaling of 60 or 70 rpm (9, 63). After completion of the first three minutes, work is increased by 16.5, 33, or 49.5 W/min (100, 200, 300 kpm/min) until maximal criteria are achieved or the patient cannot maintain the proper pedaling cadence. Normative data have been published by James and coworkers (42) and Washington and colleagues (85). The initial workload and stage increments are provided in table 3.7.

Godfrey Cycle Ergometer Protocol

The Godfrey protocol was the first 1-min protocol to be used systematically with pediatric patients (72). The protocol places patients into three different categories based on height. Workload is increased at 1 min intervals until exhaustion is reached. The cycling cadence is 60 rpm. Peak workload is compared to normal values for size and gender that have been previously published by Godfrey (31). Due to the short stage duration this test is not recommended for evaluation of submaximal physiological variables (9). Table 3.8 provides details for conducting the test.

McMaster Arm Ergometry Test

To perform the McMaster arm ergometry test (9) the patient should be seated comfortably in a wheelchair or standard chair with the axle of the ergometer at shoulder height. The distance from the chair to the ergometer should be adjusted so that the arm is fully extended when the ergometer crank arm is farthest from the patient. The patient should maintain an upright posture and be able to fully extend the arms without bending forward at the waist. For wheelchair confined patients the back may need to be supported by a pillow (8). Some patients may require additional support or strapping to minimize trunk motion. The recommended cranking rate is 50 rpm, but a slower cranking rate may be required for children with

Table 3.6   McMaster Cycle Ergometer Protocol

Time (min)	Patient height ≤119.9 cm (12.5 W/stage)	Patient height 120–139.9 cm (25 W/stage)	Patient height 140–159.9 cm (50 W/stage)	Patient height ≥160 cm (male) (50 W/stage)	Patient height ≥160 cm (female) (25 W/stage)
0 1	Initial workload 12.5 W	Initial workload 12.5 W	Initial workload 25 W	Initial workload 25 W	Initial workload 25 W
2 3	25 W	37.5 W	50 W	75 W	50 W
4 5	37.5 W	62.5 W	75 W	125 W	75 W
6 7	50 W	87.5 W	100 W	175 W	100 W
8 9	62.5 W	112.5 W	125 W	225 W	125 W
10 11	75 W	137.5 W	150 W	275 W	150 W
12 13	87.5 W	162.5 W	175 W	325 W	175 W
14 15	100 W	187.5 W	200 W	375 W	200 W
16 17	112.5 W	212.5 W	225 W	425 W	225 W
18 19	125 W	237.5 W	250 W	475 W	250 W
20 21	137.5 W	262.5 W	275 W	525 W	275 W

W = watts. W/stage = wattage increase per stage. Each stage is 2 min long.

Based on Bar-Or and Rowland 2004.

muscular or neurologic disorders. Peak oxygen consumption measured during arm ergometry is typically 20% to 30% lower than values achieved during treadmill testing (9, 64). For this type of exercise, an arm-specific ergometer is recommended on a sturdy table that is stable enough not to shift during high-intensity exercise. If an arm ergometer is not available, it is also permissible to modify a bicycle ergometer by securing it to a stout table and replacing the pedals with appropriate handles and crank arms. Details of the McMaster Arm Ergometry test are provided in table 3.9.

Multistage Versus Ramp Protocols

The use of ramping protocols for exercise testing has been increasing in popularity since the 1990s (57). However, despite recommendations from leading national organizations (American College of Sports Medicine, American Heart Association, American College of Cardiology) and many investigators for new approaches to exercise testing, most exercise laboratories continue to select the mode of exercise and the protocol based on tradition, convenience, or familiarity (13, 19, 57, 79). In the early 1980s, electronically braked cycle ergometers began to replace the mechanically braked ergometers in the testing laboratory. This allowed for the development of cycle ergometer ramped protocols that increase work in a continuous, constant manner. Whipp and coworkers first reported on the use of a ramped cycle ergometer protocol for adults in 1981 (88). A study by Cooper et al. (23) was the first to report on the use of a ramp protocol in children.

Technological advances since the 1990s have produced software that produces exercise protocols seamlessly and permits the use of treadmills as well as cycle ergometers to perform ramp tests (57). Myers et al. (59) first reported ramp testing on a treadmill in 1991. There are two types of ramp protocols: standardized (e.g., Bruce ramp, Balke ramp) and individualized (43). These types

Table 3.7   **James Cycle Ergometer Protocol**

Time (min)	BSA <0.99 (16.5 W/stage)	BSA 1.0–1.19 (33 W/stage)	BSA >1.2 (49.5 W/stage)
0 1 2	**INITIAL WORKLOAD 33 W**		
3 4 5	49.5 W	66 W	82.5 W
6 7 8	66 W	99 W	132 W
9 10 11	82.5 W	132 W	181.5 W
12 13 14	99 W	165 W	231 W
15 16 17	115.5 W	198 W	280.5 W
18 19 20	132 W	231 W	330 W
21 22 23	148.5 W	264 W	379.5 W

BSA = body surface area (m^2). W = watts. W/stage = wattage increase per stage. Each stage is 3 min long.

Based on Bar-Or and Rowland 2004; Kaplan et al. 1980; Paridon et al. 2006; Washington et al. 1988.

of protocols are very efficient in providing exercise responses in a short amount of time, thus enabling the easy acquisition of diagnostic data within 10 to 12 min (23, 29, 63). Ramp protocols offer several potential advantages over traditional cycle ergometer and treadmill protocols. They provide uniform increases in hemodynamic and physiologic responses and eliminate the large and unequal workload increases commonly seen in multistage protocols. Furthermore, it is easier to accurately determine the ventilatory anaerobic threshold (VAT) with a ramp protocol (58, 59).

In an individualized ramp test, the slope of the ramp should be based on the child's body size and fitness level and designed to produce a maximal effort within 10 min (19, 63). For the test to be adequately individualized, prior knowledge of the patient's exercise capacity would be required (57). Unfortunately, this information is often unavailable in a clinical setting. While an individualized ramp protocol may be preferred, standardized ramp

protocols may have many of the same advantages (20, 26).

Cabrera et al. studied 46 healthy boys and girls using a standard Bruce protocol and a Bruce ramp protocol (20). The authors reported no significant difference between the two protocols in peak heart rate or peak oxygen uptake or in metabolic, hemodynamic, or ventilatory variables at peak exercise. The test duration on the standard Bruce protocol was 1.2 min longer than the ramped protocol. The authors suggested that a ramped Bruce protocol may be preferable to the standard Bruce protocol for pediatric exercise testing because it features smaller stage changes in speed and grade and reduces the changes in oxygen demand between stages (20). DiBella et al., using a protocol identical to the protocol used in the Cabrera study, found similar results in healthy children and adolescents (26). The investigators concluded that similar peak heart rates and $\dot{V}O_2$ values can be acquired in normal children and adolescents using either a standard or ramped Bruce protocol. DiBella and

Table 3.8   Godfrey Cycle Ergometer Protocol

Time (min)	Patient height ≤119.9 cm (10 W/stage)	Patient height 120–149.9 cm (15 W/stage)	Patient height ≥150 cm (20 W/stage)
1	Initial workload 10 W	Initial workload 15 W	Initial workload 20 W
2	20 W	30 W	40 W
3	30 W	45 W	60 W
4	40 W	60 W	80 W
5	50 W	75 W	100 W
6	60 W	90 W	120 W
7	70 W	105 W	140 W
8	80 W	120 W	160 W
9	90 W	135 W	180 W
10	100 W	150 W	200 W
11	110 W	165 W	220 W
12	120 W	180 W	240 W
13	130 W	195 W	260 W
14	140 W	210 W	280 W
15	150 W	225 W	300 W
16	160 W	240 W	320 W
17	170 W	255 W	340 W
18	180 W	270 W	360 W
19	190 W	285 W	380 W
20	200 W	300 W	400 W

W = watts. W/stage = wattage increase per stage. Each stage is 1 min long.

Based on Bar-Or and Rowland 2004; Godfrey 1974; Rowland 1993.

coworkers were in agreement with Cabrera et al. that the peak heart rate and peak oxygen uptake data from the Bruce ramping protocol did not support the need to develop new normative data (20). Thus, laboratories wishing to transition to the ramped protocol could do so rather easily. Paridon et al. (63) have suggested that ramped work rate increments of 20 to 25 W/min in fit adolescents and 10 W/min for young children and unfit patients are appropriate. With the increasing prevalence of childhood obesity, ramping protocols specific for this population continue to be developed. For example, the Fitkids protocol (45) and the German Society of Paediatric Cardiology protocol (27) are 90 s per stage treadmill test protocols developed to measure fitness in youngsters with impaired functional capacity. The Fitkids protocol was designed to be performed in an outpatient physical therapy setting on nonclinical treadmills; it provides clinicians with $\dot{V}O_{2peak}$ prediction equations based on patient sex, body mass, and test time. The protocol starts with a 0% grade, reaching a maximal grade of 15%, avoiding premature test termination due to lower leg fatigue (45). The German Society of Paediatric Cardiology protocol (27) was designed as a transferable protocol, appropriate for both children and adults, which would allow clinicians to track functional capacity in a single patient from childhood through adulthood.

With these advantages, ramp exercise protocols can enhance diagnostic performance, provide a more accurate estimate of exercise capacity and VAT, and improve the utility of exercise tests for predicting prognosis in a clinical setting (57, 63). However, ramping exercise does not allow for the determination of steady-state submaximal data and may not be appropriate when it is most important to determine corrected QTc intervals. Table 3.10 summarizes possible uses of a ramp protocol in a clinic setting.

Table 3.9 McMaster Arm Ergometry Protocol

Time (min)	Patient height ≤119.9 cm (8 W/stage)	Patient height 120–139.9 cm (16.5 W/stage)	Patient height 140–159.9 cm (16.5 W/stage)	Patient height ≥160 cm (male) (33 W/stage)	Patient height ≥160 cm (female) (16.5 W/stage)
0 1	Initial workload 8 W	Initial workload 8 W	Initial workload 16.5 W	Initial workload 16.5 W	Initial workload 16.5 W
2 3	16 W	24.5 W	33 W	49.5 W	33 W
4 5	24 W	41 W	49.5 W	82.5 W	49.5 W
6 7	32 W	57.5 W	66 W	115.5 W	66 W
8 9	40 W	74 W	82.5 W	148.5 W	82.5 W
10 11	48 W	90.5 W	99 W	181.5 W	99 W
12 13	56 W	107 W	115.5 W	214.5 W	115.5 W
14 15	64 W	123.5 W	132 W	247.5 W	132 W
16 17	72 W	140 W	148.5 W	280.5 W	148.5 W
18 19	80 W	156.5 W	165 W	313.5 W	165 W
20 21	88 W	173 W	181.5 W	346.5 W	181.5 W

W = watts. W/stage = wattage increase per stage. Each stage is 2 min long.

Based on Bar-Or 1983; Bar-Or and Rowland 2004; Pescatello 2014.

Table 3.10 Use of Ramp Versus Multistage Exercise Protocols

Reason for test	Recommended exercise protocol type
Ischemia	Ramp or multistage
Ectopy or dysrhythmia	Ramp
QTc evaluation	Multistage
Chest pain	Ramp
Asthma	Ramp
Fitness	Ramp or multistage
Hemodynamic analysis	Ramp or multistage

Six-Minute Walk Test

The six-minute walk test (6-MWT) is a self-paced test in which the patient is encouraged to cover as much distance as possible during the allotted time. Typically the test is performed over a 30 to 50 m course that is marked in 1-yd increments. No running is allowed, and the patient may stop to rest if needed, resuming walking as soon as possible. Patients who require supplemental oxygen should be allowed to perform the test on oxygen. The 6-MWT is used to measure exercise tolerance in children who are too physiologically compromised to be evaluated by traditional testing protocols. This test is used to follow disease progression and to assess the efficacy of surgical or pharmacologic interventions. The protocol has been used to assess pediatric heart and heart-lung transplant patients pre- and postoperatively. Numerous authors have provided reference data for children and adolescents (2, 10, 30, 44, 48, 50, 65). Care must be taken when selecting reference values for use because it has been shown that a patient's height, weight, and age influence the 6 min walking distance (2, 10, 30, 48, 50, 51, 65). The validity and test-retest reliability of the 6-MWT has been shown to be strong across a wide variety of childhood diseases and neuromuscular disorders as well as in healthy children (2, 6, 25, 36, 39, 49, 50, 53, 55, 56, 60, 62, 80, 82). A recent study by Klepper and Muir (44) reported a test-retest reliability value of 0.93 using American Thoracic Society guidelines to test healthy, elementary-aged children in the United States. These guidelines also provide recommendations for test execution and standard verbal encouragement to decrease inter- and intra-tester variability (2).

The main advantages of the 6-MWT are that it does not require special equipment, it mimics the activities of daily living (including allowing the use of supplemental oxygen during the test for patients who require it), and it is relatively easy to perform (67). Patient motivation and body habitus may significantly affect performance on the test. Often these tests have been performed without any form of monitoring. However, it is recommended that heart rate and oxygen saturation be monitored by a portable oximeter before, during, and after the test (63). Because of oximeters' sensitivity to motion, however, it is important to avoid movement-induced errors in heart rate and oxygen saturation measurements. Some researchers and clinicians have questioned the prudence of having patients with cardiac impairment exercise without continuous ECG monitoring in public hospital hallways (67).

In adult patients with significant cardiac impairment, the 6-MWT has been found to correlate relatively well with peak oxygen uptake (33, 61). However, in patients with mild or moderate impairment, the 6-MWT validity, reliability, and utility are questionable, especially in those patients who can walk more than 400 m during the test (33, 61). Some studies have found the 6-MWT to correlate reasonably well with peak oxygen uptake in children with severe cardiopulmonary disease and in those with moderate lung disease (35, 36, 60). Nixon and her coworkers found the 6-MWT to be a useful alternative to traditional treadmill or cycle ergometer testing in the assessment of oxy-hemoglobin desaturation and exercise tolerance in young patients dependent on supplemental oxygen awaiting heart-lung or lung transplantation (60). The 17 patients in the study ranged from 9 to 19 years of age. Each patient completed a 6-MWT and a cycle ergometer test using the Godfrey protocol (31). The test data showed a statistically significant correlation between the 6-MWT and peak oxygen uptake and physical work capacity. There was also a significant correlation between the minimum oxyhemoglobin values obtained from the 6-MWT and from cycle ergometer testing. The minimum oxyhemoglobin saturation levels measured on the walk test were slightly lower than on the cycle ergometer test (84% vs. 86%), leading the authors to conclude that "SaO_2 measured during progressive exercise testing on a cycle ergometer may not reflect the degree of oxyhemoglobin desaturation during self-paced walking in some patients with severe cardiopulmonary disease" (60).

Maximal Test Criteria

The ability to determine if a child has given a maximal effort during an exercise test is vital. Exercise test data provide important information to the physician about future medical management of the child. Decisions about the need for surgery, additional tests to be done, or possible pharmacologic intervention are often contingent on the outcome of the exercise test. Thus it is imperative to use established criteria to determine peak oxygen uptake so that results can be compared between laboratories and between patients (73).

The determination of a maximal effort can be quite difficult. To date no single criterion has proven to show a maximal effort. Multiple criteria have been used extensively to determine maximal exercise effort but all have limitations in varying degrees. Investigators have used peak heart rate, peak respiratory exchange ratio (RER), plateau in oxygen uptake, a rating of perceived exertion greater than 17 (RPE 6-20 scale) or greater than 9 (OMNI RPE 0-10 scale), and subjective appearance of the patient (38, 72, 79). The use of these criteria has sometimes been problematic in determining maximal effort in children. Heart rate prediction formulas using age are not accurate in children because peak heart rate during exercise is relatively constant across the pediatric age range (73). Peak heart rate values show significant individual variation with standard deviations of 5 to 10 bpm often observed (24, 37, 75). The mode of exercise influences the peak heart rate achieved. Cumming et al. (24) studied nine different exercise protocols and reported peak heart rates of 197 ± 7 bpm and 195 ± 5 bpm during cycle ergometry and 204 ± 5 bpm to 198 ± 5 bpm during a running and walking treadmill protocol respectively. Therefore using a criterion of 200 bpm as an indicator of maximal exercise effort may lead to an inappropriate number of tests being considered submaximal.

The use of RER as a criterion of maximal exercise effort is not without its own set of limitations. The RER is the ratio of carbon dioxide to oxygen uptake with resting values of 0.70 to 0.85, depending on the timing or type of food consumed or the degree of insulin resistance. RER increases as exercise increases, but there is significant variation between subjects. Peak RER values also depend on the mode of exercise, with values greater than 1.0 often observed during cycle ergometry, while peak treadmill values are often less than 1.0 (0.90 to 1.05).

A plateau in oxygen uptake is considered by some to be the gold standard for a true maximal effort during exercise testing and achievement of $\dot{V}O_{2max}$. To date there is no agreement on a single definition of oxygen uptake plateau in children and adolescents. Multiple definitions have been used over the years (5, 46, 47, 70, 73, 75):

1. Increase in $\dot{V}O_2 \leq 2.1$ ml $\cdot$ kg^{-1} $\cdot$ min^{-1} with an increase in workload

2. Increase in $\dot{V}O_2$ in the last minute of <5% or <150 ml

3. Increase in $\dot{V}O_2$ less than two standard deviations below the average change in submaximal stages

Approximately 50% of pediatric subjects will achieve a plateau during exercise testing. The percentage of subjects attaining a plateau criterion is lower in active, trained, and untrained children and adolescents than in adults, although there is considerable variation, with percentages of plateau attainment in young subjects of 8% to 95% being reported (4, 5, 23, 46, 47, 69, 75). Thus the achievement of a plateau is an excellent determinant of $\dot{V}O_{2max}$ when it is identified; the downside is that it occurs less often than we would wish in children.

A study by Rivera-Brown et al. (70) demonstrated the limitations of using traditional criteria in the assessment of $\dot{V}O_{2max}$ in adolescents. The authors studied the frequency with which $\dot{V}O_{2max}$

criteria were met during treadmill running exercise to volitional exhaustion. Thirteen male subjects underwent treadmill evaluation on three different running protocols. The criteria for achievement of maximal oxygen uptake were RER ≥ 1.0, heart rate $\geq 95\%$ of predicted maximal heart rate (220 – age), and an increase in oxygen uptake ≤ 2.1 ml $\cdot$ kg^{-1} $\cdot$ min^{-1}. Achievement of the plateau criteria for the three protocols occurred in 54%, 39%, and 85% of the subjects with the highest percentage associated with an intermittent protocol. Other studies have also shown intermittent treadmill running protocols to produce the highest attainment of a $\dot{V}O_2$ plateau (46, 47, 54, 75). Oxygen uptake (66.4-71.4 ml $\cdot$ kg^{-1} $\cdot$ min^{-1}) and RER (1.1-1.2) were similar on all three tests. Essentially all subjects met the heart rate and RER criteria for maximal effort. The results of this study and of Myers et al. (59) suggest that achievement of plateau criteria is protocol dependent, but protocol variation (for a given mode of exercise) does not affect RER or peak heart rate.

Due to the inconsistency with which young patients attain a $\dot{V}O_2$ plateau, it has been recommended that this criterion not be used to define a maximal test in children (69, 70, 73). As a result of the limitations just described, it has been suggested that subjective signs and symptoms such as ataxia, blanching of the skin around the neck and shoulders, and widening of the pupils be used in conjunction with objective criteria to identify maximal patient effort (9, 69).

Scope of Pediatric Exercise Testing

There is considerable variability in exercise testing even among pediatric testing centers. This is well illustrated by a relatively recent study by Chang and colleagues (21). They conducted a survey of 200 pediatric cardiology and pediatric pulmonology programs at children's hospitals or university hospitals as well as private practice settings in the United States, with a total of 115 programs (58%) responding to the survey. Their survey revealed that 99% of cardiology programs have exercise laboratories compared to 76% of pulmonology programs. Sixty-seven percent of respondents reported collecting oxygen consumption as part of their exercise test procedures. Sixty-three percent of the programs used both cycle ergometer and treadmill exercise modalities to evaluate their patients, with 18% using the cycle ergometer more often and 45% using the treadmill more often. Thirty-one percent of respondents reported using the treadmill exclusively. Seventy-nine percent of programs using treadmills reported using the Bruce or modified Bruce protocols, with 14% stating that they used protocols developed within their own institutions.

There is wide variation in minimal age requirements for exercise testing. Nine percent of the respondents stated they would test patients 4 years of age or younger, 25% reported their minimal age for testing to be 5 years, 31% reported age 6, 16% said age 7, and 20% said their patients had to be at least 8 before exercise testing would be performed (21). Other interesting findings from the Chang study (21) revealed that 76% of programs required a physician to be present during testing. Ninety percent of programs used technicians to perform the exercise tests, while 8% reported using nurses. Fifty-nine percent of the participating programs had a dedicated pediatric testing laboratory, and 49% shared laboratory facilities with another department.

Chang and colleagues (21) also reported significant variation in case volume among the responding programs with 26% performing less than 50 tests per year, 32% performing 50 to 99 tests per year, 20% performing 100 to 199 tests per year, 8% performing 200 to 299 tests, and 15% performing more than 300 tests per year. Fifty-eight percent of the surveyed programs performed fewer than 100 exercise tests per year. Stress echocardiography was performed in 76% of the pediatric cardiology programs, with most echocardiograms performed after the patient had completed his or her exercise test and while lying supine. Lastly, the survey results also revealed a positive relationship between the measurement of metabolic parameters (oxygen consumption) with the use of cycle ergometer, as well as the laboratories' annual case volume (21).

Conclusion

Since the 1970s, exercise testing has played an ever increasing role in pediatric clinical practice. Many exercise test protocols have been developed to aid in the cardiopulmonary assessment of clinical patients. The two primary types of test ergometers are the motor-driven treadmills and cycle ergometers. Cycle ergometers are subdivided into two categories, mechanically braked and electronically braked. The choice of ergometer depends on the objective of the test, the parameters to be evaluated, the age and health status of the patient, the laboratory space available, and the experience of the laboratory staff in the use of different test modes and protocols.

There are advantages and disadvantages to each mode of exercise. The treadmill provides a familiar form of exercise for young patients. However, treadmills are more costly, noisier, require more laboratory space, and have a greater potential injury risk than cycle ergometers. Treadmill calibration is quite easy, while mechanically braked cycle ergometers require frequent calibration and electronically braked cycle ergometers require special equipment. The cycle ergometer requires less initial expense, is more portable, provides a more stable platform for ECG and blood pressure measurements during exercise, and offers a safer mode of exercise, especially for patients with neuromuscular disorders or other gait issues.

Multistage, intermittent, or ramp protocols may be used with all types of ergometers. Multistage protocols are most commonly used, but ramp protocols appear to be gaining favor. The most commonly used multistage protocols are the Bruce and Balke treadmill protocols and the McMaster and James cycle ergometer protocols. Ramp protocols avoid many of the limitations created by multistage protocols that have large and uneven increases in workload between stages. Ramp protocols also make accurate determination of VAT easier. Intermittent protocols are rather time consuming and thus are not often used in routine clinical exercise testing. Ramp protocols have been shown to be equal or superior to traditional protocols in the exercise evaluation of children.

Exercise responses vary by mode of exercise and type of protocol. Treadmill running protocols typically produce higher oxygen uptake values (by approximately 10%) than cycle ergometer exercise. Treadmill running protocols produce higher values than treadmill walking. Balke treadmill walking protocols are useful in the evaluation of very young or unfit patients. While cycle ergometry is a frequently used form of exercise testing, it is especially helpful when evaluating obese patients or those with gait or balance issues. Peak oxygen consumption and peak heart rate are independent of protocol variation within a given mode of exercise.

Numerous studies have shown the futility of continuing to use $\dot{V}O_2$ plateau as a criterion for the determination of maximal oxygen uptake. The achievement of a plateau in $\dot{V}O_2$ during exercise testing has been shown to be protocol specific. The percentage of pediatric-aged subjects able to achieve a $\dot{V}O_2$ plateau is approximately 50%, but there is significant variation between studies (approximately 30%-90%). The use of a combination of subjective and objective data should be considered in establishing criteria for an exhaustive maximal effort (and thus $\dot{V}O_{2max}$) on a progressive exercise test.

Normal Cardiovascular Responses to Progressive Exercise

Thomas W. Rowland, MD

The sequence of physiological events that unfolds as the cardiovascular system responds to a bout of endurance exercise is truly astounding—not only in the nature of its harmonious complexity but also in the manner that such responses are so finely tuned to the increasing metabolic, thermal, and biochemical demands of contracting muscle. For clinicians performing and interpreting exercise tests, it is critical to appreciate the normal mechanisms of the exercise response as they examine the functional health of this response as defined by the quantitative and qualitative changes in its component parts.

This chapter will review our current understanding of cardiovascular system responses to a bout of progressive exercise in the upright position on a cycle ergometer or treadmill in well-hydrated, athletically nontrained young people in normothermic conditions. We will examine if, and how, such responses might differ in children and adolescents compared to adults. It should be noted that the cardiovascular findings described are limited to the model just described. Adaptations to sustained submaximal exercise (cardiovascular drift), isometric exercise, exercise at altitude, or exercise in trained athletes will not be considered. The acute responses of ventilatory function during progressive exercise will be examined in chapter 11.

Reductionist's Disclaimer

The array of physiological mechanisms controlling blood circulation that must work effectively during a bout of progressive exercise might be best compared to a symphony orchestra. A multitude of components must function independently yet harmoniously (and in tune) to create a well-functioning whole. To generate effective circulatory flow, the rate of sinus node impulse generation must increase, heart muscle must increase its contraction and relaxation, systemic arterioles must dilate in response to local metabolic need, blood pressure must be sustained, and all these components must "speak" to each other—feedback is essential for the system to function. In a symphony hall each instrument produces music independently of the whole, but they must all work, quite literally, "in concert" to create a beautiful and stirring piece of music. Each musician receives feedback not only by listening to the other musicians but also by watching a conductor standing before them waving a baton to establish the pace and dynamics of the system.

There are deeper analogies that are even more expressive of the true complexity of the operation of these two systems. Performance for each, in fact, occurs at multiple levels beneath the obvious one just described. Both are examples of *hierarchical systems*. That is, they both function at multiple levels and the functional components within each level must also work effectively if the entire system—cardiovascular response to exercise or Beethoven's *Seventh Symphony*—is to succeed.

Among the musicians, there is the instrument itself. To make beautiful music a clarinet must function within its own domain as a complex system, not only in the motion of valves or in the nature of its reed but also in the physics that go into making the sound waves reverberate with appropriate timbre. In a parallel fashion, the heart muscle must contract by a complicated system of biochemical and energetic changes that force actin and myosin filaments to slide over each other.

At yet another level, the neuromuscular system of the clarinetist must be functioning at an extraordinary level of skill, encompassing electrical transmissions and biological timing devices. There are other levels of performance as well, ranging from molecular mechanisms of auditory perception in the brain to psychological determinants of motivation and attention. It is fascinating to note that the function at each of these different levels is really quite independent of the others (e.g., the pattern of electrical innervation of the clarinetist's finger muscles has nothing to do with the design and musical output of the instrument he or she is playing). Yet all must function effectively if the multiple layers of this system are to produce exquisite music as an end product. Exercise involves deeper levels of system performance as well—in electrolyte flux, which triggers electrical alterations at the cell membranes of the sinus node, or expressions of catecholamine release from autonomic nerve endings to the heart and arteriolar walls.

It is clear, then, that a breakdown in a component within a given level of the system (e.g., poor myocardial function following ischemic injury) can influence performance (e.g., depressed maximal cardiac output, low $\dot{V}O_{2max}$, and, consequently, limited endurance exercise capacity). It follows that malfunction at *any level* in the hierarchical system will negatively affect performance. If the clarinet player is momentarily distracted, the neuromuscular system is thrown off timing, the note is played prematurely, and the rest just before the cadenza is marred. During exercise, a congenital abnormality in ion channel flux can lead to a predisposition to serious dysrhythmias and sudden death.

Recognizing the complex systems of the cardiovascular response to exercise with its hierarchical structure and feedback mechanisms, physiologists have often cautioned against *reductionist* thinking—focusing on the function of single elements—in interpreting such processes. One cannot assess components of a system in isolation, they warn, since cardiovascular function during exercise or the musical performance of a symphony orchestra are both expressions of *emergent qualities*. This means that each is not just the sum of its parts but rather an entirely new "performance" that arises when all the parts work in harmony. That is, one cannot understand the whole by characterizing its individual parts. The part played by the principal clarinet—or any other instrument in the orchestra, for that matter—is not Beethoven's *Seventh*. Likewise the diffusion of oxygen across the capillary–alveolar membrane in the lung is not the finish time in a 5 km road race.

In this chapter we will examine the importance of interactions during exercise between system components—heart rate and systemic venous return, myocardial contractility and the production of stroke volume—and how total system function relies on the relationships rather than on the function of individual elements. In understanding the function of the healthy cardiovascular system as well as one handicapped by cardiac malfunction, it is necessary to remember that the whole is beyond a simplistic focus on its individual parts.

Historical Context

In 1914, Patterson and Starling first described their "law of the heart," which indicated that the stretching of myocardial fibers when left ventricular end-diastolic volume increases triggers an augmentation of myocardial contractile force. This in turn produces a rise in stroke volume and cardiac output (20). Here, it was assumed, was a physiological explanation for the rise in circulatory flow in response to exercise: The rise in venous blood return to the heart caused an increase in left ventricular filling volume, with a subsequent expansion of stroke volume as myocardial contractility increased. The rise in stroke volume in parallel to that of the heart rate, it was concluded, was then responsible for the enhancement of cardiac output and circulatory flow during exercise.

Not all were in agreement, however. Rushmer and Smith argued that these early research models involved "greatly distorted conditions [which] seriously retarded understanding cardiac responses to exercise" (40). This contention was not unfounded because Starling's experimental heart-lung model avoided the influences of venous resistance, intrathoracic pressure, heart rate, and neurohumoral factors.

By the 1950s it was clear from studies in both animals and humans that the proposed scenario of cardiovascular responses to exercise based on the Starling law was inaccurate. Stroke volume, in fact, was observed to change little during exercise, and left ventricular end-diastolic volume remained relatively stable with increasing exercise intensities. It appeared that the primary factor in the circulatory response to exercise was a fall in peripheral vascular resistance from arteriolar dilatation within the exercising muscle, not the function of the heart itself.

"The primary cause of augmented cardiac output is believed to be the local vasodilatation in the skeletal muscle," wrote Guyton in 1967 (10). "The heart has relatively little effect on the normal regulation of cardiac output." In their review article published in the same year, Bevegard and Shepherd were in agreement: "The heart serves as a force-feed pump designed to discharge whatever volume it receives by increasing its rate or stroke volume. Unless there is dilatation of resistance vessels in some systemic vascular bed, mediated by local, humoral, or nervous mechanisms, an increased rate will not result in an increase in cardiac output" (5).

Subsequent studies have borne this out. Ross et al., for example, performed atrial pacing on 17 patients ages 6 to 41 who demonstrated no evidence of myocardial dysfunction (24). As the paced rate rose to 190 bpm, stroke volume fell to one-half of the original values while cardiac output did not change.

The identification of the fall in peripheral resistance as the primary mover in the circulatory responses to exercise is consistent with a model based on Poiseuille's law. Poiseuille was a French physiologist who determined using small glass tubes that the volume of blood flow in a system is directly proportional to the pressure gradient across the tube, a viscosity coefficient, and the fourth power of the diameter of the tube. Put more simply, this law can be expressed as $Q \sim P/R$, where, during the circulatory responses to exercise, the increase in rate of blood flow (Q) is regulated by the decline in peripheral vascular resistance (R), while the pumping action of the heart is responsible for maintaining a pressure head (P).

Empirical Evidence

Recent advances have permitted the noninvasive study of circulatory responses at high exercise intensities by Doppler and two-dimensional echocardiography, acetylene and carbon dioxide rebreathing techniques, and changes in thoracic bioimpedance. Observations from such studies have generally supported the conclusions from earlier reports that identified a peripheral rather than central cardiovascular control of these responses.

Empirical observations have indicated that the following occur during the course of a progressive bout of upright exhaustive endurance exercise:

1. Total peripheral vascular resistance (TPR) declines by approximately 60%.

2. Stroke volume rises in the initial phase of upright exercise but demonstrates little subsequent change as exercise intensity increases.

3. Left ventricular filling pressure, as indicated by end-diastolic dimension, remains constant or gradually decreases.

4. Inotropic (contractile) function is augmented together with lusitropic function (diastolic relaxation).

Peripheral Vascular Resistance

During the course of a progressive bout of upright exhaustive endurance exercise, TPR declines by approximately 60%. TPR during progressive exercise has been calculated as mean arterial pressure divided by cardiac output. Typically, values demonstrate a greater rate of fall in the early work stages followed by a more gradual decline at high exercise intensities. In a review of eight reports in young subjects, the average value of TPR fell from 18.8 ± 2.0 units at upright rest to 7.1 ± 1.8 units at maximal exercise, a 62% reduction (25). In direct comparisons, both values of TPR and the magnitude of decline with exercise have been independent of sex (18) and athletic training (16).

What mechanism is responsible for this arteriolar vasodilatation within exercising muscle that triggers increased blood flow? The magnitude of circulatory blood flow during endurance exercise is tightly coupled to oxygen uptake, leading Guyton to propose in 1967 that the muscles "autoregulate" blood flow by means of arteriolar dilatation with respect to their oxygen demands (10). While this concept remains popular, pinpointing the specific means by which the contracting cells communicate their metabolic requirements to the mechanisms altering arteriolar tone has remained elusive. A number of agents that can cause arteriolar dilatation are produced locally in the muscle cell in response to increased metabolic activity—acetylcholine, potassium ions, hydrogen ions, inorganic phosphate, carbon dioxide, adenosine—but blocking studies have failed to identify any single principal determinant.

It is possible that mechanisms other than production of local vasodilatory agents may be important in the exercise-induced fall in peripheral resistance. Cell-to-cell conduction of vasodilatation from the precapillary level upstream to the major arterioles could occur. Others have suggested that the red blood cell might act as an oxygen sensor, or that flow-mediated dilatation via nitric oxide or prostaglandins in response to wall stress might play a role (42).

It should be recognized that the exercise-induced decline in TPR reflects the net sum of changes in arteriolar tone occurring among all vascular beds throughout the body. While active vasodilatation occurs in arterioles that supply exercising muscle, diminished blood flow from vasoconstriction occurs in vessels that supply the kidney, mesentery, skin, and inactive muscle. Endurance exercise, then, affects a *redistribution* of blood flow through regional changes in arteriolar tone. As a result, the 20% of cardiac output delivered to skeletal muscle at rest increases to 80% with exercise.

Stroke Volume

During the course of a progressive bout of upright exhaustive endurance exercise, stroke volume rises in the initial phase but demonstrates little subsequent change as exercise intensity increases. In the initial stages of a progressive exercise test performed in the upright position by a healthy subject, stroke volume rises by about 25% above resting value, then changes little (plateaus) at higher work intensities (figure 4.1). This pattern has been consistently observed by numerous measurement techniques, including indirect Fick (CO_2 rebreathing) (3), direct Fick (41), thermodilution

(11), thoracic bioimpedance (6), dye dilution (9), radionuclide angiography (23), and Doppler echocardiography (17). It is observed similarly in males and females, children and adults, fit and unfit youth, and athletically trained children as well (34, 36).

Evidence indicates that the initial increase in stroke volume reflects the mobilization of blood that was sequestered in the legs by gravity when the subject assumed the upright position on the treadmill or cycle ergometer (4). In standing or sitting, blood volume in the dependent lower extremities increases by approximately 500 to 1,000 ml in the adult subject. Central blood volume, cardiac output, and stroke volume are subsequently reduced, the latter falling by 20% to 40%. At the onset of upright exercise, this dependent blood volume is mobilized by both arteriolar vasodilatation and the pumping action of the contracting muscles of the lower limbs (43). As a result, central volume is restored, and cardiac output and stroke volume resume levels approximating those when the subject was supine.

This process was documented experimentally in 10 healthy adolescent males (mean age 15.3 ± 0.5 yr) using Doppler echocardiography (34). As indicated in figure 4.2, when these subjects moved from a supine position to a sitting position on the cycle ergometer, stroke volume fell by 25%. At the onset of exercise, stroke volume rose to approximate that when supine, then remained stable to the point of subject exhaustion.

As would be expected, subjects performing exercise in the supine position typically do not demonstrate any initial rise in stroke volume (4).

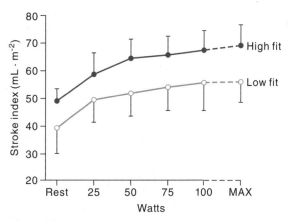

Figure 4.1 Pattern of stroke volume response to progressive cycle exercise in high- and low-fit untrained 12-year-old boys.

Reprinted, by permission, from T. Rowland et al., 1999, "Physiological determinants of maximal aerobic power in healthy 12-year old boys," *Pediatric Exercise Science* 11: 317-326.

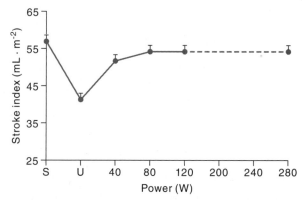

Figure 4.2 Stroke volume index at rest supine (S) and values when assuming the sitting upright position (U), followed by progressive cycle exercise in adolescent males.

Reprinted, by permission, from T. Rowland and V. Unnithan, 2013, "Stroke volume dynamics during progressive exercise in healthy adolescents," *Pediatric Exercise Science* 25: 173-185.

Likewise, a "flat" stroke volume response has usually been observed in any exercise that is not influenced by gravity, including prone simulated swimming (27), astronauts in zero-gravity conditions (1), arm exercise (15), and upright exercise in a swimming pool (7).

These observations imply that after an early rise related to orthostatic changes in central blood volume, stroke volume remains stable in the course of an upright bout of progressive exercise. In this initial stage, stroke volume rises as a manifestation of the Starling law, with augmented cardiac filling volume as blood is mobilized from the legs. After this, no significant change in stroke volume is observed, and increases in cardiac output result solely from a rising heart rate.

Left Ventricular Filling Pressure

During the course of a progressive bout of upright exhaustive endurance exercise, left ventricular filling pressure, as indicated by end-diastolic dimension, remains constant or gradually decreases. Left ventricular end-diastolic dimension can be estimated by two-dimensional directed M-mode echocardiographic measurements during exercise. Using this technique, four studies in children have revealed an identical pattern: a slight increase at the onset of upright exercise (consistent with the initial refilling of the ventricle), followed by a slight gradual decline to the point of maximal exercise (16, 17, 26, 29) (figure 4.3). This pattern is identical in young athletes and untrained youth (16) and mimics that described in adult subjects as well (17, 22). Typically, these changes in end-diastolic volume are sufficiently minor as to escape statistical significance.

The stability of left ventricular diastolic size during the course of a progressive test can be assumed to reflect a constancy of ventricular filling pressure. Thus, the stability of stroke volume production (after the initial refilling rise) is accompanied by an unchanging ventricular preload as work intensity rises.

Inotropic and Lusitropic Function

During the course of a progressive bout of upright exhaustive endurance exercise, inotropic (contractile) function is augmented together with lusitropic function (diastolic relaxation). While left ventricular end-diastolic dimension and stroke volume remain essentially stable, a progressive rise in both myocardial systolic and diastolic performance is observed as work intensity increases.

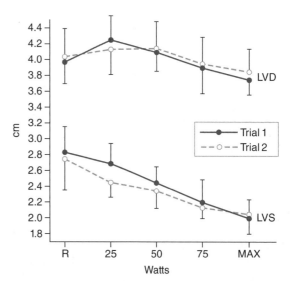

Figure 4.3   Left ventricular systolic and diastolic dimensional changes during maximal upright cycle exercise in 10- to 12-year old boys.

Reprinted, by permission, from T.W. Rowland and J.W. Blum, 2000, "Cardiac dynamics during upright cycle exercise in boys," *American Journal of Human Biology* 12: 749-757.

Augmented myocardial inotropic velocity and force are observed in the course of progressive exercise by many markers of systolic function (33). Aortic flow velocity and systolic ejection rate (stroke volume/ejection time) increase typically by a factor of 1.5 at maximal exercise, and ventricular ejection and shortening fractions rise by 1.3. As previously noted, left ventricular end-diastolic dimension remains stable; therefore, all increases in ejection and shortening fractions occur by a reduction in end-systolic dimension. Longitudinal myocardial shortening velocity, as estimated by tissue Doppler imaging, increases more than twofold from rest to the point of subject exhaustion. As a result, systolic ejection time normally declines from approximately 0.25 s at rest to 0.18 s at peak exercise. At maximal exercise the left ventricle of an adolescent ejects blood at a remarkable rate of 400 ml/s.

The relative contributions of the possible triggers of increases in inotropic function with exercise are difficult to decipher. Many factors are known to augment myocardial contractility, and all are at work in the course of a progressive exercise test. In the initial stages, contractility increases according to the Starling law as the left ventricle expands from blood mobilized from the dependent legs. Later, increases in heart rate, a decline in afterload from diminished peripheral vascular resistance, sympathetic nervous stimulation, and a rise in circulating catecholamines—all recognized independent determinants of myocardial inotropy—contribute to the enhanced contractility that results from exercise.

This intensification of contractile function during progressive exercise must, by necessity, be matched by augmented filling (diastolic function) of the ventricle. Filling of the left ventricle is dictated by the pressure gradient across the mitral valve, reflecting the balance of "upstream" pressure (a function of left atrial size, pressure, and compliance) and "downstream" pressure (created by myocardial relaxation properties, measured by tissue Doppler E' as the velocity of myocardial relaxation, as well as a "suction effect" of ventricular contraction). The transmitral pressure gradient during a progressive exercise test, as estimated by Doppler echocardiographic-determined peak diastolic blood velocity (E wave), increases fourfold (32). Concomitant with this rise in ventricular filling gradient, tissue Doppler E' typically increases by a factor of 2.5, indicating the primary role of a change in "downstream" factors in augmenting ventricular filling during exercise. On the other hand, the ratio of E to E', considered a marker of "upstream" factors, remains unchanged or declines gradually during progressive exercise.

Synthesis

In summary, with increasing work intensities, heart rate and cardiac output rise as arteriolar dilatation in contracting muscle reduces peripheral vascular resistance. Stroke volume (after an initial small "refilling" phase) and left ventricular end-diastolic dimension remain essentially stable, while systolic and diastolic myocardial function are steadily enhanced. Any depiction of normal cardiovascular response to progressive exercise must be consistent with each of these observations.

During the course of progressive exercise, systemic venous return to the heart increases three to four times over that at rest, yet left ventricular filling (end-diastolic) volume per beat remains remarkably constant. This implies that the rise in heart rate must be precisely matched to the volume of systemic venous return. The tachycardia of exercise serves, then, to maintain a constant end-diastolic ventricular size and little change in atrial pressure.

The most likely mechanism for this matching of heart rate with volume of venous return is the Bainbridge reflex, whereby a rise in atrial pressure

triggers sympathetic stimulation to the sinus node (2). This reflex, although historically clouded in controversy, has been clearly documented in both humans and animals. Alternatively, some have suggested that the close association of heart rate with systemic venous return might reflect sympathetic reflexes originating in the contracting muscle (8). By whatever means, this matching acts to "defend" left ventricular end-diastolic dimension, preventing chamber enlargement which, by the law of LaPlace, would result in an increase in wall tension and heart work efficiency (14).

The empirical observations demonstrate an apparent paradox. How can a progressive rise in ventricular contractile force (with diminution of left ventricular systolic dimension) be commensurate with a stable stroke volume? The answer lies in the matter of *time*. The increased myocardial contractility during progressive exercise ejects the same volume of blood (i.e., stroke volume) in a shorter ejection period. Augmented inotropic (and lusitropic) function act, then, to maintain rather than increase stroke volume as work intensity increases and ejection time declines (figure 4.4). This effect of enhanced myocardial function also serves to preserve a sufficient diastolic time period, critical for not only ventricular filling but

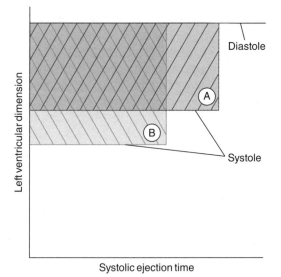

Figure 4.4 As systolic ejection time shortens from low-intensity exercise (stroke volume indicated by area of rectangle A) to higher intensity (rectangle B), the left ventricle empties more completely (systolic dimension decreases and shortening fraction rises in B compared to A). However, the stroke volume does not change (the area of rectangle A equals that of rectangle B).

also myocardial perfusion from coronary artery flow, as the heart rate rises.

In terms of ventricular work, certain features—stroke volume, ventricular diastolic size, and preload—are no different from those in the resting state. Exercise augments ventricular work in its demands for increased systolic and diastolic function as well as frequency of contraction, these factors contributing to the rise in myocardial oxygen uptake during exercise.

This picture of the circulatory responses to endurance exercise is that of a series of interactions between functional cardiovascular elements. The data are consistent with a model of arteriolar vasodilatation in contracting muscle, responding to cellular metabolic needs, as the principal controller of the rate of circulatory flow with exercise. Heart rate must rise to (a) match systemic venous return, resulting in a stable ventricular diastolic size and filling volume, and (b) increase cardiac output to sustain blood pressure. At the same time, myocardial contractility must increase to both maintain stroke volume and accommodate sufficient diastolic time for coronary perfusion. The circulatory response to progressive exercise, then, can only be understood as a complex series of critically important interactions that define the rise in blood flow to satisfy the metabolic needs of contracting muscle.

All available evidence indicates that this basic scenario of the physiological events surrounding the normal circulatory adjustments to exercise is identical in children and young adults. The former exhibit certain quantitative differences, which will be outlined in the chapters that follow, but the underlying patterns of cardiovascular response to progressive exercise are similar.

Physiological Basis of Cardiovascular Fitness

Cardiovascular fitness is interpreted in various ways by both the general public and the scientific community. Often the term is used to refer to one's capacity to perform endurance exercise, either as time per distance (1-mi run, 5-km road race) or time duration (treadmill endurance time). We assume that such performance is linked to the capacity of the heart and lungs to supply enough oxygen to satisfy the aerobic metabolic demands typical of endurance events.

Right Ventricular Responses

In assessing the cardiac responses to exercise, attention has traditionally focused on the left ventricle as the pump (a) responsible for systemic perfusion and (b) most accessible to standard diagnostic techniques. Function of the right ventricle, it has been assumed, must reflect, at least over time, that of the left. Recently, however, evidence has suggested that the circulatory demands endurance exercise places on the right ventricle may differ from those on the left. Studies of adult ultra-endurance athletes have indicated greater evidence of fatigue and myocardial stress in the right ventricle post-event compared to the left (19). Such effects, it has been postulated, might result in long-term right ventricular dysfunction and could play a role in the risk of arrhythmia-based sudden death in athletes.

The right ventricle has unique anatomic and contractile features. Compared to the left ventricle, the right is a thin-walled, crescent-shaped structure composed of an inlet trabeculated portion and a smooth-walled outlet. The right ventricle contracts in a peristaltic motion, unlike the left, whose contraction is characterized as a helical twist, or wringing out, of the myocardium. During exercise, the reduction in resistance is lower in pulmonary arteries than in systemic arteries. This poses a potential problem for the thin-walled right ventricle, which is expected to tolerate afterload less effectively than the left. Supporting this concern, LaGerche et al. found greater levels of stress in the right ventricular wall than in the left ventricular wall in adult subjects during progressive exercise (12).

Other studies, however, have failed to reveal any dysfunctional responses of the right ventricle during a standard bout of progressive exercise (13). In a study of college-aged men, Rowland et al. found that the pattern of stroke volume change, the increase in tricuspid diastolic inflow gradient, and the rise in systolic and diastolic myocardial velocities in the right ventricle mimicked those observed in the left ventricle (39).

Similar studies have not yet been performed in children. Future investigations will help reveal if the relatively greater afterload faced by the right ventricle during exercise has long-term negative implications or clinical significance.

In the exercise testing laboratory, cardiovascular fitness is usually identified in terms of physiological measures, most specifically maximal oxygen uptake ($\dot{V}O_{2max}$), the highest oxygen uptake achieved by a subject during a progressive cycle or treadmill test, expressed relative to body mass. Here, in accordance with the Fick equation, oxygen uptake is interpreted as a surrogate of maximal cardiac output. $\dot{V}O_{2max}$ obtained during a laboratory exercise test thus provides information about what can be termed "true" cardiovascular fitness, or the maximal capacity of the heart to generate cardiac output ($\dot{Q}_{max}$).

$\dot{V}O_{2max}$ per kg is tightly linked to endurance performance, but the interpretation of finish times in endurance events as indicative of true cardiovascular fitness is fraught with difficulty. Performance on a weight-bearing endurance event, such as running, is related not only to true cardiovascular fitness ($\dot{Q}_{max}$) but also to body composition because body fat serves as a metabolically inert load that must be transported. Also, a number of other less well-defined factors, such as motivation, strength, and anaerobic fitness, presumably contribute. In one study of sixth grade students, Rowland et al. found that $\dot{V}O_{2max}$ and body fat content contributed equally to performance on a 1-mi (1.6 km) run, but together the two factors explained only about 60% of the variance in run times (30).

Body composition affects the physiological definition of cardiovascular fitness ($\dot{V}O_{2max}$ per kg) as well, since body fat is an inert component of the denominator. Excess body fat decreases $\dot{V}O_{2max}$ per kg, while a lean child will exhibit an inflated value; both effects are independent of the true cardiovascular fitness in the numerator.

Given a group of children of similar size and body composition, what physiological factors define true cardiovascular fitness? Why is one child "in shape," another "out of shape"? Within the group, such variation in fitness will be identified during exercise testing by $\dot{V}O_{2max}$ (in absolute terms, since size and body composition are matched). $\dot{V}O_{2max}$ is the product of maximal heart rate times stroke volume times arterial venous difference (i.e., oxygen extraction in the skeletal muscle). Maximal heart rate and arterial venous oxygen difference are independent of cardiovascular fitness; that is, they are similar in high- and low-fit youth. The principal factor defining true cardiovascular fitness is the capacity for the heart to generate stroke volume at peak exercise (31, 38).

What defines this capacity? The first clue comes from an examination of the trajectory of stroke volume during the course of a progressive exercise test in youths of different levels of fitness ($\dot{V}O_{2max}$). Such data are presented in figure 4.1, which illustrates the patterns of stroke volume response to progressive cycle exercise in a study of nontrained 12-year-old boys of high and low fitness ($\dot{V}O_{2max}$ values of 54.8 ± 1.2 and 38.8 ± 2.5 ml · kg^{-1} · min^{-1} in the two groups, respectively) (31). It is evident that the *pattern* of stroke volume response is identical, with the expected small initial rise and subsequent plateau. The curves are parallel yet displaced upward with increasing fitness, reflecting larger values of stroke volume. A similar pattern of stroke volume response has been described in highly trained child cyclists ($\dot{V}O_{2max}$ 60.0 ± 6.0 ml · kg^{-1} · min^{-1}) with parallel displacement above that of the highly fit children (35).

These data indicate that any influences that differentiate maximal stroke volume (and thereby defining their level of cardiovascular fitness) are not apparent *during* exercise but rather in the resting, pre-exercise state. Fitness-related differences in stroke volume at maximal exercise are expressed similarly at rest. Therefore one must examine factors that determine individual differences in *resting* stroke volume (body size being considered) when searching for physiological explanations for cardiovascular fitness.

Resting stroke volume is most directly related to variations in preload (left ventricular end-diastolic dimension) (35). Individual differences in true cardiovascular fitness, then, are best explained by factors that influence cardiac end-diastolic size, such as plasma volume and resting heart rate (21). Differences in these factors among untrained individuals presumably reflect genetic effects for the most part.

In the process of endurance athletic training (and improving $\dot{V}O_{2max}$), an enlarging resting stroke volume with increased ventricular size results from increases in plasma volume and resting bradycardia that occur as cardiovascular fitness improves to match the augmented metabolic capacity (aerobic enzyme activity, mitochondrial density) of exercising muscle. The traditional idea that volume overload and augmented cardiac output incurred during repeated training sessions triggers ventricular enlargement by stretching myocardial fibers appears unlikely because, as noted earlier, left ventricular end-diastolic size does not increase during a bout of endurance exercise.

Interestingly, myocardial systolic and diastolic function in youngsters, both at rest and during exercise, appear to be independent of true cardiovascular fitness. Echocardiographic markers of inotropic and lusitropic function outlined earlier in this chapter are similar during progressive exercise regardless of fitness level (28, 35, 37). Contrary to the popular conception, then, the contractile health of the heart muscle in healthy youths is no different whether the fitness level is high or low.

Conclusion

The normal circulatory responses to the increased metabolic demands of endurance exercise are met by a finely coordinated interplay of cardiac, vascular, and neuro-adrenal adaptations. The principal factors that regulate the rate of blood flow are peripheral—manifested as a fall in arteriolar vascular resistance in response to local vasodilatory humoral responses within the skeletal muscle. The function of the cardiac pump is to propel forward blood flow to maintain system pressure. Ventricular diastolic volume (preload) and stroke volume remain constant during a progressive exercise test. Concomitantly, both systolic and diastolic function are augmented in parallel, preserving constancy of stroke volume as ventricular ejection and filling times shorten.

Exercise Electrocardiography

Thomas W. Rowland, MD

In providing for the body's circulatory demands, cardiac mechanical function is contingent on the directives of an intricate, finely tuned electrical system. In effect, the heart is an *electric* pump. From the spontaneously generated action potentials in the sinus node, the electrical depolarization propagated through the heart triggers cellular calcium release to the contractile apparatus, resulting in myocardial contraction.

Not surprisingly, abnormalities in the electrical system of the heart can interfere with normal cardiac function. Similarly, the impact of myocardial dysfunction on the heart's performance can be reflected in changes in its electrical properties (such as abnormal heart rhythms or changes in depolarization or repolarization patterns).

These electrical signals are generated at the sinoatrial (SA) node, an elliptical body situated in the epicardial junction of the superior vena cava and right atrium. Automaticity (ability to create an electrical impulse) and rhythmicity (at a regular tempo) of the SA node result from the balance of sodium, potassium, and calcium ion currents across cell membranes, which cause spontaneous depolarization.

The generated electrical impulse spreads through the atria, triggering atrial contraction, and then traverses the atrioventricular (AV) node, the sole normal pathway to the ventricles. Here the impulse slows, permitting sequential electrical activation and myocardial contraction of the atria followed by the ventricles. Without this "brake" at the AV node, cardiac function would be reduced because the atria and ventricles would contract simultaneously. Specialized conduction tissues (the right and left bundle branches and the Purkinje fibers) then transmit impulses to the ventricular myocardium. In the atria and ventricles, the passage of electrical current generates electrophysiological and biochemical changes that result in electromechanical coupling, converting cellular transmembrane ion flux to actin–myosin activation and myocyte contraction.

The electrocardiogram (ECG) accurately reflects the activity of the cardiac conduction system as the course of this electrical activity in the heart is transmitted to the surface of the body. Standard positioning of leads provides insights into the origin and course of these impulses in both the frontal and transverse planes of the thorax. The standard ECG provides information about the anatomic position of the cardiac chambers, average electrical vectors in both the atria and ventricles, and the duration of atrioventricular conduction (PR interval), ventricular conduction (QRS duration), and ventricular repolarization (QT interval). The ECG tracing indicates abnormalities of impulse generation (arrhythmias), the location and extent of ischemic myocardial damage (ST changes), the size and hypertrophy of cardiac chambers, and the effects of serum electrolyte abnormalities and drugs on the heart.

Historically, clinical exercise testing was introduced to detect electrocardiographic evidence of myocardial ischemia (ST-T wave changes) in adults experiencing angina pectoris. Today electrocardiographic monitoring during exercise is used to assess coronary artery disease in adult patients. In pediatric populations, in which ischemic heart disease is rare, other clinically important information can be obtained from the electrocardiographic responses to a progressive exercise test, including

- identifying a "maximal" exercise effort,
- sinus node and atrioventricular node function (heart rate response and conduction),
- effects of heart block,
- provocation of arrhythmias,
- ischemic ST-T wave changes (aortic stenosis), and
- risk stratification (long QT syndrome, ventricular pre-excitation).

Life Span and Heartbeats

The remarkable robustness of the cardiac electrical system is underscored by the observation that the cardiac pacemaker at the SA node and the conduction pathways are called upon to generate and transmit an electrical impulse through the heart about once a second for 80 yr—that's a total of 2,522,880,000 discharges—without respite. This feat might be considered banal only when compared to the sinus node of the hummingbird, which fires 20 times a second. The average hummingbird, though, if it survives the first year of life, usually lives for only about 4 yr, during which time the node discharges around 2.5 billion times, interestingly identical to the figure for the human heart.

In fact, the resting heart rate (f) across the range of all mammals can be expressed relative to body mass (M) as $f \sim M^{-0.25}$ suggesting that we and our animal brethren are all born with the same total number of heartbeats in our life's "bank," and that the rate we "spend" them is related to our body mass. All this appears to be an expression of metabolic rate—and life span—that are both associated with body mass by a similar mathematical relationship. Fortunately, human beings are outliers in these data, which predict that we would be out of heartbeats by the age of 20 to 25 (47).

Effects of Exercise on the Cardiac Conduction System

Electrical activity and mechanical pumping are both *intrinsic properties* of the heart. That is, removed from the thorax and placed in an appropriate nutritive medium, both will continue to function. *In vivo,* however, the cellular transmembrane ion shifts that generate an electric current in the heart are influenced by extrinsic factors, specifically changes in autonomic innervation.

Sympathetic preganglionic nerve fibers arise from autonomic nuclei within the medulla oblongata of the brain stem. These communicate with postganglionic neurons at the cervical level of the spinal cord. These in turn send nerve fibers that exert their influence via norepinephrine on nerve endings at the sinus and atrioventricular nodes and on atrial and ventricular muscle fibers. Parasympathetic fibers, also originating in the brain stem, innervate the sinus and atrioventricular nodes via branches from the vagus nerve, but they have little direct effect on the atrial or ventricular myocardium.

The stimulatory effects of sympathetic nervous activity increase the rate of sinus node firing and augment atrial and ventricular contractility and relaxation properties via beta-receptor activation in the heart. Increases in parasympathetic innervation, on the other hand, decelerate the sinus node and slow conduction through the atrioventricular node.

A progressive bout of endurance exercise is marked by a steady rise in sympathetic nervous activity and parasympathetic withdrawal. This change of balance in autonomic tone is responsible for changes in the electrical conduction system observed during exercise. The rate of sinus node discharge increases (which itself contributes to augmented myocardial contractile function), while electrical conduction is accelerated throughout the electrical system. The velocity of impulses propagated through the AV node is enhanced, as is electrical transmission through the ventricles. Sympathetic stimulation also serves to increase the rates of depolarization and repolarization, or electrical recovery, in the ventricles.

We do not know precisely which triggers cause this enhancement of sympathetic nervous influence on the heart, but the effect is probably multifactorial. Increased right atrial size and pressure from augmented systemic venous return (Bainbridge reflex), reflexes from metabo- and mechanoreceptors in exercising muscle, and central command from cortical centers may all play a role.

These changes in the electrical system of the heart during exercise are revealed by changes in the surface electrocardiogram:

- Heart rate
- P wave amplitude
- PR interval
- QRS duration
- QRS amplitude
- J-point depression
- T wave amplitude
- QT interval

Heart Rate

At rest, the discharge rate of the sinus node is largely under parasympathetic control, often evidenced by *sinus arrhythmia,* or vagal-induced phasic variations in RR intervals corresponding to respirations. In a study by Marcus et al., when children (with sinus node dysfunction) were administered atropine, a parasympathetic blocking agent, the heart rate at rest rose from an average of 89 bpm to 128 bpm (31). This magnitude of rise is similar to that observed in such blocking studies in adults, suggesting a similar degree of vagal influence on the resting heart rate in both age groups.

Commencement of exercise initiates a progressive vagal withdrawal and augmentation of sympathetic nervous activity. This increased sympathetic drive, supplemented by similar effects of circulating catecholamines from adrenal output, speeds the rate of sinus node discharges (shortening of RR interval on the ECG) by accelerating the rate of spontaneous depolarization. Heart rate rises and any sinus arrhythmia seen at rest is abolished. At peak exercise, autonomic influence on the sinus node is entirely sympathetic; administration of atropine (in studies of adults) causes no change in peak heart rate.

P Wave Amplitude and PR Interval

P wave voltages often increase during a bout of progressive exercise in children (14). The progressive rise in sympathetic cardiac influence during a standard exercise test speeds electrical transmission through the atria and atrioventricular node. Consequently, the PR interval on the ECG shortens, typically from 0.12 to 0.16 s at rest to about 0.10 to 0.14 s at peak exercise.

QRS Duration and Amplitude

The QRS duration typically remains stable or shortens slightly during the course of a progressive exercise test (5). In a lead normally characterized by a prominent R wave, such as V_5, the amplitude of this wave usually decreases during exercise in children. Thapar et al. reported an average decrease in R wave amplitude of 5 mm among 70 black children aged 7 to 14 during maximal cycle testing (55). The study by Paridon and Bricker suggested, however, that variability in R wave amplitude responses to exercise can be expected among young subjects (37). They found an overall average decrease in R wave in lead V_5 during maximal treadmill tests in 82 healthy children aged 4 to 18. However, 23 of 61 boys and 3 of 21 girls demonstrated increased or unchanged amplitudes. In the boys under age 10, 72% showed no change or increased R voltage in V_5 compared to 21% over age 10.

The differences in R wave amplitude response to exercise in healthy youths appear to have no prognostic significance. In adults, however, the story may be different. Decreases in R wave voltage on the ECG with exercise are expected in healthy subjects, while either unchanged R wave amplitude or an increase in voltage is considered an ischemic response and a predictor of coronary artery disease.

J-Point Depression

The J point is defined as the junction of the end of the QRS complex with the beginning of the ST segment. Experience in pediatric testing laboratories indicates that J-point depression, followed by an upslope of voltage reaching the baseline before 0.08 s, is a benign, common finding during progressive exercise. J-point depression must be differentiated from an ischemic ST segment response, in which ST depression is flat for more than 0.08 s, or the upslope to baseline is delayed (see later section in this chapter).

T Wave Amplitude

The height of the T wave is often increased during exercise (5, 41, 56). For instance, among 300 children aged 12 to 18 undergoing maximal treadmill testing, Thompson et al. described increased T wave amplitude in 48 (16%) (56). Riopel et al. reported an average rise in T wave amplitude in V_5 from 4.8 ± 2.5 mm at rest to 7.3 ± 2.8 mm at maximal treadmill exercise in 288 healthy children aged 4 to 21 (41). However, the study of Thapar et al. revealed that average T wave amplitude in lead V_5 *decreased* from 5.2 to 3.9 mm during the initial stages of a maximal cycle test and then returned to resting values at peak exercise (55).

QT Interval

The increased velocity of ventricular myocardial repolarization is reflected in a shortening of the QT interval during a standard exercise test. Assessment of a normal QT shortening during exercise and in recovery poses several challenges. Precise measurement of the QT interval is often made difficult by the asymptotic return of the terminal portion of the T wave to baseline. Moreover, during exercise, the T and P waves often merge at high heart rates, obscuring the end of the T wave. For this reason some have used the so-called tangent method, in which the termination of the T wave is defined by the intersection of a tangent drawn from the downslope of the T wave to a baseline (such as a Q-Q line) (figure 5.1).

Values of the QT interval must be adjusted for heart rate, but the optimal means for doing so remains controversial. Traditionally, absolute values of QT have been adjusted for heart rate by Bazett's formula, where the corrected QT interval (QTc) is calculated by dividing the absolute QT interval by the square root of the RR interval. The normal value of QTc at rest is generally considered to be 0.40 ± 0.06 s with some differences by sex.

Absolute QT interval decreases during a maximal exercise test and lengthens in recovery. In a study of 18 children aged 4 to 15, mean values fell (determined by the tangent method) from 0.39 ± 0.02 pre-exercise to 0.23 ± 0.01 at peak exercise (56). Normative values for QTc response to exercise have usually been established during the course of recovery, when measurement is facilitated by a slower heart rate and less merging of the T wave and subsequent P wave. It is generally believed that in a healthy subject, QTc is not substantially affected by exercise (range 0.36-0.44 s).

Berger et al. reported QT interval findings in lead II in 94 boys and girls aged 8 to 17 in the supine position serially during recovery from a maximal upright cycle test (7). Values were estimated using the tangential method. At 1 and 2 min postexercise, QTc by the Bazett formula was similar to that measured before exercise, but values subsequently rose by about 0.02 ms to a plateau.

These issues bear significance in the diagnosis of long QT syndrome, in which lengthening of QTc can occur during exercise and recovery. This topic will be addressed in a later section of this chapter.

ECG Setup and Monitoring

With the availability of modern ECG systems, monitoring of the full 12-lead tracing has become routine practice during clinical exercise testing. Electrodes for recording the precordial leads (V_1-V_6) are applied in the standard configuration over the anterior and left lateral chest. However, the limb leads are placed on the torso to reduce motion artifact. Usually those on the upper extremities are located bilaterally in the infraclavicular fossae, while those on the legs are applied to each side of the lower rib cage. Some have found, however, that applying the limb leads to the back (over the scapulae and flanks) provides clearer ECG tracings at high work intensities. This may be particularly true in obese subjects.

Modern self-adhesive electrodes and cable systems have significantly reduced the motion-induced artifact that was often encountered in ECG monitoring during maximal exercise testing in the past. Still, careful preparation of the skin with alcohol and light abrasion at each electrode site remains important in assuring clear, artifact-free tracings. The lead-wire box is normally attached at the subject's waist or back. In treadmill testing of very young children, it is often advisable to attach the box to the handrails rather than to the exercising subject. In some postpubertal subjects it may be necessary to shave chest hair to permit the usual electrode placement. These subjects are usually appreciative if the testing staff is willing to alter the normal precordial lead placement to avoid this difficulty.

Even with these precautions, artifact sometimes occurs during exercise, which makes it difficult to interpret the ECG properly. This can be caused by such things as breast tissue, obesity, clothing, or a malfunctioning electrode. Changing electrodes or electrode positions may solve the problem. If

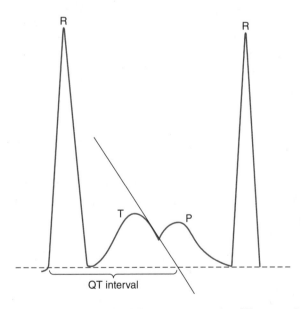

Figure 5.1 A method for estimating the QT interval at rapid heart rates when T and P waves are superimposed. The termination of the T wave is provided by a projection of the tangent of the its steepest downslope to the Q-Q baseline.

an electrode needs to be changed or moved once an exercise test is underway, this can be safely done while the subject grasps the handrails or is supported by a staff person rather than by restarting the test.

It is important to carefully interpret the baseline ECG obtained at rest before beginning the exercise test. Unusual heart rate, arrhythmias, conduction delays, ST segment abnormalities, and alterations in intervals need to be identified in order to properly assess changes that occur in response to exercise. A shift of the QRS axis to the right (with a negative QRS complex in lead I) is the most common difference between the resting ECG using the setup just described for standing or sitting and the standard supine ECG tracing. The QRS waveform may also be different. Thus the normalcy of QRS axis and standard voltage criteria for ventricular hypertrophy cannot be interpreted from the pre-exercise ECG.

Modern ECG monitoring systems permit real-time presentation of all 12 leads plus a rhythm tracing, usually of lead II. Hard-copy printouts are typically obtained at 2 or 3 min intervals during exercise and recovery as well as on demand by the testing staff when particular abnormal events occur, such as observed arrhythmias or the occurrence of patient symptoms (dizziness, chest pain, palpitations).

Measuring Heart Rate

ECG monitoring systems are generally very accurate in measuring heart rate electronically during exercise. Occasionally, erroneous values are evident due to motion artifact, unusual voltages, or frequent ectopy. In such cases, the rate can be estimated by manually measuring RR intervals on the ECG printout in seconds (1 mm = 0.04 s) and calculating heart rate as 60/RR interval. The accuracy of this approach assumes a constant paper printing speed of 25 mm per s.

Heart rate is the cardiovascular variable most easily measured during exercise. Consequently, abundant experimental and observational information is available for the expected normal responses according to age, sex, protocol, and influencing variables.

Pre-Exercise Value

Many factors influence pre-exercise heart rate. When assessed in the basal condition, resting heart rate decreases with age across the pediatric years, from an average of approximately 85 bpm at age 4 to 60 bpm at age 18. This decline in heart rate directly parallels the decline of size-related metabolic rate.

Low resting rates are observed in highly trained child endurance athletes; low rates can also be caused by certain drugs (particularly beta-blocking agents), hypothyroidism, and sinus node dysfunction. High resting heart rates occur in hot climate conditions. Low (and high) heart rates at rest may sometimes be explained as expressions of normal biological variability within the population in both autonomic influence and intrinsic resting sinus node discharge frequency. Immediately before an exercise test, however, the effects of these influences are often obscured by pretest anxiety and anticipatory tachycardia. For this reason it is not unusual for pretest (not "resting") rates to reach 90 to 100 bpm.

Exercise Response

Interestingly, the effects of pretest anxiety on heart rate appear to be abolished immediately once an exercise test is underway. This is evidenced by the tight linear coupling of heart rate normally witnessed with increasing metabolic demand ($\dot{V}O_2$) as workload rises, at least to moderate-to-high intensities. At a level of approximately 75% $\dot{V}O_{2max}$ a tapering of heart rate is often seen, both in children and adults. For example, in a study by Rowland and Cunningham, all 11- to 13-year-old subjects demonstrated a tapering of heart rate above 60% $\dot{V}O_{2max}$ during a progressive treadmill walking test (44). In a third of the children a plateau was observed, defined as less than a three-beat increase in the final stage.

This tapering of heart rate at high work intensities has not been adequately explained (8). Some have felt that the "break point" of the heart rate rise occurs concomitantly with the ventilatory anaerobic or lactate threshold, but why a rise in anaerobic metabolism or a fall in blood pH should trigger a deceleration of heart rate with increasing work is not clear. It has been suggested that the heart rate tapering phenomenon may be an artifact of the testing protocol rather than a true physiological phenomenon (8).

The *rate of rise* of heart rate during a standard progressive test is most obviously dictated by the exercise protocol selected. In addition, the heart rate at a given workload declines as a child ages. Thus, frequent high load increments in a young subject will elicit a faster heart rate response than smaller load changes in an older person. An unexpected rapid heart rate rise during progressive exercise can be a manifestation of low stroke volume production, either as an expression of low cardiovascular fitness in an otherwise healthy person or as an indication of heart disease. An unusually slow increase in heart rate as work intensity rises is typically observed in trained young endurance athletes. Other possibilities include subjects with a high degree of vagal tone, sinus node dysfunction, and those taking certain drugs, such as beta antagonists, calcium channel blockers, and migraine medications (ergot alkaloids, serotonin receptor agonists).

Maximal Heart Rate

Maximal heart rate is useful in assessing the degree of the subject's exhaustive effort as well as detecting sinus node dysfunction. An assessment of peak heart rate at the limits of exercise is thus especially important in evaluating those patients with a history of syncope, seizures, or dizziness, particularly during physical activities. Expected values are influenced by testing modality, type of exercise, and body position. Average maximal heart rate achieved by children during treadmill running is generally 195 to 205 bpm, with values about 5 bpm less for walking. Mean maximal heart rate during upright cycling in groups of children is typically 185 to 195 bpm. Studies indicate an average maximal heart rate in children during supine cycling of 172 bpm (20) and with progressive exercise on a rowing ergometer of 190 bpm (62).

These "normative" published values for maximal heart rate in children reflect group means, and it is important to realize that each obscures a wide inter-individual variability. For example, in utilizing several treadmill and cycle protocols, Cumming and Langford reported standard deviations of values for maximal heart rate ranging from 5 to 7 bpm (21). In the treadmill walking and running tests described by Sheehan et al. such standard deviations were 12 and 9 bpm, respectively (48). In this author's laboratory, convincingly true exhaustive efforts have been witnessed at a heart rate as low as 170 bpm and as high as 225 bpm. That such inter-individual variability is not simply a reflection of differences in subject motivation is indicated by the similar variability observed during testing of elite-level child athletes (46).

Most studies have indicated that maximal heart rate during progressive testing of healthy children is not significantly affected by level of aerobic fitness or sex. Moreover, as indicated in both cross-sectional and longitudinal studies, values remain constant throughout the pediatric years. Paridon and Bricker described a consistent mean maximal heart rate of 197 bpm during treadmill testing in a cross-sectional study of children aged 4 to 18 (37). Rowland and Cunningham tested 9 girls and 10 boys annually for 5 yr starting at age 9 on a treadmill walking protocol (unpublished data). Maximal heart rate was stable, within the range of 200 to 203 bpm. Bailey et al. performed a similar 8-yr longitudinal study of 51 boys beginning at age 8 (3). Maximal heart rate on a treadmill running protocol averaged 196 bpm. Mean values between the annual tests did not differ by more than 3 bpm. Given this stability of maximal heart rate during the growing years, formulae used in testing of adults to predict maximal heart rate, such as 220 – age, are inappropriate for testing children, at least until the mid-teen years.

Limited information suggests that maximal heart rate during exercise testing does not demonstrate any degree of circadian rhythmicity. When Reilly and Brooks had 15 adult males perform cycle exercise tests spread out over six separate days, circadian variation in heart rate was evident at rest but was abolished during exercise (40). In a study of 14 healthy adolescent males, no difference was observed in maximal or submaximal heart rates between morning and afternoon testing sessions (45).

Depressed Maximal Value

Considering 95th percent confidence limits from the studies just described, the lower value of an acceptable peak heart response to a progressive upright exercise test might be defined as 175 bpm for treadmill testing and 170 bpm on the cycle ergometer.

The most common cause of failure to achieve these rates is lack of adequate subject effort, evident from an absence of subjective signs of fatigue (hyperpnea, discomfort, flushing) and a low RER value (i.e., <1.00). The effects of previously mentioned drugs that can elicit a bradycardic response to exercise should also be recognized. Low peak heart rates are observed in patients whose sympathetic response to exercise is typically impaired, such as those with a heart transplant, Down syndrome, anorexia nervosa, or congestive heart failure.

A limited peak heart rate as an expression of sinus node dysfunction is infrequent in children and adolescents. However, such "chronotropic incompetence" is not uncommonly observed in patients following cardiac surgery, particularly those procedures involving cardiopulmonary bypass. The exact etiology for this depression of maximal heart rate, however, remains uncertain. Damage to the sinus node itself as a result of bypass cannulation as well as disruption of myocardial sympathetic innervation from surgical manipulation may be contributing factors. Typically the reduction in peak heart rate is not profound, usually no less than approximately 85% of predicted. The extent to which this limitation contributes to reductions of aerobic fitness in these patients is unclear.

Sinus node dysfunction has been described as a consequence of myocarditis and in association with long QT syndrome, as well as in children with unoperated congenital heart disease. Cases of symptomatic, idiopathic sinus node disease have been described, some which have been congenital and familial (6).

Heart Rate Recovery

Heart rate recovery postexercise is faster in individuals with a higher level of aerobic fitness, and heart rate at a certain time point after exercise is directly correlated with age or body size and peak exercise heart rate (41, 59). For example, Washington et al. cited average heart rates at 1 min of recovery of 133, 138, and 148 bpm in boys grouped as body surface area <1.00 m^2, 1.00 to 1.19 m^2, and >1.2 m^2, respectively (59). Heart recovery has generally been found to be faster in boys than girls, but this may be explained by the higher levels of aerobic fitness in young males. Given the large number of variables influencing recovery heart rate, the use of norms to predict aerobic fitness in children is problematic.

Simhaee et al. (50) and Lin et al. (30) have reported that metabolic risk factors such as waist circumference and serum lipid levels are inversely correlated with heart rate recovery rate in children and adolescents. Similar findings have been described in adult populations (34), in which heart recovery rate following exercise

has been linked to future mortality (18). Future investigations may thus indicate a more precise role and the utility of measuring recovery heart rate dynamics in the exercise testing laboratory as predictors of health outcomes.

Identifying Heart Block

First-degree heart block (delayed AV conduction) and Mobitz type I second-degree heart block (intermittent AV conduction) are usually manifestations of a high degree of parasympathetic influence on the AV node. Such delays are not uncommon in highly trained endurance athletes and individuals who possess a high level of vagal tone. These forms of heart block are normally abolished during the vagal withdrawal and the augmented sympathetic activity of progressive exercise.

Third-degree, or complete heart block, is characterized by a complete lack of transmission of electrical impulses through the AV node. In this case, the automaticity of an "escape" pacemaker high in the ventricles initiates the depolarization of the Purkinje fibers and myocardium, albeit at a rate slower than that of the sinus node, which continues to activate the atria. The rate of response of the ventricular pacemaker to exercise, although under sympathetic control, is also less than that of the sinus node.

There is considerable variability in maximal heart rate that can be generated by the ventricular pacemaker in patients with complete heart block. In published reports the range has generally been from 50 to 145 bpm, a peak rate that appears to be independent of that at rest. This variability presumably reflects different sensitivities of the ventricular pacemaker to autonomic stimulation. No relationship has been observed between the incidence of syncope and maximal achievable ventricular heart rate in patients with congenital complete heart block (27).

Patients with complete heart block often exhibit ventricular ectopy during exercise testing. Winkler et al. reported multiple premature ventricular contractions, ventricular couplets, or ventricular tachycardia in half of their patients during exercise (63). Both the explanation for this exercise-induced ectopy and its clinical significance are unclear. No clear-cut association has been established between the appearance of ventricular ectopy and clinical outcome in patients with complete heart block but no structural disease (49). Sudden death in conjunction with ventricular ectopy can occur in those who have associated serious cardiac abnormalities (63).

Complete right or left bundle branch block, as indicated on the ECG by a widened QRS complex (≥0.12 s), can rarely be induced by exercise. Such cases have been observed in children with healthy hearts and in those with significant cardiac disease (10). Importantly, the sudden appearance of bundle branch block during an exercise test can be confused with the advent of ventricular tachycardia. The former can be identified by the persistence of a normal P-QRS relationship.

Detecting Arrhythmias

Supraventricular and ventricular arrhythmias differ in their electrophysiological mechanisms as well as in their response to sympathomimetic stimulation. Consequently, these arrhythmias are usually dissimilar in their response to exercise as well as in their clinical implications.

Supraventricular Arrhythmias

Premature atrial beats, generated outside the sinus node but above the atrioventricular node, are benign. If present on the pre-exercise ECG, they usually disappear with exercise (43); when stimulated by exercise in an asymptomatic patient, they are also of no clinical concern (13). However, a patient with atrial premature beats during exercise who presents with a history of syncope or unexplained tachycardia might require further testing because such ectopy could reflect a potential for episodes of supraventricular tachycardia (58).

Supraventricular tachycardia (SVT) usually reflects a reentry circuit created by dual atrial–ventricular conduction pathways either within the AV node or by accessory extranodal pathways (ventricular pre-excitation, Wolff-Parkinson-White [WPW] syndrome). Clinical experience indicates that SVT is seldom triggered by exercise, even in those with a past history of this tachyarrhythmia. In one study, SVT developed during exercise testing in only 2 patients out of 56 referred for evaluation of palpitations (12). Draper et al. reported findings on exercise testing in 53 patients aged 5 to 18 who all had previous documented episodes of SVT (22). Thirty had normal pre-exercise ECGs, and 23 demonstrated evidence of ventricular pre-

excitation. Only three (6%) developed SVT during their exercise test.

Strasberg et al. described not a single case of SVT induced by treadmill exercise among 54 patients aged 12 to 64 who had pre-excitation on the resting ECG (52). These authors concluded that this lack of exercise-provoked SVT "probably reflected the lack of exercise-induced premature complexes with the critical timing for engaging only one of the two pathways." These findings indicate that a progressive exercise test should not be expected to be a highly useful means for triggering SVT, even in those with previously documented episodes.

In the unusual event of SVT occurring during an exercise test, diagnosis may be challenging. The QRS complex will remain normal in most cases, and associated alterations (or absence) of P waves may be difficult to identify on the ECG at high workloads. The usual initial indications of exercise-induced SVT are either an unexpected, abrupt rise in heart rate or failure of the heart rate to change as workload is increased or decreased.

Ventricular Arrhythmias

Exercise testing for assessment of ventricular arrhythmias is most frequently conducted on youth referred for

- isolated ectopy (premature ventricular contractions, or PVC's) identified on an ECG performed after an irregular heart rhythm has been noted on a routine physical examination, or
- syncope, seizures, or dizziness during sports play or other physical activities.

In these cases concern is raised that, in contrast to usually benign supraventricular arrhythmias, such children might carry a risk for important, life-threatening tachyarrhythmias such as hemodynamically unstable ventricular tachycardia and/or ventricular fibrillation.

Rhythm disturbances arising in the ventricles are often sensitive to electrical excitation triggered by sympathetic stimulation. Thus, ventricular arrhythmias, as compared to those supraventricular in origin, are more readily triggered by exercise. Consequently exercise testing is a better provocative test for assessing vulnerability to ventricular than supraventricular tachyarrhythmias.

Isolated, unifocal PVC's on a resting ECG occur with a frequency of approximately 0.3 to 2.2% of the pediatric population. PVC's tend to disappear as a child grows (15, 25). In an asymptomatic child without underlying heart disease, an unremarkable family history, and an otherwise normal ECG, this ventricular ectopy is almost always benign. Still, the detection of PVC's on a resting ECG raises two concerns: a) is this ectopy a reflection of unrecognized heart disease, and are these PVC's at rest harbingers of more serious, life threatening ventricular rhythms, such as ventricular tachycardia or fibrillation. Exercise testing provides useful information in addressing these issues.

Benign PVC's characteristically disappear during an exercise test as sinus tachycardia suppresses the ventricular ectopic focus (15, 29, 61). This effect of exercise is usually observed regardless of frequency of the ectopy (15) or its appearance in a bigeminal or trigeminal pattern (65). Jacobsen et al., for example, reported that among 17 healthy children with unifocal PVC's, ectopy was abolished by light exercise in all (25). This is not universally the case, however. Cagdas et al. described their experience in exercise testing of 149 healthy children with PVC's, mostly unifocal and isolated (15). The frequency of ectopy was diminished or PVC's were abolished during exercise in 62%, increased in 8%, and remained unchanged in 31%. In follow-up, ventricular ectopy at rest had disappeared in half of the subjects.

Suppression of PVC's during exercise testing may occur at low heart rates and is sometimes even abolished by the tachycardia of pretest anxiety. However, in some subjects, high work intensities associated with a heart rate of 170 to 180 bpm may be required to eliminate ventricular ectopy (43).

The interpretation of isolated, unifocal PVC's at rest that are not abolished by exercise is unclear (61). As noted above, such persistence is not inconsistent with benign ventricular ectopy. Still, decisions regarding the need for further evaluation of such individuals needs to take in consideration any history of pertinent symptomatology and presence or absence of underlying cardiac disease.

Occasional, isolated PVC's are commonly observed in healthy subjects during an exercise test, and these can be considered benign. However, the appearance of increasing or complex ectopy (couplets, triplets) in the course of an exercise test is of concern, since exercise-induced ventricular arrhythmias may signal a risk for ventricular tachycardia or fibrillation.

Complex ventricular ectopy characterized by multi-focality (varying QRS morphology) or occurring in pairs (couplets), triplets, or short bursts of ventricular tachycardia suggest greater electrical instability of the ventricle which might eventuate in more dangerous ventricular tachyarrhythmias (i.e., sustained ventricular tachycardia), particularly during the sympathetic stimulation of exercise. Nonetheless, the resolution of complex ventricular ectopy during an exercise test in an asymptomatic patient without underlying heart disease remains compatible with a benign outcome (29).

In summary, isolated, unifocal PVC's (as well as complex ventricular ectopy) at rest that disappear during exercise in a healthy, asymptomatic child are most commonly benign, but, as Walsh et al. have pointed out, this "is not universally true" (58). Ventricular ectopy that is not abolished or is triggered by exercise may deserve further evaluation.

Ventricular tachycardia (VT), defined as three or more consecutive beats of ventricular origin at a rate >120 bpm, is rare in the pediatric population. Most commonly observed in patients with significant cardiac disease (particularly those characterized by ventricular hypertrophy), it can also be hereditary (catecholaminergic polymorphic ventricular tachycardia), related to electrical disturbances (long QT syndrome), or idiopathic in apparently healthy youth.

Exercise testing is often an effective means of triggering ventricular tachycardia in predisposed individuals, particularly those symptomatic from their tachyarrhythmia.

Among 25 patients with recognized ventricular tachycardia, Rocchini et al. found that 73% of those who were symptomatic demonstrated an increase in their tachyarrhythmia with exercise testing (42). However 9 of the 10 asymptomatic children had partial or complete suppression of ventricular tachycardia with exercise.

Bricker et al. reviewed results of 2,761 exercise tests over a 10-year period in patients referred to the Texas Children's Hospital (11). Ventricular tachycardia was observed during exercise in 14 patients and during post-exercise recovery in 8 cases. Of these, 17 had a cardiac abnormality (long QT interval, arrhythmogenic right ventricle, congenital heart disease), while 5 had a normal heart.

In patients both with and without heart disease who experience syncope from ventricular tachycardia during exertion, VT can be demonstrated during exercise testing with a high frequency (17).

However, in some such cases the exercise test will be normal. For instance, patients with long QT syndrome, in whom VT and syncope are frequently related to exercise, exercise testing often fails to demonstrate VT. As Wren has pointed out, then, "a negative test in the presence of exercise-related symptoms or in a high risk patient should not be reassuring" (64).

Ventricular tachycardia during exercise testing is characterized by the sudden appearance of a regular tachycardia with widened, bizarre QRS complex. This rhythm disturbance must be distinguished from other causes of wide-QRS complex rhythms that can be triggered by exercise, including exercise-induced complete bundle branch block, supraventricular tachycardia with aberrant ventricular conduction, and pre-excited tachycardia in WPW syndrome (58).

Detecting Ischemia

The increased cardiac metabolic requirements (myocardial $\dot{V}O_2$) associated with a bout of progressive exercise provide an opportunity to detect abnormalities in the myocardial oxygen supply:demand resulting from an insufficiency of coronary blood flow. In adults with atherosclerotic coronary artery disease this is manifest symptomatically by angina pectoris and electrocardiographically by depression of ST-T waves, particularly over the left lateral precordial leads (V_5 and V_6). As classically defined, such ischemic changes are characterized by a depression of the ST segment of 1 mm (0.1 mV) or more, lasting for greater than 0.08 s or with an upslope that does not reach the baseline before 0.08 s. These ischemic changes need to be differentiated from benign J-point depression, in which the J point (junction of the end of the QRS complex with the ST segment) falls but is followed by an upsloping ST segment that reaches the baseline in less than 0.08 seconds (figure 5.2).

By these identifying characteristics, the appearance of such ischemic changes during an exercise challenge carries predictability of significant coronary artery disease in adults with a moderately high level of both specificity and sensitivity. It is important to note that these particular ECG criteria as markers of a diminished coronary-supply-to-metabolic-demand ratio have not been validated in children and adolescents. Similarly, issues of sensitivity and specificity of such ischemic ECG changes in heart disease in pediatric subjects have not been addressed. Consequently, the applicability of the ECG criteria for ischemia routinely used in adult subjects to children and adolescents remains uncertain.

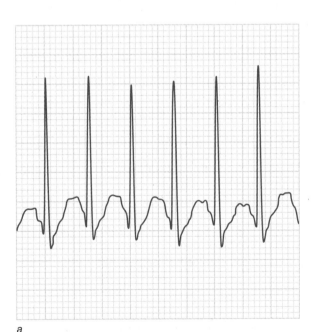
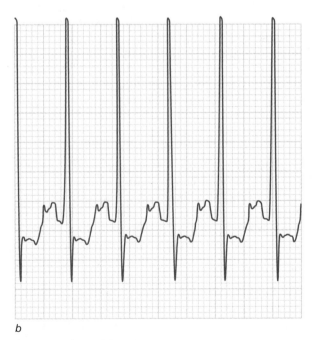

a   b

Figure 5.2 *(a)* Normal depression of the J point and upward sloping of the ST segment with exercise. *(b)* An ischemic pattern of depressed J point, flattening of the ST segment, and delayed return to baseline.

Reprinted, by permission, from J.T. Bricker 1993, Pediatric exercise electrocardiography. In *Pediatric laboratory exercise testing*, edited by T.W. Rowland (Champaign, IL: Human Kinetics), 51.

The number of cardiovascular diseases in children and adolescents that carry the potential for compromising coronary circulation is few; moreover, these anomalies are, in general, very uncommon in the general pediatric population. Still, attention to ST-T wave changes during exercise testing in young subjects is important, since a) although rare, the abnormalities which can limit coronary blood flow during physical activities in this age group characteristically pose a risk of sudden cardiac death, and b) chest pain, a symptom potentially reflecting (although rarely) such anomalies, is a common cause for referral of a child or adolescent for clinical exercise testing.

Aortic Stenosis

In one sense, patients with significant aortic outflow obstruction have the same potential for impaired coronary artery flow as adults with atherosclerotic coronary disease; the point of obstruction being simply upstream (i.e., just before the coronary ostia). During exercise any such impairment of coronary artery flow is accentuated, since the obstruction is fixed, while the demand of a hypertrophied myocardium is increased. Thus, it might be predicted that with a combination of sufficient outflow gradient and exercise demand, ischemic changes would result that would be manifest on the ECG.

Forty years ago, in fact, a series of studies indicated that this was the case. Children with aortic stenosis whose resting gradient exceeded 50 mmHg (by cardiac catheterization) were found to demonstrate ischemic ECG changes during exercise testing, typically with ST depression of 2 mm (16, 23, 26). These data showed that, lacking such changes, one could confidently predict a gradient <50 mmHg. However, ECG changes were found to have no predictive value following surgical or balloon intervention.

Clinical decisions regarding intervention for relief of obstruction in patients with aortic outflow obstruction are now made largely via echocardiographic criteria. However, it should be recognized that ischemic ST changes on the ECG at rest or during exercise in patients with aortic stenosis obstruction has been recognized as a risk factor for sudden death with this condition.

Congenital Coronary Artery Anomalies

Congenital abnormalities of the origins or course of the coronary arteries may limit myocardial

False Positive Tests

A number of conditions have been recognized to produce ST-T wave changes during an exercise test, which simulate a true ischemic effect. "False positives" can be caused by nonischemic heart or pericardial disease, hyperventilation, drugs, electrolyte abnormalities, anemia, pectus excavatum, and mitral valve prolapse. ST changes in patients with ventricular repolarization abnormalities, such as those with Wolff-Parkinson-White syndrome or bundle branch block cannot be interpreted as ischemic in origin. Unexplained ST changes with exercise mimicking an ischemic effect have also been reported in healthy individuals.

perfusion and lead to risk of sudden death, particularly during sports play. This appears to be particularly true when the left main coronary artery arises from the right aortic sinus, with a subsequent intramural course in the wall of the aorta. Other congenital coronary arteries anomalies that may pose clinical risk include coronary hypoplasia or stenosis, vessels buried within the heart muscle (myocardial bridging), and the right main coronary artery arising from the left sinus of Valsalva with an intramural course.

The frequency of such coronary abnormalities is low in the pediatric population (0.1%-0.3%), but anomalous origin of a coronary artery is a recognized cause of unexpected sudden death in young athletes. These tragedies are difficult to prevent, since the presence of a coronary artery anomaly is typically occult and not easy to diagnose. While chest pain, dizziness, or syncope can serve as premonitory symptoms, patients are usually asymptomatic, and not infrequently the first indication of a coronary artery anomaly is sudden death, often during physical activities.

Following the experience of atherosclerotic coronary disease in adults, it would be expected that identification of a patient with an anomalous coronary artery—and particularly one at risk for sudden death—would be facilitated by recognizing ischemic ST changes during a progressive exercise test. Unfortunately, this may not be the case. Basso et al. described 27 cases of sudden death during sports play from a coronary anomaly, with a mean age of death of 16 ± 5 years (4). Of these, previous ECGs had been obtained in 9, and 6 had undergone an exercise stress test, all of which were unremarkable. Based on these findings, the authors concluded that "neither routine 12-lead ECG nor exercise stress tests were particularly informative for the diagnosis of congenital coronary artery anomaly."

Supporting this conclusion, Osaki et al. reported their experience with 31 children (mean age 6) with an anomalous coronary artery (36). Of the 13 who underwent exercise testing, only 1 displayed ischemic ST changes. In a review of the literature, among 18 patients with congenital coronary anomalies <35 years old who had exercise stress tests, ischemic ECG changes were observed in 2 (4).

Kawasaki Disease

Kawasaki disease is an acute, idiopathic vasculitis affecting young children, characterized by fever, conjunctivitis, rash, lymphadenopathy, oral erythema, and desquamation of the skin of fingers and toes. While generally benign, inflammatory involvement of the coronary arteries may eventuate in coronary ectasia, aneurysms, stenosis, and risk of sudden death in a very small minority (~0.08%).

In the older child with a past history of Kawasaki disease, there remains question of possible long-term risks associated with coronary artery involvement, even in the absence of aneurysms. Some health caretakers have elected to refer patients (particularly athletes) with earlier Kawasaki disease for exercise testing for assessment of possible ischemic changes, even if there is no echocardiographic evidence of residual alterations in coronary artery diameter. Recommendations from the American Heart Association for care of patients who do exhibit coronary artery aneurysms include annual exercise testing (19).

While such an approach is intuitively rational, it is surprising that evidence of ischemic ECG changes during exercise appears to be of limited value in recognizing coronary artery abnormalities in patients with a history of Kawasaki disease. Even with significant coronary artery involvement, no ST changes have usually been observed (24, 38, 39). However, Allen et al. reported two patients with large aneurysms who developed ST changes with exercise testing (1), and 5 of the 13 patients tested by Kato et al. with aneurysms or stenosis (by angiography) exhibited ischemic changes (28). The predictive value of ST changes with exercise in patients with coronary artery

abnormalities for adverse outcomes following Kawasaki disease is uncertain.

Evaluation of Prolonged QT Interval

Long QT syndrome encompasses a family of inherited cardiac electrical disturbances characterized by a delay in ventricular repolarization. Patients with this syndrome are susceptible to episodes of ventricular tachyarrhythmia (torsades de pointes) and sudden death, often precipitated by sympathetic stimulation (exercise, particularly swimming, as well as emotional distress). At least 12 genetic types have been identified, each with its own phenotypic expression, but genetic testing is negative in about 25% of cases.

Exercise testing can be useful in discriminating patients with long QT syndrome from normal individuals who have a borderline long QT interval on the ECG. While the QT interval corrected for heart rate by the Bazett formula typically changes little with exercise in healthy subjects, the QTc may lengthen, especially in recovery, in those with long QT syndrome. The sensitivity of this finding, however, is influenced by the genetic type of long QT syndrome.

Most information regarding the effects of exercise testing in patients with long QT syndrome has been gathered in types 1 and 2, the most common variants. In adults, QTc is prolonged with exercise in type 1 but not type 2. Takenaka et al. found that among patients with type 1 average QTc (in lead V_5) increased from 0.51 ± 0.14 s at rest to 0.60 ± 0.05 s at peak exercise, while no significant change was observed in QTc in comparison groups of patients with type 2 and healthy control subjects (54). It has been reported that in type 2 patients QTc may lengthen initially during the early stages of progressive exercise but return to pre-exercise values at high heart rates and into the recovery phase (35). QTc in type 4 patients may prolong with exercise (32).

In children, prolongation of QTc during recovery from exercise compared to pre-exercise values has been described in patients with long QT syndrome compared to normal youth (2, 53, 60). Aziz et al. compared QT responses to maximal cycle exercise in 50 youth with long QT syndrome with those of 108 healthy children (2). In the normal subjects, mean values for QTc at peak exercise were similar to those at rest (0.41 s) and remained stable during recovery until five minutes post-exercise when a rise to 0.43 s was recorded (figure 5.3). In patients with type 1 long QT syndrome the average QTc at rest of 0.45 s increased to 0.49 s at peak exercise and remained unchanged during the course of 9 min of recovery. Exercise triggered a shortening of QTc in type 2 long QT syndrome in recovery, with a return to pre-exercise values after 5 min.

These limited data suggest that lengthening of QTc in early recovery from exercise compared to pre-exercise values is valuable in identifying patients with type 1 long QT syndrome but not necessarily type 2. The behavior of the QT interval on the ECG with exercise in children with other types of long QT syndrome is still incompletely characterized.

Risk Stratification With Ventricular Pre-Excitation

While ventricular pre-excitation generally carries a favorable prognosis, these patients carry a very small risk for malignant tachyarrhythmias and sudden death. This can occur when, in the course of atrial fibrillation or flutter, impulses are conducted at a high frequency through the acces-

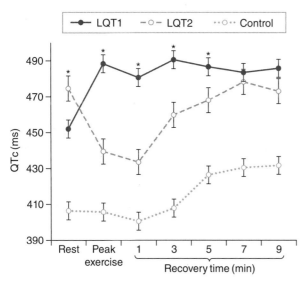

Figure 5.3 Corrected QT interval (QTc) before and during exercise and in recovery in pediatric patients with types 1 and 2 long QT syndrome compared to healthy control subjects. P-value <0.001 for LQT versus controls at all points. *P-value <0.001 for LQT1 versus LQT2.

Adapted, by permission, P.F. Aziz et al., 2011, "Genotype- and mutation site-specific adaptation during exercise, recovery, and postural changes in children with long QT syndrome," *Circulation: Arrhythmia and Electrophysiology* 4: 867-873.

sory pathway and transmitted to the ventricles, ultimately degenerating to ventricular fibrillation. The risk of such an event is greatest in those who possess an accessory pathway with a short antegrade refractory period.

Published data suggest that disappearance of ventricular pre-excitation on the ECG during the course of an exercise test can identify those patients at low risk (9, 51, 52). That is, the augmented sympathetic activity and vagolysis during exercise are more likely to abolish transmission of electrical impulses through those accessory pathways possessing a long refractory period.

Spar et al. evaluated 76 patients <21 years of age with ventricular pre-excitation by both exercise testing and electrophysiological study (51). Eleven (14%) demonstrated sudden loss of pre-excitation with exercise, 18 (24%) showed gradual loss, and 47 (62%) persistence. With atrial pacing, the mean value for cycle length with 1:1 conduction in the accessory pathway were 375 ± 135 ms, 325 ± 96 ms, and 296 ± 52 ms in the three groups, respectively. Of those with conduction <270 ms, none were in the sudden loss group, 5 of 18 were in the gradual loss group, and 18 of 47 were in the no loss group.

In the 17 children with Wolff-Parkinson-White syndrome described by Bricker et al., 4 showed total disappearance of pre-excitation with treadmill exercise (9). Disappearance of delta wave with exercise was related with a long anterograde refractory period of the accessory pathway (360-390 ms). In the study of Strasberg et al. in 36 patients aged 12 to 64, complete normalization of the QRS complex occurred in 50% (52). However, Moltedo et al. described loss of pre-excitation in only 8% of 50 youth with Wolff-Parkinson-White syndrome with exercise testing (33).

In summary, total and sudden disappearance of ventricular pre-excitation during an exercise test is generally indicative of low risk but only occurs in a minority of patients. Persistence of pre-excitation on the ECG during exercise is similarly consistent with a benign course, but risk of a tragic event associated with rapid accessory pathway conduction during atrial flutter and fibrillation cannot be ascertained.

Conclusion

One hundred years after exercise testing was first devised to examine electrocardiographic changes in adult patients with coronary artery disease, the

ECG remains a central element in the clinical exercise laboratory. Exercise places demands on myocardial metabolism, augments sympathetic influence on sinus node discharge rate, and promotes automaticity of ectopic cardiac foci. Changes in cardiac electrical conduction expressed on the ECG as manifestations of these effects can provide valuable information in patients with primary electrical conduction disorders as well as those with a propensity for arrhythmias and limitations of coronary perfusion.

Blood Pressure Response to Dynamic Exercise

Bruce Alpert, MD, and Ranjit Philip, MD

Blood pressure (BP) is routinely measured during exercise testing on either a cycle ergometer or a treadmill. Dynamic exercise causes increases in cardiac output (CO) and heart rate with simultaneous dilation of the systemic vascular bed. Because BP is the product of CO and peripheral vascular resistance, the rise or fall of BP may help us to assess the integrity of myocardial contractility, chronotropic competence, and dilation of peripheral arterioles. Hence, during exercise testing, blood pressure has significant clinical importance. Well-studied cardiac lesions include left-sided obstructive lesions such as coarctation of the aorta and aortic stenosis. The aims of this chapter are the following:

- Review the basic physiology of exercise BP.
- Discuss the technical aspects of measurement of BP.
- Review the normal BP response and BP data from published studies of healthy children.
- Note racial, gender, age, and obesity-related differences in maximal exercise BP.
- Evaluate the prognostic value of exercise BP testing.
- Briefly discuss BP response in special conditions including congenital heart disease (CHD). This topic will be discussed in detail in another chapter.
- Discuss the importance of change in BP (delta BP) versus absolute values in the interpretation of results.

Basic Physiology of Exercise Blood Pressure

Blood pressures at rest and during exercise are lower in children than in adults (53). The circulatory changes that occur from a resting state to exercise are complex. At rest, skeletal muscle receives about 15% to 20% of the CO, although it constitutes almost 40% of the total body mass (13). Peripheral muscle perfusion increases to 80% to 90% of CO during exercise, and the peripheral oxygen extraction increases. From rest to maximal exercise CO increases three- to fivefold. Because BP is the product of CO and peripheral resistance, the response of the peripheral vascular bed is also of importance. During exercise, there is a dramatic dilation of the peripheral vascular bed, which results in a reduction of resistance. The increase in CO with exercise results in an increase in systolic (S) BP. The vasodilation causes the diastolic (D) BP to remain largely unchanged. However, with isometric exercise (constant muscle length against force or tension such as weightlifting) both SBP and DBP increase.

Increasing exercise intensity calls for increased oxygen consumption and CO, so BP rises with each progressive stage of an exercise protocol. Arterial baroreflex mechanisms maintain BP at a higher level during exercise. The withdrawal of parasympathetic nervous activity is initially responsible for the rapid increase in heart rate, which enables adequate CO. The high BP attained during moderate to severe exercise is achieved by a summation of metabolic reflexes and vasoconstrictive mechanisms (55). The reduction in cardiac output and SBP during recovery is achieved by a withdrawal of sympathetic output and an increase in vagal tone. These produce reductions in both heart rate and peripheral resistance. Resting levels are achieved within minutes (49, 56). Supine SBP returns to normal values quickly during recovery and may actually be below baseline levels for a few hours postexercise.

In summary, SBP should show a progressive increase during exercise and a progressive decline after exercise. An impaired BP response, that is, a decrease in BP or a failure to increase SBP with exercise, indicates abnormal function of these regulators and may imply clinically significant left ventricular dysfunction or a left-sided obstructive lesion (36, 60). Although important, the drop in SBP may not be consistently present in patients with impaired cardiac function. Occasionally, SBP can drop in children in the absence of severe cardiac disease (7).

Technical Aspects of Blood Pressure Measurement

The technical aspects of exercise BP measurement are critical. It is important to have accurate and reliable measurements in order to use the data for clinical decision making. It is generally recommended that BP be measured at the following points:

- At rest before beginning the exercise test.
- Frequently during the exercise test to evaluate BP elevation or to detect impending hypotension. This is usually done every 3 min, with more frequent measurements if symptoms of hypotension are present.
- During the recovery period to ensure that SBP returns to approximately baseline values.

BP Measurements by Auscultation

Mercury sphygmomanometers were formerly used in association with auscultation to measure exercise BP. Due to concerns about environmental hazards, mercury devices have been replaced with other technology (9). A calibrated aneroid device is an acceptable alternative; calibration should be performed at least yearly (10-12). Environmental noise from equipment (treadmills, monitors, fans,

Does It Matter When BP Is Taken to Document Response to Exercise?

Systolic BP rises with increasing dynamic work as a result of increasing cardiac output, whereas diastolic pressure usually remains about the same or is moderately decreased because of vasodilation of the vascular bed. The average rise in systolic BP during a progressive exercise test is about 10 mmHg/MET.

After maximum exercise there is usually a decline in systolic BP, which normally reaches resting levels in children in about 6 min, then often remains lower than pre-exercise levels for several hours. Systolic BP at maximum exertion or at immediate cessation of exertion is considered a clinically useful first approximation of the heart's inotropic capacity.

However, a common experience in clinical practice is that the so-called max BP recorded to indicate the BP at peak exercise, due to technical reasons, is recorded well into recovery. The initiation of BP measurement occurs when the child indicates the need to terminate the test (indicating peak exercise). Unfortunately, by the time the automated cuff takes the measurement it is usually well into recovery and hence may be a lower value than at maximum exertion. With normal chronotropic and parasympathetic responses, the BP can fall precipitously, and thus one should not delay measurement of BP at either maximum exertion or immediate cessation of exertion. To avoid a delay in reading, we recommend a pre-emptive BP measurement when it is presumed that peak or maximum exertion is nearing, based on the subject's heart rate and his or her perceived exertion rating. This would most likely be the best prediction of BP at maximum exertion.

air conditioners, etc.) as well as the subject's respiration make auscultation of Korotkoff sounds difficult. If a treadmill is used as the exercise stimulus, arm motion may be a major variable as well because of the normal pendulum-like reciprocal motion. Whether using a treadmill or a cycle ergometer, the BP cuff must be wrapped and usually taped in place to avoid slippage. Arm motion can be reduced in various ways. If the child is undergoing testing on a treadmill, the arm can be steadied by the person measuring the BP (by lightly taping the stethoscope head over the brachial artery before the study begins, the tester will have a hand free to reduce arm motion). The use of a cycle ergometer makes the measurement of BP much easier; the observer does not have to chase the child up and down the treadmill to obtain the BP, and the arm may be supported by a cradle or by the tester during measurement.

Good vision, hearing, and hand-eye coordination are necessary attributes of the BP recorder; retraining is periodically needed (51). The appropriate cuff size must be chosen based on the patient's arm size. A small-sized cuff for the arm circumference will lead to a falsely elevated BP measurement. We recommend that the cuff bladder completely encircle the limb; the cuff width is optimal at greater than or equal to 40% of the limb circumference measured at the midpoint (62).

Automated Blood Pressure Systems

Automated devices have become the norm for measuring blood pressure. Oscillometry is the method most widely used in automated devices. This method is less susceptible to external noise as no transducer is required over the brachial artery. Unfortunately, oscillometric devices are prone to artifact from movement and vibration, especially during treadmill testing. These devices detect the maximal pulse amplitude, which correlates with mean arterial BP and the proprietary software calculates SBP and DBP. It usually underestimates DBP and the level of inaccuracy in DBP increases with increasing exercise (8).

Several automated systems have been commercially available for measuring exercise blood pressure. In one study (3), Alpert et al. compared an automated system (Critikon 1165; no longer manufactured) to a single observer (a second-year fellow in pediatric cardiology). There were excellent correlations between the automated system and the observer for SBP and acceptable values for the comparisons of DBP. Another study done to evaluate the use of a motion-tolerant BP monitor (CardioDyne NBP 2000, Luxtec, Worcester, MA) against standard manual auscultation at rest and during exercise revealed that SBP and DBP measured by each technique were not significantly different from each other and were highly correlated both at rest and during exercise (42). Some laboratories prefer to use an automated system because it has no terminal digit preference and is consistent from day to day. The amount of change from one observer to another during serial testing of the same subject may exceed the inaccuracies of the automated system. Algorithms for measuring Korotkoff sounds that involve the differential subtraction of sounds recorded by microphones facing the artery and an external-facing microphone may be more accurate than the human ear. Because some frequency components of the Korotkoff sounds are at the lower end of the human ear frequency response, the automated system may have another inherent advantage.

Normal Blood Pressure Response to Dynamic Exercise in Healthy Children

The normal response of BP in progressive exercise testing is that SBP increases while DBP is maintained or slightly decreased. Systolic BP usually reaches pre-exercise levels within 6 min of recovery after maximal exercise. It often remains lower than baseline for a few hours. A sudden cessation of exercise can sometimes lead to a steep drop in SBP in healthy children. This is because of peripheral venous pooling and a hindered immediate post-exercise increase in systemic vascular resistance to counteract the decline in cardiac output (17).

Systolic BP is dependent on age, body surface area (BSA), sex, race, and physical fitness, which should be taken into account during routine exercise testing. Several studies have included the measurement of exercise BP. It is of interest that Godfrey, in his classic monograph from 1974 (27), did not mention BP response to exercise in any study of healthy subjects or of patients with either pulmonary or cardiac disease.

Age and Body Surface Area

Blood pressures at rest and during exercise are lower in children than in adults. A child with a BSA of 1.25 m^2 attains a SBP of around 140 mmHg at maximum exercise, while a child with a BSA of 1.75 m^2 achieves a SBP of around 160 mmHg (60).

Riopel et al. were the first to report SBP and DBP data from large groups of healthy children (53). They studied 279 healthy 4- to 21-year-olds (119 white males, 50 black males, 66 white females, 44 black females) during treadmill stress testing. The children were divided into four groups by BSA. The study found that the largest increase in SBP occurred during the first minute of exercise, with a more gradual increase occurring with each subsequent minute (figure 6.1, tables 6.1 and 6.2). There were progressively higher SBP values with increasing BSA, whereas DBP remained unchanged or decreased slightly.

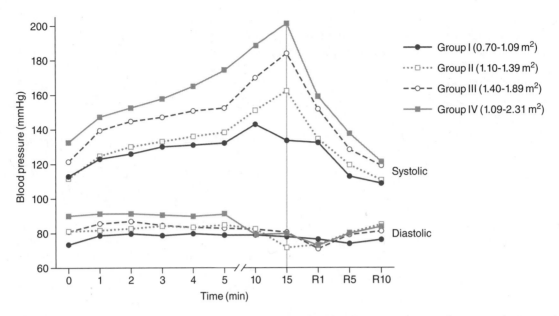

Figure 6.1 Mean (±1 standard deviation) systolic and diastolic blood pressures at each minute during and after exercise in the four groups (I-IV). R_1 through R_{10} indicate values obtained during the 10 min postexercise period.

Data from Riopel, Taylor, and Hohen 1979.

Table 6.1   Systolic Blood Pressure (mmHg ± 1 Standard Deviation)

	Group I		Group II		Group III		Group IV	
	WHITE	**BLACK**	**WHITE**	**BLACK**	**WHITE**	**BLACK**	**WHITE**	**BLACK**
BOYS (NO.)	28	15	41	19	44	19	6	6
Pre-ex	107 ± 9	121 ± 12	108 ± 12	113 ± 8	118 ± 13	124 ± 11	128 ± 5	119 ± 10
1 min ex	121 ± 12	129 ± 16	121 ± 14 †	131 ± 9	137 ± 17 ‡	139 ± 11	150 ± 10	146 ± 14
Peak ex	142 ± 16	145 ± 15	145 ± 15	154 ± 17	171 ± 22 ‡	181 ± 16 §	206 ± 9	200 ± 22
GIRLS (NO.)	19	10	21	11	25	22	1	1
Pre-ex	110 ± 9	113 ± 13	106 ± 12	112 ± 11	112 ± 10 *	118 ± 7	124	100
1 min ex	123 ± 11	123 ± 13	122 ± 14	128 ± 15	126 ± 14 §	141 ± 17	150	122
Peak ex	147 ± 12	145 ± 14	145 ± 13	154 ± 24	155 ± 14	161 ± 20	184	140

*p < 0.05. †p < 0.02. ‡p < 0.01. §p < 0.005. No. = number of subjects. Pre-ex = before exercise. Peak ex = peak exercise. 1 min ex = 1 minute of exercise.

Reprinted from *American Journal of Cardiology*, Vol 44, D.A. Riopel, B.A. Taylor, and R.A. Hohen, "Blood pressure, heart rate, pressure-rate-product and electrocardiographic changes in healthy children during treadmill exercise," pg. 701. Copyright 1979, with permission of Elsevier.

Table 6.2 Diastolic Blood Pressure (mmHg ± 1 Standard Deviation)

	Group I		Group II		Group III		Group IV	
	WHITE	BLACK	WHITE	BLACK	WHITE	BLACK	WHITE	BLACK
BOYS (NO.)	28	15	41	19	44	19	6	6
Pre-ex	74 ± 7	79 ± 9	75 ± 11	79 ± 11	79 ± 9	85 ± 10	85 ± 9	80 ± 9
1 min ex	76 ± 6	82 ± 9	77 ± 9	81 ± 8	83 ± 12	80 ± 4	86 ± 9	85 ± 10
Peak ex	85 ± 7	85 ± 7	84 ± 9	87 ± 9	77 ± 12	84 ± 11	95 ± 8	87 ± 11
GIRLS (NO.)	19	10	21	11	25	22	1	1
Pre-ex	71 ± 9	77 ± 5	76 ± 7	74 ± 8	78 ± 6	84 ± 7	92	96
1 min ex	74 ± 9	82 ± 4	79 ± 7	83 ± 12	81 ± 9	89 ± 11	96	92
Peak ex	79 ± 7	83 ± 4	84 ± 6	85 ± 11	79 ± 10	84 ± 12	110	100

*$p < 0.05$. †$p < 0.02$. ‡$p < 0.01$. §$p < 0.005$. No. = number of subjects. Pre-ex = before exercise. Peak ex = peak exercise. 1 min ex = 1 minute of exercise.

Reprinted from *American Journal of Cardiology*, Vol 44, D.A. Riopel, B.A. Taylor, and R.A. Hohen, "Blood pressure, heart rate, pressure-rate-product and electrocardiographic changes in healthy children during treadmill exercise," pg. 701. Copyright 1979, with permission of Elsevier.

Pulse pressure (SBP minus DBP) was greater in large subjects at maximal exercise.

James et al. studied 149 healthy subjects (95% white), 5 to 33 years of age. Ninety males and 59 females underwent maximal cycle ergometer stress testing (37). The subjects were stratified by BSA divisions (<1 m^2, 1-1.19 m^2, >1.2 m^2). Maximal SBP was positively correlated with body size and age. Maximal DBP increased up to 14% above the value obtained pre-exercise while sitting. These results are in contrast to data obtained by Riopel et al. (53). The maximal SBP for the upper 2.5% of James' total population ranged from 234 to 254 mmHg. No stated complications occurred in the subjects whose SBP exceeded 230 mmHg. The maximal SBP value was related to both power output and resting SBP. James et al. speculated that there might have been differences in stroke volume, systemic vascular resistance, and duration of left ventricular ejection that interacted to produce these high levels of exercise SBP.

Alpert et al. tested 405 healthy children (221 whites and 184 blacks) aged 6 to 15 (4). The protocol was performed on a mechanically braked cycle ergometer, with 3 min stages, to maximal voluntary effort. Statistical analyses yielded regression lines with confidence bands for the 5th, 25th, 50th, 75th, and 95th percentiles of SBP (figure 6.2). Alpert et

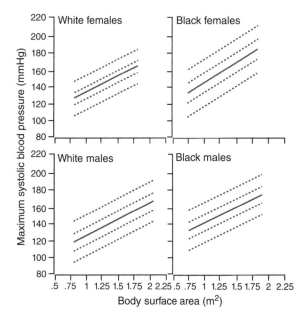

Figure 6.2 Nomograms of maximal systolic blood pressure (mmHg) against body surface area. Solid line represents 50th percentile of systolic blood pressure. Top line represents 95th percentile confidence band with dashed lines below representing 75th, 25th, and 5th percentile confidence bands.

Reprinted from *Journal of Pediatrics*, Vol 101, B.S. Alpert et al., "Responses to ergometry exercise in a healthy biracial population of children." pg. 541. Copyright 1982, with permission of Elsevier.

al. did not report DBP values because of technical problems with measuring DBP during exercise on a mechanical ergometer in a noisy environment and because the authors lacked confidence in the accuracy of the measurements of the fourth and fifth Korotkoff sounds (muffling and disappearance, respectively). These nomogram plots allow the estimation of the maximal SBP value (which occurs either during exercise or immediately postexercise) in relation to the other exercise variables, such as heart rate, maximal work rate, and peak working capacity index (workload per kilogram of body weight).

The effort that the subject or patient puts forth is often judged by variables such as heart rate, work rate, or oxygen consumption. If a patient with aortic stenosis achieves a 25th percentile SBP response with 75th percentile heart rate and work capacity, this may imply, for example, that the patient has a gradient severe enough to limit exercise BP significantly. If a patient postoperative for coarctation of the aorta has a SBP response in excess of the 95th percentile but only a 5th percentile working capacity, the clinician may consider recatheterization and possibly balloon angioplasty or reoperation. Previous investigators (37, 67) expressed their data as mean and standard deviation, making prediction of the expected individual responses cumbersome.

Figure 6.2 shows that the SBP response varies directly with body size (m^2 of surface area). The nomograms demonstrate that maximal SBP values greater than 220 mmHg are above the normal range for subjects with BSA less than 2 m^2, confirming the treadmill data reported by Riopel et al. (53). James et al. (37) used a cycle ergometer and reported SBP values that were higher than those of Alpert et al. (4). No explanation for this difference could be found; Alpert speculated that methodologic differences or the use of mechanically braked versus electronically braked ergometers might explain this.

Washington et al. (67) reported data from cycle ergometer exercise using the protocol of James et al. (37) on 151 white children (70 girls and 81 boys) aged 7.5 to 12.75. They used an automated apparatus to measure BP, a technique previously validated by Alpert et al. (3) in a study of 121 children. They used the divisions of BSA that had been described by James et al. (37). The data show a trend of increasing mean maximal SBP from smaller to larger BSA (122 to 130 to 139 mmHg in males and 126 to 131 to 142 mmHg in females),

but no statistically significant differences were found. Maximal SBP values did not exceed 183 mmHg in any child. The studies were performed at intermediate altitude (1,600 m) in Denver, CO; some of the values, such as $\dot{V}O_{2max}$ (maximal oxygen consumption), may have been limited because of this. The effect of altitude on exercise BP could be tested either by having children at altitude (such as Denver) breathe 21% oxygen at 1 atm (760 mmHg) through a face mask or by having children at sea level breathe 18% oxygen during exercise. The values of maximal DBP were invariably higher than the values obtained at rest for each child. This is in contrast to the report by Riopel et al. (53) that a majority of children had decreased DBP values. Washington considered a decrease in DBP during exercise "an abnormal response." This finding may relate to differences between treadmill and cycle ergometer exercise physiology or to technical differences in the measurement of DBP. In addition, Riopel's study was performed at sea level and Washington's at 1,600 m. The degree of hypoxia reached in Washington's subjects may have led to a difference in vasodilation during maximal exercise from what occurs at sea level. The Denver children may have undergone a relative vasoconstriction or a failure to vasodilate maximally, leading to the difference observed.

Another study by Ahmad et al. (1) examined multiple parameters including exercise endurance time, heart rate, BP, and metabolic variables between ages 5 and 18. With increasing age, the maximum SBP and DBP increased in males and females. After age 13, boys had significantly higher maximum SBP and DBP than girls.

Sex

Riopel et al. showed that, within a given age group, boys have a higher peak SBP than girls (53), probably because of higher maximal stroke volume. Wanne and Haapoja (66) performed maximal cycle ergometer studies on 497 healthy 9- to 18-year-old children. They reported data for SBP at various submaximal heart rates, up to 170, and into recovery. Boys consistently demonstrated higher SBP values than girls, and postpubertal boys had the highest SBP values of any group. Diastolic BP decreased significantly in every group tested, in sharp contrast to the findings of James et al. (37) and Washington et al. (67). Maximal SBP values in excess of 200 mmHg were found in 22 subjects. Only two subjects demonstrated elevated resting SBP. Three of the 15-year-old males had maximal SBP values of 240 mmHg at submaximal exercise; no complications were noted in these subjects.

Race

There has been variability in the reported data regarding racial differences in BP responses. Riopel et al. observed few racial differences in SBP in 288 healthy children undergoing treadmill exercise (53). A series of studies from the Medical College of Georgia investigated differences between blacks and whites in responses to cycle ergometer testing. In the initial study, Alpert et al. (2) described racial differences among the 405 children discussed previously. They found no significant SBP differences at rest between blacks and whites, but when the groups were compared by BSA, blacks had significantly higher maximal SBP values. These differences remained significant when the data were expressed as changes in SBP from rest to exercise, or the percentage of this change. No published data, to our knowledge, have determined whether exercise BP responses of black patients with cardiac disease differ from those of their white counterparts.

Treiber et al. (63) replicated the finding by Alpert et al. Black children aged 4 to 6 had higher exercise SBPs than did whites. Treiber et al. discussed these findings with respect to prediction of essential hypertension. Arensman et al. (9) found that, in a population of 10-year-old boys, whites demonstrated higher cardiac output responses to exercise (supine) and blacks had greater systemic vascular resistance. The height of exercise SBP has been shown in adults to be highly predictive as a marker for later-onset essential hypertension. Mechanisms of exercise response in a biracial population of children must be considered.

It seems important that clinicians testing children with diseases should be able to compare the data to those of healthy children of the same race (as well as possibly gender, stage of development, altitude, etc.). Guo et al. (31) studied 294 healthy Asian children aged 5 to 14 using the Bruce protocol on a treadmill. The mean SBP of boys at maximal exercise increased by 38.6% compared to resting level; the girls increased by 34.4%. After maximal exercise, mean SBP reached resting level in 6 min, but DBP varied. We are unaware of data for large populations of Hispanics and other groups for use in clinical pediatrics.

Exclude Athletes From Sport Participation Based on a High BP Response?

Systolic blood pressure rises with increasing dynamic work as a result of increasing cardiac output, whereas diastolic blood pressure usually remains about the same or may be heard to zero in some normal subjects. An exaggerated systolic blood pressure response to exercise has been defined as a maximal value of ≥210 mmHg for men and ≥190 mmHg for women. A rise in diastolic blood pressure during exercise of >10 mmHg above the resting value or an absolute value of 90 mmHg also is considered abnormal and could predict an increased likelihood of CAD. Recommended relative indications for exercise test termination are a systolic or diastolic blood pressure of >250 and >115 mmHg, respectively. An exaggerated systolic blood pressure response to exercise could indicate an increased risk for future hypertension, left ventricular hypertrophy, and cardiovascular events. However, there is no evidence to suggest that a high blood pressure response to exercise in an athlete is an indication of an increased risk for sudden cardiac death. Systolic blood pressure at maximum exertion or at immediate cessation of exertion is considered a clinically useful first approximation of the heart's inotropic capacity. The absolute value should not be used as a reason to exclude athletes from sport participation.

Athletes

In a review of studies of normotensive and hypertensive youth, Dlin (18) summarized data on children who had been trained by either dynamic or static exercise. He noted that highly trained adolescents had higher systolic BP response values than untrained youth. Because training produces lower heart rates both at rest and during exercise, Dlin hypothesized that the higher BP occurred because athletes can achieve higher work rates at equivalent heart rate values than untrained individuals and that these higher work rates lead to higher SBP. By increasing physical fitness, SBP max increases. Thus, the maximal delta SBP (i.e., the difference between SBP max and resting SBP) reaches greater values in athletes (61).

Turmel et al. (64) evaluated the BP response to exercise in 44 endurance athletes during intense training and following 3 wk without training. Other parameters such as 24 h ambulatory BP monitoring, a 24 h Holter assessment, and sampling of blood were also obtained. Of the 11 athletes who had an exaggerated systolic BP response to exercise during training, seven of them had a similar response during the nontraining period. Athletes with an exaggerated BP response during the training period had higher SBP values on 24 h ambulatory BP monitoring. Those with an exaggerated BP response during the nontraining period had higher levels of markers of heart rate variability (SDNN and pNN50) and lower apolipoprotein-A1 levels. It was postulated that these findings may indicate the first marker of vascular abnormalities.

Thus, the rate at which the BP rises during exercise is more important than an absolute high value of maximal BP. In fact, an exaggerated maximal BP can contradictorily be a sign of fitness in athletes. How then can we differentiate the high maximal BP seen in a person with lower cardiovascular fitness? The BP rises rather slowly in physically fit persons. This is in contrast to those with lower levels of fitness or deconditioned persons in whom there is an early rapid rise in BP (14).

Obese Adolescents

Among adults, blood pressure response to exercise is usually higher in obese subjects than in lean subjects (17). The same observation has also been made in prepubertal children (12).

When to Terminate Exercise Testing Based on Blood Pressure Response

Data are available for healthy children and adolescents for both SBP and DBP, as shown in tables 6.1 and 6.2. No investigator has found an exaggeration of peak SBP in an otherwise healthy exercising subject to be predictive of complications. Therefore we do not recommend that an exercise test be stopped because an arbitrary value of SBP has been reached. There is no definite evidence of an intrinsic danger of a SBP value of 230, 240, or 250 mmHg. We do, however, believe that SBP must remain within a range that the tester can measure. If the SBP exceeds a measurable value, or if symptoms develop, then there is adequate justification to stop the test. Of note, the American Heart Association states

that clinical judgment should always be used, and test termination is usually indicated if there is severe hypertension; i.e., SBP > 250 mmHg and DBP > 120 mmHg (23). A drop in SBP of more than 10 mmHg from baseline BP despite an increase in workload, when accompanied by other evidence of ischemia, is an indication to suspend the test.

The normal DBP response to exercise in children is not widely agreed upon. The investigations described previously found varying and conflicting trends. It is highly unusual, however, for children exercising on a treadmill to have significant increases in DBP.

Prognostic Value of Exercise BP Testing

There are several areas of ongoing research into the clinical application of BP responses to exercise, including the prediction of future cardiovascular risk such as hypertension and heart failure.

Future Hypertension and Cardiovascular Disease Risk

Early detection of hypertension can prevent critical damage to target organs. There is evidence that in adolescents and young adults, an exaggerated exercise BP response is a predictor of future sustained hypertension (19, 35, 71). An exercise SBP >195 mmHg predicts a two- to threefold increased risk of future sustained hypertension in normotensive volunteers. Even in children, an exaggerated BP response is a prognostic factor for hypertension (40).

Mahoney and colleagues found that the rise in SBP and the exercise DBP correlate better with left ventricular hypertrophy (LVH) than does resting BP (43). This is significant because LVH is an independent risk factor for future cardiovascular disease (38).

As previously mentioned, during maximal exercise, trained athletes often achieve greater SBP increases than untrained persons, which does not necessarily put them at an increased risk for developing hypertension. These data support the concept of evaluating the SBP and DBP with relatively low submaximal exercise workload to ascertain an exaggerated BP response (39). When testing at these lower levels of exercise, there are additional advantages of necessitating minimum cooperation from the subjects. This is helpful during testing because it curbs the effect of confounders such as level of conditioning and the impact of exercise duration on the test results.

Several studies in adults have illustrated that the DBP response to exercise is a similar or greater predictor than SBP for future hypertension, in particular with DBP max values greater than 90/100 mmHg or an increase > 10 mmHg (39-41).

Risk of Heart Failure

The intolerance to exercise observed in patients with heart failure is correlated with the prognosis (42). Guimarães and colleagues studied 26 children with stable chronic heart failure with an ejection fraction of <45% due to idiopathic dilated cardiomyopathy (30). Due to the depressed myocardial contractility and possibly medications that can affect chronotropic response, children with heart failure had significantly lower SBP values than did healthy subjects. The depressed BP response to exercise has been previously reported in numerous adult studies with heart failure (65, 68).

Special Conditions

Specific research has been performed for various congenital heart disease diagnoses, particularly left-sided obstructive lesions. In addition, we will discuss more complex lesions, including patients with tetralogy of Fallot, single ventricle physiology, and transplants.

Congenital Heart Disease

Children with CHD often have impairment of their functional capacity that may occur preoperatively or postoperatively, as well as in the long term. Multiple factors contribute to their poor functional capacity, including the disease process itself, related to its treatment or to the secondary detraining that occurs from being less active due to the condition (60). Lunt et al. established that in comparison to healthy adolescents, adolescent patients with CHD were less likely to reach minimum exercise requirements and to execute vigorous exercises (41).

The types of CHD may be categorized as obstructive (pressure overload such as aortic stenosis and coarctation of the aorta), left-to-right shunts, cyanotic heart disease (tetralogy of Fallot), and volume overload (such as aortic regurgitation and mitral regurgitation).

Aortic Stenosis

In aortic stenosis, the left ventricular ejection time is prolonged at rest and also during exercise, adjusting for changes in cardiac rate. It is widely believed that as the aortic valve (or sub-valve or supra-valve) stenosis becomes more severe, the SBP response to exercise falls significantly. In some patients at peak exercise, SBP may rise only by 10 to 20 mmHg above the resting level. In rare cases exercise SBP may fall below the resting value, reflecting serious impairment of left ventricular stroke output.

Alpert et al. (4) published data from 29 patients (19 males and 10 females) with aortic stenosis who were 11.8 ± 3.9 (SD) years old. The maximal SBP values in these patients in response to treadmill exercise were compared to data from 116 control subjects of comparable age (12.2 ± 3.6) and sex distribution (68 males, 48 females). Only one patient's exercise test was terminated prior to maximal voluntary effort because of ischemia. The increase from resting to maximal exercise (delta SBP) was 30.3 mmHg in the patients with aortic stenosis, compared to the delta SBP in controls of 43.1 mmHg ($p < 0.001$). A delta value of exercise SBP greater than 35 mmHg was of excellent predictive value in patients with aortic stenosis. If a patient's SBP rose by more than 35 mmHg during exercise, then the patient had only a 10% chance of having a catheterization gradient that exceeded 50 mmHg (i.e., the patient had a 10% chance of having moderate or severe aortic stenosis). Thus, the exercise responses of the patients with aortic stenosis were lower than those of the control subjects; the authors speculated that "blood pressure measurement during exercise may increase the clinician's ability to select for catheterization only those patients with aortic stenosis who are likely to require surgery."

Two reports from Cincinnati, authored by Whitmer et al. (70) and James et al. (38), also addressed the patient with aortic stenosis, both pre- and postoperatively. Whitmer et al. reported results from cycle ergometer testing in 23 patients (19 males, 4 females) within 6 mo before and 3 to 30 mo after surgical intervention for valvular or discrete subvalvular aortic stenosis. The patients were aged 5 to 19 (mean age 10.7). In 16 patients with both pre- and postoperative SBP responses to exercise, a significant increase ($p < 0.025$) was noted, from a mean (± SD) of 121 ± 22 mmHg to 143 ± 33 mmHg. James et al. (38) studied only preoperative patients with valvular aortic stenosis or subvalvular aortic stenosis with varying severities.

Sixty-five patients aged 4 to 24 (mean age of 12; 56 with valvular and nine with subvalvular aortic stenosis) underwent cycle ergometer testing. For patients with the most severe aortic or subaortic stenosis, peak exercise SBP was lower than that of the control subjects ($p < 0.03$), and it tended to be the lowest in patients with the highest resting left ventricular-to-aortic pressure gradients. Systolic BP decreased during exercise to levels below resting values in 32% of patients whose resting gradient was greater than 70 mmHg.

Severe aortic valve stenosis, defined as a mean Doppler gradient >40 mmHg or a peak Doppler gradient >70 mmHg in a symptomatic patient, is a contraindication to exercise testing (25). It is reasonable to consider exercise testing in moderate aortic stenosis (mean Doppler gradient of 25-40 mmHg or a peak Doppler gradient of 40-70 mmHg) (50). It is particularly useful in providing advice for sport participation and prognostic information when there is discordant clinical and echocardiographic data. In the absence of symptoms and the presence of a normal exercise test, the current sport participation guidelines for this subset of patients is low static and low–moderate dynamic exercise (e.g., golf, bowling, baseball, softball, volleyball). In summary, patients with aortic stenosis have a suboptimal or a blunted BP response to exercise.

Coarctation of Aorta

After coarctectomy, some patients have an elevated SBP during submaximal exercise and develop a peak SBP difference between the upper and lower extremities. After coarctation repair, abnormally high SBPs commonly occur with exercise, even in patients who have a normal BP at rest. There are many postulated reasons for this. From a mechanical perspective, occult coarctation can lead to an elevated BP response. Alterations in baroreceptor function (29) and hyperresponsiveness of the renin-angiotensin system (54) usually persist even after repair. In addition, subjects after coarctation repair are thought to have an abnormal response to sympathomimetic agents (33), and they may have structural alterations in their central and peripheral arteries that lead to a persistence in upper body peripheral vascular resistance. In children with coarctation of the aorta, the left ventricle responds to the increased afterload initially with a hyperdynamic systolic function, increased left ventricular mass with time, and subnormal left ventricular wall stress. These maladaptive changes tend to persist after

coarctation repair even in normotensive patients with no significant resting SBP gradient (32). It is theorized that there is increased aortic stiffness after coarctation repair. The cumulative effect of aortic stiffening along with possible mild aortic narrowing after repair results in left ventricular pressure overload. This leads to persistence of left ventricular hyperdynamic function and hypertrophy (32).

Two articles of clinical importance advocated for exercise BP measurement in the routine evaluation of patients with coarctation. Connor (15) and Connor and Baker (16) described the use of the arm–leg SBP gradient as a measure of significant coarctation or residual (recurrent) coarctation. Connor noted that an arm–leg gradient immediately following exercise that exceeded 35 mmHg suggested that recatheterization was indicated. The exercise study, by increasing cardiac output and reducing systemic vascular resistance, was able to unmask a fixed obstruction at the level of the coarctation or coarctation repair.

The concept of unmasking an obstruction is important. Even when an angiogram, computed tomography (CT), or magnetic resonance imaging (MRI) does not suggest that a significant coarctation is present, a patient may exhibit an exaggerated SBP in response to maximal dynamic exercise. The physician may also wish to consider whether studies to define the presence or absence of cerebral aneurysms (Berry aneurysms) are indicated. We do not routinely perform these studies. To date, no data on healthy individuals exist (to the best of our knowledge) that could help determine whether any arbitrary SBP level is dangerous to any specific patient. In addition, the vascular changes similar or identical to those in essential hypertension may already be present and may progress relentlessly despite adequate resection of the coarctation. The landmark study by Maron et al. (45) demonstrated without a doubt that hypertension in coarctation can have devastating consequences.

At the same time as Connor's papers appeared, Freed et al. (24) reported data on exercise-induced systolic hypertension in 30 patients aged 6 to 30 after repair of coarctation. The study used a Bruce treadmill protocol to voluntary exhaustion as the end point. Twenty patients with mild cardiac disease served as the control population. As expected, the patients with coarctation had higher postexercise SBP than did the control patients ($p < 0.001$). The arm–leg SBP difference in the coarctation patients increased from 10 mmHg at rest to

69 mmHg after exercise ($p < 0.001$). The arm SBP value correlated highly ($r = .91$) with the arm–leg gradient; the authors suggested that the residual obstruction at the coarctation resection site formed a mechanical impediment to flow and thus caused the postexercise SBP elevations in the arm.

The influence of different surgical procedures on hypertension after repair of coarctation has also been studied extensively. Connor and Baker (16) compared the arm–leg SBP gradient postexercise for a group of patients whose coarctations had been repaired using a Dacron patch angioplasty with that of patients treated with end–end anastomosis (before the era of absorbable sutures). The mean arm–leg gradient postexercise was 6.8 mmHg in the former group and 36.1 mmHg in the latter, despite very small BP differences between the groups at rest. Markel et al. (44) studied 28 children who had undergone repair of coarctation by end–end anastomosis or by patch angioplasty. Maximal SBP after exercise was equivalent to control for patients with no arm–leg gradient either pre- or postexercise. They concluded that patients with no arm–leg gradient at rest and systolic hypertension were very unlikely to have operable recoarctation and should receive antihypertensive therapy or be limited from strenuous physical activity. Smith and his coworkers (59) compared BP at rest and with exercise in 50 patients who were post-coarctectomy (end–end anastomosis or synthetic patch aortoplasty). The SBP mean was higher in the patients with end–end anastomosis than in those with patch aortoplasty. The arm–leg pressure difference was also greater in the former group. These data suggest that the patch aortoplasty was superior to end–end anastomosis in that group's experience.

The technique of subclavian flap angioplasty was in use during this era, but too few patients were old enough to exercise to allow comparison of surgical data. In assessing series such as this, the clinician should be careful to define whether age at operation, age at exercise testing, and time since surgery are known so he or she can interpret the results most meaningfully.

More recently, to research the development of hypertension in children after repair of coarctation by either end–end anastomosis or construction of a subclavian flap, Giordano and colleagues (26) measured resting and exercise hemodynamics as well as 24 h ambulatory BP. Both groups were followed for a mean of 13 yr after surgery. The subclavian flap group had a higher incidence of late hypertension as evidenced by 24 h SBP and DBP levels and SBP levels during exercise in comparison to the end–end anastomosis group. The authors opined that due to the greater resection of abnormal aortic tissue when repaired by end–end anastomosis, the residual aortic stiffness was lower.

From the studies presented, it appears that the SBP level as well as the arm–leg gradient is of use for clinicians seeking to define whether a residual or functional recoarctation is present. The severity of narrowing in coarctation can be assessed by blood pressure differences between the arms and legs. This correlates well with invasive hemodynamic evaluations (28). Therefore, the success of surgical repair in coarctation is assessed based on the absence or presence of these gradients and the level of difference in gradients. However, exercise arm–leg gradients may not always correlate with narrowing of the reanastomosed region measured by imaging (39). Increased cardiac output during exercise leads to an increase in arm BP, while vasodilation in the working leg muscles leads to a small decrease in leg BP (22). Thus, an arm–leg gradient of 40 to 50 mmHg may reflect physiological circulatory adaptations to exercise. In summary, significantly elevated exercise gradients in patients after coarctation repair can indicate residual narrowing but usually are a reflection of the hypertensive response of the arms, which is a known entity seen long-term in these patients. The persistent hypertensive response affirms the idea that arterial dysfunction of the precoarctation vascular bed is not influenced by surgery (32).

Volume Overload Conditions

There are limited studies on the effect of volume overload conditions (aortic and mitral valvular regurgitation, left-to-right shunt, and chronic anemia) on exercise BP. In chronic aortic regurgitation, the left ventricle (LV) is subject to excess preload and afterload. The more the regurgitation, the more the volume overload on the LV and hence the more excess preload. The regurgitation also increases the LV end-diastolic volume. This in turn increases the LV wall stress that increases afterload. In addition, systolic hypertension created from the increased stroke volume that is ejected into the high-impedance aorta further increases LV afterload. The excess preload and afterload in moderate to severe aortic regurgitation eventually leads to worsening LV dilation and systolic dysfunction.

Alpert et al. (6) studied exercise responses in 137 patients with either left ventricular pressure (aortic stenosis or coarctation) or volume overload (mitral or aortic regurgitation). There were 70 patients with aortic stenosis, 25 with coarctation, 20 with aortic regurgitation, and 22 with mitral regurgitation. They compared these patients to 405 healthy control children. The patients with aortic regurgitation had maximal SBP that exceeded the value predicted from the control data. These patients' values were very close to those obtained in patients with coarctation; the values exceeded those for both patients with aortic stenosis and patients with mitral regurgitation. The physiology of patients with coarctation and aortic regurgitation differed greatly, but their SBP responses exceeded the normal response significantly. In the small number of subjects with mitral regurgitation, BP responses did not differ from those of the control subjects.

Tetralogy of Fallot

Progressive arterial desaturation occurs at low levels of work in children with cyanotic heart disease. After "corrective" surgery, maximal work capacity, heart rate, and stroke volume persist below normal (52). Cardiac output is low for the amount of oxygen consumed. The BP response to exercise is usually blunted.

Hirschfeld et al. (34) studied 28 patients aged 7 to 30. To test the adequacy of the repair from a hemodynamic standpoint, a cardiac catheterization was done. This was compared with their exercise performance as evidenced by a graded exercise test. Patients were evaluated 2 to 9 yr after surgery. At catheterization, 23 of the 28 patients had acceptable surgical results. Maximal working capacity, maximal heart rate, and exercise BP were diminished when compared to normal values despite the satisfactory hemodynamics obtained by cardiac catheterization. In a study of 135 patients performing 279 exercise studies, Wessel et al. (69) reported the SBP in 40 selected patients post-intracardiac repair. They reported a mean value for maximal SBP of 137.5 ± 18 (SD) mmHg, with a range from 115 to 175 mmHg. These data are not compared with control subjects from the same laboratory but appear lower than the normal values obtained in other cycle ergometer studies. More recently, Sarubbi et al. confirmed the preceding findings, as well (57).

Ebstein's Anomaly

In Ebstein's anomaly, due to the significant cardiomegaly, there are respiratory limitations as well. Due to the cardiorespiratory constraints, there is decreased exercise tolerance. Barber et al. (10) exercise tested 14 patients with unrepaired Ebstein's anomaly. These patients had a suboptimal or blunted BP response. They were all deconditioned, with a significantly increased heart rate at rest. In addition, they had lower levels of exercise time, total work, maximal oxygen uptake, and oxygen saturation.

Dextro-Transposition of the Great Arteries

Mathews et al. (47) reported the results of treadmill exercise in 21 patients following repair of d-transposition of the great vessels. The children were aged 4 to 15, with a mean age of 9, and were asymptomatic. All had undergone the Mustard operation. The patients' data were compared to those of 61 control youngsters whose mean age was 14, with a range from age 9 to 20. There was no difference between patients and control subjects for maximal SBP or DBP response. The values for maximal SBP were, however, all in the lower range of normal.

A more recent prospective follow-up study looked at long-term exercise capacity after a neonatal arterial switch operation. Fifty-four of the 56 patients (96.4%) had normal exercise capacity. There was no difference in the HR and SBP response to exercise in comparison to controls.

Single Ventricle

Numerous studies have included exercise data both pre- and post-Fontan operation. Driscoll and his coworkers (20) from the Mayo Clinic reported results from 81 patients pre-Fontan, 33 with tricuspid atresia, 38 with univentricular heart, and 10 with other complex congenital heart disease. They compared these results to 29 patients post-Fontan. The SBP responses were reduced ($p < 0.05$) compared to those of the control subjects. The SBP values were expressed as a percentage of predicted values. Surprisingly, the pre-Fontan group achieved 90% and the post-Fontan group achieved 85% of predicted values. There were similar reductions in exercise DBP both pre- and post-Fontan. The Mayo group (72) reported pre- and post-Fontan data for 20 patients. The SBP and DBP were within the normal range at rest and with exercise both pre- and postoperatively. The DBP showed a significant increase following the Fontan operation.

Sickle Cell Disease

Children with sickle cell disease have lower exercise values of SBP and work rate. This was shown by Alpert et al., who compared data from 47 children with sickle cell disease aged 5 to 18 to data from 170 healthy, black, age-matched controls (5). Patients who demonstrated ischemia on the exercise electrocardiogram, and males in general, had the lowest exercise SBP responses. The degree of impairment was directly correlated to the hemoglobin value.

In a more recent study, McConnell et al. reported SBP responses from 43 patients with sickle cell anemia (48). Systolic BP responses in sickle cell anemia patients who showed ST segment depression to cycle ergometer exercise were higher than those in sickle cell anemia patients who did not. No comparisons with healthy subjects were performed. Only patients with hemoglobin values below 8.5 g/dl had ST segment depression. The mechanism of the comparatively higher SBP in the presence of myocardial ischemia is unknown.

Transplant

Blood pressure response to exercise stress testing has been studied in heart and renal transplant recipients. Ehrman et al. studied a group of orthotopic heart transplant ($n = 28$) and heart surgery ($n = 19$) patients with similar resting ejection fractions and left ventricular end-diastolic pressures (21). They were exercised to a symptom-limited maximum workload to assess differences in cardiovascular and gas exchange responses. Testing was performed at a mean of 3 and 6 mo after surgery, respectively ($p < 0.05$). Transplant patients had a higher resting SBP and DBP ($p < 0.01$) and a significantly higher heart rate ($p < 0.01$) at rest in the supine and standing positions and during min 2 through 7 of supine recovery. No significant differences were found for SBP during recovery or peak heart rate. They concluded that SBP response is more appropriate than heart rate for assessing recovery of the denervated heart after maximal exercise.

Giordano et al. (26) assessed exercise tolerance and BP response to treadmill exercise in children after renal transplantation. At comparable workloads in comparison to healthy children, the patients had reduced exercise tolerance, increased HR, and increased maximum SBP. Those off antihypertensive medications had a higher maximum SBP and HR.

Interpretation of Results

Several studies have addressed the SBP and DBP levels achieved during maximal dynamic exercise. There is debate over whether an absolute SBP value can be too high and whether the DBP should change. Thus, it is necessary to use normal values for changes in SBP related to variables such as cardiac output, stroke volume, workload, oxygen consumption, heart rate, and respiratory exchange ratio so that the magnitude of BP response may be judged with respect to physiological changes needed to perform work.

In general, children with a high exercise BP usually have a higher baseline BP. Thus it is important to note the baseline BP when making interpretations of the SBP response because the exercise BP prognostic value may be due to tracking effect. Using delta SBP between rest and exercise may be of more benefit, especially in hypertensive patients with higher resting SBP but with an absolute increase in exercise BP similar to that in normotensive subjects.

In summary, significant increases in the SBP and DBP during effort, low or falling levels of SBP during effort, low amplitude of delta SBP, and slow recovery of the SBP are considered abnormal BP responses and appear to be of significant prognostic value for future outcomes, such as hypertension, status of heart transplant and heart failure patients, and cardiovascular events.

Conclusion

Exercise BP is an important variable in decisions relating to many clinical conditions and can be of prognostic significance. Several studies have addressed the SBP and DBP levels achieved during maximal dynamic exercise. Systolic BP rises with increasing dynamic work as a result of increasing CO, whereas DBP usually remains the same or moderately declines. There is a debate over whether a SBP value can be too high and whether DBP should change. Blood pressure must perfuse the exercising muscle; values that are too high or too low may reduce the efficiency of work physiology. Children with persistently elevated resting BPs continue to have higher BPs during exercise. It is not clear whether a higher exercise BP in an otherwise normal child will lead to a higher chance of hypertension in the future. This opens a new vista of research in this cohort of children. Follow-up studies in these children with a higher-than-expected exercise BP would be of value.

Maximal Oxygen Uptake

Ali M. McManus, PhD, and Neil Armstrong, PhD, DSc

Maximal oxygen uptake ($\dot{V}O_{2max}$) is the highest rate at which a child can consume oxygen during exercise. It provides a composite measure of the pulmonary, cardiovascular, and hematological components of oxygen delivery and the mechanisms of oxygen utilization in the exercising muscles. It has long been regarded as the best single measure of aerobic fitness (21). Although $\dot{V}O_{2max}$ provides an elegant assessment of a person's ability to perform aerobic exercise, it does not fully describe all aspects of aerobic fitness. Rapid shifts in skeletal muscle metabolism from rest to exercise of varying intensities, similar to the physical activity patterns noted in children (64), are best described by the pulmonary oxygen uptake kinetic response, which provides a superior measure of the integrated response of oxygen delivery and the metabolic requirements of rapid changes in exercise intensity (19). Similarly, a child's ability to sustain submaximal aerobic exercise, also a useful indicator of aerobic fitness, is better described by blood lactate accumulation than $\dot{V}O_{2max}$ (13). Regardless of these limitations, interest in the $\dot{V}O_{2max}$ of children has not waned since the first laboratory investigations of boys in the 1930s and girls and boys in the 1950s (20, 86). Indeed, it "has become the most researched variable in paediatric exercise science" (14, p. 269).

In the clinical setting it is very common to conduct exercise tests without measuring $\dot{V}O_{2max}$, and the results are routinely used to evaluate exercise tolerance and maladaptive responses to exercise such as cardiac arrhythmias. Usually an exercise test is terminated when the child reaches his or her volitional maximal tolerance (e.g., time to exhaustion) or when clinical symptoms or electrocardiographic findings indicate abnormal function such as arrhythmia. The additional assessment of $\dot{V}O_{2max}$ during a clinical exercise test provides both diagnostic and prognostic advantages. $\dot{V}O_{2max}$ provides a physiological marker of exercise tolerance and overcomes some of the difficulties of determining whether voluntary maximal effort reflects true aerobic fitness or motivation. $\dot{V}O_{2max}$ is a primary marker for determining the response of the cardiorespiratory system to exercise intervention (19) and therefore offers a robust measure for tracking the response to pharmacological or surgical intervention. $\dot{V}O_{2max}$ has been used as a predictor of mortality and of hospital admissions (47, 72, 81) and can also be used as a proxy for cardiac functional reserve (78), thus improving our ability to judge the severity of functional impairment. Exercise testing with $\dot{V}O_{2max}$ assessment therefore improves our ability to assess exercise tolerance, diagnose a problem, determine whether exercise, pharmacological, or surgical intervention is needed, and monitor the effectiveness of interventions and the progression of disease.

Despite the proven usefulness of the $\dot{V}O_{2max}$ test in assessing dysfunction (106), achieving quality and consistency of data in children remains a problem. Various methodological considerations are often overlooked during $\dot{V}O_{2max}$ testing in children and adolescents. $\dot{V}O_{2max}$ varies with age, maturity, and sex. When coupled with the influence of body size and composition (3, 5) and additional issues related to measurement, it is easy to see how tests can be misinterpreted.

In this chapter we will provide an overview of the basic principles of the physiological responses to aerobic exercise in children and adolescents. For the purpose of this chapter we define children as those 12 years and under and adolescents as 13- to 18-year-olds. We will discuss the measurement of $\dot{V}O_{2max}$, highlighting methodological issues pertinent to children and adolescents, and we will discuss the development of $\dot{V}O_{2max}$ in relation to sex, age, maturity, body size, and body composition. Finally, we will discuss what is considered normal. Examples of $\dot{V}O_{2max}$ data across a range of disorders are provided for reference.

Exercise Metabolism in Children and Adolescents

Skeletal muscle metabolism during exercise can be evaluated from the pulmonary oxygen uptake ($p\dot{V}O_2$) kinetic response. The $p\dot{V}O_2$ kinetic response is triphasic, with phase I depicting an immediate increase in $p\dot{V}O_2$ at the onset of exercise that signals a rise in cardiac output ($\dot{Q}$) and is independent of muscle $\dot{V}O_2$. Phase II follows with an exponential increase in $p\dot{V}O_2$, which provides a very close reflection (within about 10%) of the muscle $\dot{V}O_2$ kinetics and is described by a time constant (τ). The smaller the τ, the greater the aerobic contribution to adenosine triphosphate (ATP) resynthesis, a marker of enhanced aerobic fitness. Within the moderate intensity exercise domain, $p\dot{V}O_2$ then attains steady state (phase III); however, during heavy exercise, a slow component manifests at phase III, and this reflects a loss of muscle efficiency and ensuing fatigue (18).

Children have a faster phase II $p\dot{V}O_2$ kinetic response than adults (42), which supports the argument that during childhood there is a greater reliance on aerobic metabolism (15). Boys show a faster phase II τ and truncated slow component than girls during exercise above the ventilatory (or anaerobic) threshold (43). Obese children have also shown a slower phase II τ and elongated slow component (58). Essentially, a slower phase II response and extended slow component make rapid and frequent transitions between exercises of varying intensities energetically more challenging. The oxygen uptake kinetic response provides insight into aspects of exercise metabolism not available from a $\dot{V}O_{2max}$ test and may have considerable clinical relevance. Much is to be gained from the transient kinetic response to exercise, and those interested can find more comprehensive descriptions of oxygen uptake kinetic assessment in youth elsewhere (15).

Physiological Responses to Aerobic Exercise

When a child or adolescent undertakes an acute bout of aerobic exercise, a series of physiological adjustments are made to ensure that oxygen is delivered for use at the working muscles, heat generated by the muscle is removed, and an adequate supply of blood is maintained to the brain and heart (69). Oxygen uptake ($\dot{V}O_2$) by the muscles has conventionally been described using the Fick principle, where $\dot{V}O_{2max}$ is the product of maximal cardiac output ($\dot{Q}_{max}$) and maximal arterial-venous oxygen difference (a-vO_2 diff$_{max}$).

The relative contributions of the Fick components of $\dot{V}O_{2max}$ have yet to be clarified in young people; however, when components of the Fick equation are compromised as they are in disease states, exercise capacity and $\dot{V}O_{2max}$ will diminish. We will consider this assertion by reviewing the response of each of these components during exercise in the child and adolescent and providing examples of health conditions that result in compromised function.

Cardiac output at maximum reflects the delivery of oxygen to the working muscle and is the product of maximum heart rate (HR_{max}) and maximum stroke volume (SV_{max}), as illustrated in figure 7.1. Data consistently demonstrate that, in healthy children and adolescents, HR_{max} at $\dot{V}O_{2max}$ is independent of age, sex, and aerobic fitness (5, 24, 105). Exceptions include youngsters with complete heart block, those using medication such as β blockers, and those with chronotropic incompetence where a diminished HR response to exercise results, although this does not always limit functional capacity (24, 82). During a progressive exercise test to maximum, an almost linear rise in HR is expected until about 75% of $\dot{V}O_{2max}$. At that point HR levels off to a value at or before $\dot{V}O_{2max}$ that is dependent upon the ergometer (29) and the testing protocol used (7). An incremental treadmill running protocol generates a HR_{max} of 200 ± 7 bpm, whereas with an incremental cycle ergometer protocol a HR_{max} of 195 ± 7 bpm would be expected in healthy children and adolescents.

Stroke volume, therefore, is the principal factor governing alterations in $\dot{Q}_{max}$ during exercise. Accurate measurement of stroke volume (SV) and $\dot{Q}$ during exercise, particularly maximum exercise, is one of the most challenging measures in exercise physiology (112) (see chapter 8). The direct Fick method has been the gold standard for assessing $\dot{Q}$ in adults, but since this method requires catheterization, it is a high-risk procedure and would generally only be used in children whose treatment protocol required catheterization (112). Noninvasive approaches include acetylene or carbon dioxide rebreathing, Doppler echocardiography, and impedance cardiography. Reliability of the Doppler echocardiographic method is good, with the

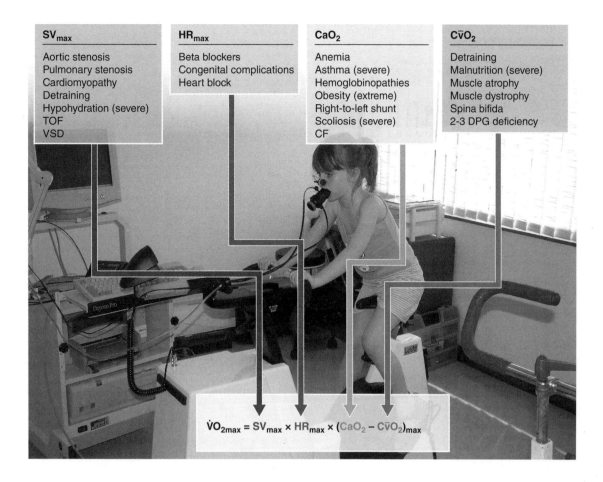

SV$_{max}$

Aortic stenosis
Pulmonary stenosis
Cardiomyopathy
Detraining
Hypohydration (severe)
TOF
VSD

HR$_{max}$

Beta blockers
Congenital complications
Heart block

CaO$_2$

Anemia
Asthma (severe)
Hemoglobinopathies
Obesity (extreme)
Right-to-left shunt
Scoliosis (severe)
CF

C$\bar{v}$O$_2$

Detraining
Malnutrition (severe)
Muscle atrophy
Muscle dystrophy
Spina bifida
2-3 DPG deficiency

$$\dot{V}O_{2max} = SV_{max} \times HR_{max} \times (CaO_2 - C\bar{v}O_2)_{max}$$

Figure 7.1 Maximal oxygen uptake described by the Fick equation, with examples of health conditions that result in compromised function.

CF = cystic fibrosis. TOF = tetralogy of Fallot. VSD = ventricular septal defect.

Equations are reprinted from Bar-Or and Rowland 2004.

coefficient of variation ranging from 5.2% to 8.1% for $\dot{Q}$ (75, 91, 92) and about 8.5% for SV (91). The only study to investigate reliability coefficients for maximal $\dot{Q}$ and SV using impedance cardiography reported a coefficient of variation of 9.3% for both (109). Gas rebreathing has the lowest reliability, with the coefficient of variation for $\dot{Q}$ around 12% (74). The Doppler technique offers the most reliable method for the assessment of $\dot{Q}$ and SV, but it requires highly skilled personnel (94). Although less reliable than Doppler, impedance cardiography may offer an acceptable assessment of $\dot{Q}$ and SV without the expertise that Doppler requires.

Cardiac index (CI), which is $\dot{Q}$ normalized to body surface area, is between 8 and 12 l · min^{-1} · m^{-2}) at maximum, with differences generally dependent on sex and training status (table 7.1). Values for maximal stroke index (SI$_{max}$ – SV normalized to body surface area in ml · m^{-2}) are also provided in table 7.1. SI$_{max}$ is normally between 45 and 60 ml · m^{-2}, although this is somewhat higher in boys than in girls and is also dependent on training status. The 30% to 40% increase in SV from rest to maximum is largely dependent upon an increase in end-diastolic volume or a decrease in end-systolic volume.

Myocardial contractility is unlikely to play a role in the healthy child because myocardial contractile capacity is not altered by aerobic fitness (100). In contrast, a recent study of asymptomatic children with repaired tetralogy of Fallot demonstrates reduced right ventricular strain rate, which was associated with a reduced $\dot{V}O_{2max}$ (44). There are various diseases that affect myocardial function

Table 7.1   Stroke Index, Cardiac Index, and Arterial-Venous Oxygen Difference at Maximal Exercise on a Cycle Ergometer

Reference	Sex	Age (yr)	Method	HR_{max} (bpm)	SVI_{max} (ml·m^{-2})	CI_{max} (l·min^{-1}·m^{-2})	a-vO_2 diff$_{max}$ (ml·100 ml^{-1})	$\dot{V}O_{2max}$ (ml·kg^{-1}·min^{-1})	$\dot{V}O_{2max}$ (l·min^{-1})
Cumming 1977 (33)	F	11.8 ± 3.1	Catheterization	174 ± 11	46 ± 3[†]	8.6 ± 1.8[†]	—	—	—
	M	12.6 ± 3.5	Catheterization	170 ± 17	56 ± 13	10.1 ± 1.8	—	—	—
Rowland et al. 1997 (90)	M	10.9 ± 1.3	Doppler	193 ± 10	59 ± 11	11.33 ± 2.32	13.9 ± 3.0	50.9 ± 8.3	1.9 ± 0.3
Rowland et al. 2000 (93)	F	11.7 ± 0.5	Doppler	198 ± 9	55 ± 9[†]	10.9 ± 1.7[†]	12.3 ± 1.9	40.4 ± 5.8[†]	1.84 ± 0.31
	M	12.0 ± 0.4	Doppler	199 ± 11	62 ± 9	12.3 ± 2.2	12.2 ± 1.7	47.1 ± 6.1	1.98 ± 0.28
Rowland et al. 2011 (99)	M	14.6 ± 0.8	Doppler	188 ± 7	59 ± 7[†]	11.1 ± 1.52[†]	18.1 ± 2.8	57.4 ± 4.8[†]	—
	M	15.3 ± 0.5	Doppler	195 ± 11	46 ± 10	9.02 ± 2.05	17.5 ± 3.7	44.4 ± 6.6	—
	F	14.6 ± 0.7	Doppler	189 ± 12	50 ± 5[†]	9.53 ± 0.99[†]	18.1 ± 2.7	43.5 ± 3.4[†]	—
	F	15.0 ± 0.6	Doppler	191 ± 9	41 ± 4	7.8 ± 1.03	17.6 ± 3.1	36.0 ± 5.1	—
Obert et al. 2003 (75) (pretraining control group)	F	10.4 ± 0.3	Doppler	202 ± 7	46 ± 6[†]	9.4 ± 1.2[†]	13.1 ± 2.8[†]	42.4 ± 5.6[†]	—
	M	10.5 ± 0.3	Doppler	202 ± 7	49 ± 5	9.7 ± 0.8	15.6 ± 1.5	51.5 ± 6.3	—
Winsley et al. 2009 (113)	F	10.2 ± 0.3	Impedance cardiography	192 ± 11	45 ± 6	8.7 ± 1.1	12.6 ± 1.6[†]	—	1.23 ± 0.08[†]
	M	10.1 ± 0.5	Impedance cardiography	195 ± 11	47 ± 8	8.9 ± 1.4	14.8 ± 2.1	—	1.41 ± 0.18

F = females. M = males. a-vO_2 diff$_{max}$ = maximum arterial-venous oxygen difference. CI_{max} = maximum cardiac index. HR_{max} = maximum heart rate. $\dot{V}O_{2max}$ = maximum oxygen uptake. SVI_{max} = maximum stroke index.

[†] = significant differences noted.

and will result in a reduction in SV. These include disorders that create an outflow obstruction, such as aortic stenosis, pulmonary stenosis, and tetralogy of Fallot; disorders that are associated with deficient contractility, such as cardiomyopathy and obesity; and disorders that result in deficient forward stroke volume, such as ventricular septal defect (VSD) and tetralogy of Fallot (24) (see figure 7.1). A number of studies have explored whether cardiac function is impaired in the obese youngster and whether this may relate to a suboptimal $\dot{V}O_{2max}$. Much of the data suggest obese and healthy weight youngsters have similar CI_{max} and global LV systolic function (39, 54, 76, 89). More recent use of speckle-tracking echocardiography has shown lower longitudinal left ventricular systolic strain in obese youngsters, but the effect on aerobic fitness remains unknown (54, 76).

Arterial-venous oxygen difference reflects the difference between arterial oxygen content (CaO_2) and venous oxygen content ($C\bar{v}O_2$) and provides a marker of oxygen extraction by the exercising muscle. Arterial oxygen content is primarily an outcome of hemoglobin concentration, which shows sex-divergent changes with advancing age. Values for hemoglobin prior to puberty are similar in girls and boys, at about $135 \text{ g} \cdot l^{-1}$ (35). Hemoglobin concentration continues to increase in boys until it reaches about $152 \text{ g} \cdot l^{-1}$ by age 16 years (35). Values are lower in girls, with hemoglobin concentration plateauing around $137 \text{ g} \cdot l^{-1}$ (35). Arterial oxygen content also relies on the maintenance of partial pressure of oxygen (P_aO_2), which is dependent upon alveolar ventilation, and pulmonary diffusion capacity. Therefore conditions that limit the oxygen-carrying capacity such as anemia, hemoglobinopathies, cyanotic heart disease, respiratory disorders (e.g., severe asthma and cystic fibrosis), and chest wall disorders (e.g., scoliosis and extreme obesity) can all result in arterial desaturation (24) (see figure 7.1).

The venous oxygen content reflects the ability to extract oxygen as it flows through the muscle and is determined by the amount of blood flowing through the muscle and capillary density. The changes in blood flow during exercise represent the increase in cardiac output and the redistribution of that output to the working muscle. There are also changes at the peripheral vascular bed with a decrease in vascular resistance and an increase in vasodilation, which further augments blood flow through the muscle (46). We can assume that a dense and open capillary bed increases capillary blood flow, reduces diffusion distances, and reduces the transit time of blood flowing through the exercising muscle, thus aiding oxygen extraction and therefore lowering venous oxygen content during exercise. Conversely, if muscle blood flow is disrupted, as in disorders such as muscle atrophy and dystrophy or in severe malnutrition, the oxygen content in venous blood remains high, indicating poor levels of oxygen extraction (24) (see figure 7.1).

Arterial-venous oxygen difference can be directly assessed using cardiac catheterization (53), but more commonly it is estimated by solving the Fick equation (a-vO_2 diff$_{max}$ = $\dot{V}O_{2max}$ / $\dot{Q}_{max}$) (113). Values at rest and during exercise differ little by age, with a-vO_2 diff increasing linearly with increases in exercise intensity and showing a plateau near maximal exercise (56, 93). Data on young people's a-vO_2 diff at $\dot{V}O_{2max}$ are provided in table 7.1, falling between 12.0 and $18.0 \text{ ml} \cdot 100 \text{ ml}^{-1}$. Since a-v$O_2$ diff$_{max}$ is independent of aerobic fitness, it is unlikely to limit $\dot{V}O_{2max}$ in the healthy child.

Measuring Maximal Oxygen Uptake in Children

The assessment of $\dot{V}O_{2max}$ requires measurement of gas exchange at the mouth, most commonly using a commercially available automated gas analysis system or metabolic cart. These devices are primarily designed for adults, so careful consideration of the type of system and related equipment is important. Key output variables from metabolic carts are the fraction of expired oxygen, the fraction of expired carbon dioxide, and the volume of inspired or expired air (figure 7.2). Metabolic carts therefore consist of O_2 and CO_2 gas analyzers and a turbine or pneumotachograph to measure flow and volume. Calibration of the gas analyzers with known concentrations of calibration gas, as well as calibration of the volume sensor, must occur before each test to ensure stable and accurate data.

Metabolic carts either use a mixing chamber, which stores sampled gas for periodic sampling, or use breath-by-breath technology where every breath is sampled. If the primary outcome is $\dot{V}O_{2max}$, then a mixing chamber system may be the more appropriate choice. However, the size of the mixing chamber is important, and using a large mixing chamber for a child with a small tidal volume may result in considerable measurement error (13). Breath-by-breath systems allow continuous measurement of gas concentrations and volume, with immediate reporting.

Gas sampling can either be set to a breath average in a breath-by-breath system (e.g., eight-breath average) or reported per increment of time in breath-by-breath and mixing chamber systems (e.g., 15 s average). Rather than assuming the manufacturer's default, sample intervals during a $\dot{V}O_{2max}$ test in children should be no larger than 30 s so that the data are not overly smoothed (68). Conversely, the sample rate must be of sufficient magnitude to ensure that the true physiological signal is distinguishable from the noise generated from large inter-breath variations notable in the child's normal pattern of breathing (83). Whatever the chosen sample rate, it should always be reported to allow cross-study comparisons.

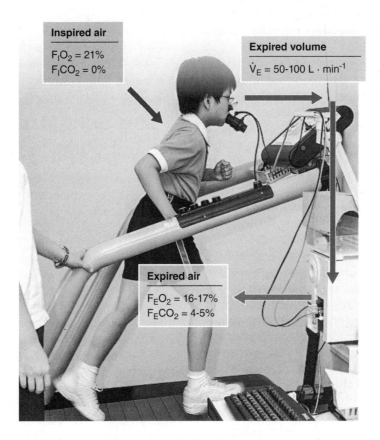

Figure 7.2 An illustration of the key components needed for the measurement of oxygen uptake using a metabolic cart.

Equations and labels are reprinted from Myers 1996.

Modern gas analysis systems use a small, lightweight turbine or a pneumotachograph, and these are connected to the child via a mouthpiece (with nose clip) or face mask (figure 7.3). Since most commercial gas analysis systems come with adult-sized mouthpieces and face masks, it is imperative that child-sized mouthpieces and face masks be purchased to ensure the comfort of the child and the quality of data obtained. The combined dead space of the turbine or pneumotachograph plus mouthpiece or face mask should be considered, and the experimental setup should result in a dead space of about 60 ml for children with a body surface area >1.0 m² or 35 ml in the smaller child (103). While many children are comfortable with a mouthpiece and nose clip, this is impractical for those with very small noses, and for some children the face mask is easier to use (65). Nevertheless it is very important to avoid leakage with masks, and leakage is common in children with small faces and noses.

Ergometer Choice

The $\dot{V}O_{2max}$ test can be conducted on a variety of ergometers, most commonly the cycle ergometer or treadmill. The highest values for $\dot{V}O_{2max}$ are obtained during treadmill running, and since walking and running are natural forms of movement for ambulatory children, the treadmill test is often optimal. There is a risk of falling in the younger child, and even though a treadmill with adjustable or additional handrails is often recommended for the pediatric exercise laboratory, holding onto the handrails reduces the increase in HR and $\dot{V}O_2$ and prolongs the test (25, 88). If young children or those at a greater risk of falling are being tested, it is worth considering a treadmill with a safety harness for support.

If a cycle ergometer is preferred, this must have fully adjustable handlebars, seat, pedal cranks, and a child-sized saddle should be fitted (13). Conducting the $\dot{V}O_{2max}$ on a cycle ergometer is

Figure 7.3 A turbine with a mouthpiece and a pneumotachograph with a face mask.

recommended for children with neuromuscular diseases that affect ambulation or for those who have difficulty adapting to treadmill belt motion (88). Cycle ergometers are robust and easily calibrated, and they allow other cardiac and vascular measures to be made during exercise. Cycling does, however, require considerable quadriceps strength during a test to maximum, which can result in substantial local muscle fatigue in the child with smaller quadriceps muscle mass. This can alter peripheral hemodynamics, notably venous return, cardiac output, and HR values (29).

Supine cycle ergometry is also an important modality for children and adolescents with orthopedic or neurological conditions that make upright cycling or treadmill exercise difficult. Supine cycling is also used when $\dot{V}O_{2max}$ assessment is coupled with other measurements such as echocardiographic assessment of cardiac function (115). A recent study compared maximum values for HR, respiratory exchange ratio (RER), and $\dot{V}O_2$ during upright and supine cycle ergometer exercise in 80 adolescents aged 13 years (62). Neither HR nor RER were significantly different in the girls or boys at maximum effort. Significant differences were apparent in $\dot{V}O_{2max}$, but these differences were small in both boys (upright 2.24 ± 0.80 l · min^{-1} versus supine 2.19 ± 0.79 l · min^{-1}) and girls (upright 1.89 ± 0.48 l · min^{-1} versus supine 1.79 ± 0.47 l · min^{-1}). As the authors conclude, these data show that maximal effort is possible using supine cycle ergometry, and although a significantly lower $\dot{V}O_{2max}$ was apparent using the supine exercise test, the difference may be too small to be clinically relevant. Rowland et al. (96) also report a lower $\dot{V}O_2$ during supine cycling (1.60 ± 0.09 l · min^{-1}) compared to upright cycling (1.78 ± 0.17 l · min^{-1}) at 100 W in a group of 10- to 15-year-old boys. Interestingly, HR was lower during supine cycling at higher exercise intensities (above 75 W), but no differences were apparent in CI or stroke index (SI) at 75 W. Similar values for left ventricular end-diastolic and end-systolic dimensions, as well as changes in systemic vascular resistance, were reported for the two body positions (96). A supine cycle ergometer may therefore be acceptable for clinical exercise testing when cardiac function is a priority.

Protocol Choice

Even though $\dot{V}O_{2max}$ is protocol independent (46), it is important that some basic principles are adhered to when designing or choosing a protocol. Safety is the paramount consideration, and the child should be subjected to the least possible discomfort during the test. Contraindications must be ruled out prior to testing; see the American Heart Association Council on Cardiovascular Disease in the Young, Committee on Atherosclerosis, Hypertension, and Obesity in Youth statement on clinical stress testing in the pediatric age group (79). Staffing during a maximal test should include at least two testers. The child and the parent or guardian should be given clear guidelines about presenting to the test, which includes not eating a meal within 2 h of the test and not engaging in vigorous exercise the day before or the day of the test (117). Parents should also make sure that the child wears appropriate clothing (e.g., T-shirt, shorts, and running shoes). Children need time to habituate to the test ergometer, and upon arrival in the laboratory they should have the opportunity to try out walking or running on the treadmill or cycling on a cycle ergometer. The test must begin with a low-intensity warm-up, and the test duration should ideally be about 8 to 12 min. If subsidiary measures such as blood lactate are required, the length of each stage within a test and the workload increments should be carefully considered. For instance, the steady-state lactate stage length should be 3 min, and workload increments should not exceed 4 ml · kg^{-1} · min^{-1} in $\dot{V}O_2$. A discontinuous protocol will be necessary when the subsidiary measure is taken with the child stationary, which will lengthen the total testing time. It is important that the length of rest periods between stages be standardized; 1 min is recommended. There should be a simple signal to indicate when the test should be terminated that both the child and the testers understand, and this should be followed by an appropriate low-intensity recovery (117). Finally, objective and subjective end points for test termination are needed. (Please refer to the test termination section for details.)

The most commonly used protocols in the pediatric exercise laboratory are incremental and ramp protocols. An incremental test should use stages of equal duration, such as 2 or 3 min per stage, and they can be continuous or discontinuous (e.g., a 1 min break between stages). Similar increments of intensity per stage should be used, such as an increase of 2 km $\cdot$ h^{-1} in speed per stage. In a healthy population of children ≥ 8 yr, most incremental treadmill tests begin with a walking stage at 4 or 5 km $\cdot$ h^{-1}, then speed is increased to jogging, which is usually 8 km$\cdot$h^{-1}. Further increases in speed are usually in 1 or 2 km $\cdot$ h^{-1} increments per stage until about 10 or 12 km $\cdot$ h^{-1}, at which point the treadmill gradient is increased by about 2% per stage until voluntary exhaustion. In a clinical setting, treadmill speeds may be as low as 2 km $\cdot$ h^{-1} with increases of just 0.5 km $\cdot$ h^{-1} per stage until 5 or 6 km $\cdot$ h^{-1}, then increases in gradient until termination of the test.

A ramp protocol uses continuous increases in exercise intensity until maximum. Ramp protocols on cycle ergometers are popular because the test is quick and power output is easy to measure in watts (W). A ramp cycle ergometer protocol normally begins with a light stage, such as 3 min at 10 W, followed by 10 W increments per minute, with cadence between 70 and 80 revolutions per minute (rpm). Again in the clinical setting this protocol may be adjusted to begin with an unloaded stage, followed by smaller increments per stage, such as 5 W $\cdot$ min^{-1}.

The most commonly used treadmill protocol in a clinical setting has been the Bruce protocol. The Bruce protocol uses 3 min stages starting with walking at 2.7 km $\cdot$ h^{-1} at an elevation of 10%. Speed then increases to 4.0 km $\cdot$ h^{-1} followed by 5.5 km $\cdot$ h^{-1}, 6.8 km $\cdot$ h^{-1}, 8.0 km $\cdot$ h^{-1}, 8.8 km $\cdot$ h^{-1}, and 9.7 km $\cdot$ h^{-1}, each accompanied by a 2% increase in gradient up to 22%. The advantage of the Bruce protocol is that it can be used across age groups and provides submaximal responses (34). It should be noted that the practice of reporting submaximal $\dot{V}O_2$ at exercise intensities above the ventilatory (or anaerobic) threshold has been criticized because of evidence of a p$\dot{V}O_2$ slow component emerging in both prepubertal children and adolescents (43). The disadvantages are that the Bruce protocol uses large and inconsistent workload increments. Reasons for using the Bruce protocol are often convenience rather than a consideration of the child or the data. Rowland notes reasons such as "The adult laboratory is using it," "It's convenient because it's in our automated testing equipment," and "It provides some uniformity across laboratories, and we're able to compare it with published normal values" (88, p. 2). Although $\dot{V}O_{2max}$ is generally resilient to protocol changes, when large increments in workload are used with children, peripheral fatigue often occurs before central fatigue, and the Bruce protocol can limit the achievement of a true cardiorespiratory maximum (117).

Test Termination

Children with a variety of clinical conditions can safely exercise to maximum, and in a clinical laboratory a list of clinical symptoms will always guide test termination. It is important that test termination be based on the same criteria for all children who can attain a maximum effort. Traditionally a plateau in $\dot{V}O_2$ despite further increases in exercise intensity has been used as the standard marker of $\dot{V}O_{2max}$ (51). In 1952 Åstrand (20) reported that only a minority of children and adolescents terminate a progressive exercise test to exhaustion with a plateau in $\dot{V}O_2$. It has since become clear that a plateau in oxygen uptake is seldom apparent in children (89), and the term $\dot{V}O_{2peak}$ (or peak $\dot{V}O_2$) has been adopted to denote a maximum test in the absence of a plateau. The term "peak" can lead to some confusion in the clinical setting because it is often used when adult cardiac patients cannot reach maximum. Subsequent work in children and adolescents has shown that $\dot{V}O_{2peak}$ is reliable, with a coefficient of variation of about 4% (103). Supramaximal testing has also shown that when strict criteria to delineate a maximum effort are applied, $\dot{V}O_{2peak}$ does in fact represent a maximal value (7, 23). With ramp protocols the usual physiological markers (such as HR$_{max}$ or maximal RER) are not as robust. Barker et al. (23) found the 1.0 RER maximal criteria, as well as the 85% age-predicted HR$_{max}$ and 195 bpm HR$_{max}$ criteria, underestimate $\dot{V}O_{2max}$. A supramaximal check may be indicated to verify $\dot{V}O_{2max}$ findings, and a follow-up test at 105% of the maximal power output achieved during the ramp test has been used to provide confirmation of whether a maximal effort has been elicited (23). This approach to verifying $\dot{V}O_{2max}$ with a supramaximal test about 10 min after the initial test has been used successfully in children with various disorders. For example, a follow-up supramaximal test at 110% of the maximal speed achieved in the treadmill $\dot{V}O_{2max}$ test verified the initial maximum $\dot{V}O_2$ value in children with spina bifida (36). Likewise, in children with an expiratory flow limitation, a follow-up supramaximal test at 105% of the maximal power output achieved during a ramp cycle ergometer test was used to verify the initial maximum effort (84). These data show that when appropriate exercise protocols and termination data are used, $\dot{V}O_{2peak}$ does in fact represent a maximal value, and the terms $\dot{V}O_{2peak}$ and $\dot{V}O_{2max}$ can therefore be used interchangeably in the pediatric literature.

Physiological indicators that are used to help confirm maximal effort include HR, RER, and blood lactate (7, 97). Typically HR_{max} values of about 200 bpm for treadmill exercise and 195 bpm for cycle ergometer exercise are expected in children and adolescents. The maximal achievable heart rate during exhaustive exercise has been used extensively as a marker of exertion in normal children, but limitations, particularly in children with congenital heart disease, must be recognized. The heart rate response in children with a heart defect varies considerably, depending on the particular defect, largely as a result of chronotropic insufficiency. Children with simple shunt lesions such as atrial septal defects and ventricular septal defects have been found to generally have near-normal cardiovascular responses (61, 114). In contrast, most cyanotic disorders (e.g., transposition of the great arteries, Fontan patients) show a high incidence of chronotropic impairment with a resultant low HR_{max}, which reduces the value of HR as a criterion for confirming maximum effort (40).

The increase in RER during a maximal exercise test to values of about 1.00 reflects an increasing anaerobic metabolic contribution and is indicative of a near maximal effort. Maximal RER, like HR, is ergometer dependent and is also protocol dependent. RER values of 0.99 to 1.00 offer an acceptable marker of maximum for treadmill exercise, but these can be boosted considerably using a supramaximal bout of exercise (13). Higher values of at least 1.06 would be indicative of a maximal effort for cycle ergometer exercise (97). Blood lactate accumulation has been commonly used in adults as a physiological indicator of maximum, but in children, post-maximum blood lactate accumulation has shown poor reliability (87). This is most likely because blood lactate accumulation is both ergometer and protocol dependent, and values are dependent on the timing of the postexercise

sample (7, 29). It is therefore not recommended as a marker of maximal effort in youth (14).

Subjective measures are also important for determining maximum effort, and rating of perceived exertion is commonly used in the pediatric exercise laboratory. Perceived exertion is the ability to sense your own physical effort (73) and helps facilitate continuation or discontinuation of a test. A variety of scales have been used in clinical and nonclinical settings (50, 57, 85), and a perceptual measure of fatigue provides an important addition to the test termination decision. With experience, subjective fatigue is quickly recognizable during exercise tests with children. Profuse sweating, an inability to maintain the desired exercise intensity, dyspnea, and an unsteady gait are all indicators that the child is approaching or has reached his or her maximum effort. However, this marker should be used only as an aid to other more objective criteria and never alone.

Developmental Patterns in Maximal Oxygen Uptake

Treadmill- and cycle ergometer–determined $\dot{V}O_{2max}$ values have been extensively documented in children and adolescents (4, 55). We will focus on data for youngsters over the age of 8 years because the younger child (<8 yr) finds maximal testing more difficult to manage, and these data are less robust (24).

Age, Growth, and Maturation

Absolute $\dot{V}O_{2max}$ ($l \cdot min^{-1}$) increases progressively with age. Both cross-sectional and longitudinal data show a distinct developmental pattern in absolute $\dot{V}O_{2max}$ in boys and girls. Using data from studies that provide absolute $\dot{V}O_{2max}$ data for girls and boys across a range of chronological ages, figure 7.4 shows a near linear increase in $\dot{V}O_{2max}$ in boys with age (figure 7.4a). An increase with age in girls is also apparent (figure 7.4b), albeit smaller than boys, until about age 14 when values tend to plateau. These data mimic the pattern illustrated by Armstrong and Welsman (4) who analyzed more than 5,000 treadmill-determined and 5,000 cycle ergometer–derived $\dot{V}O_{2max}$ values in children and adolescents aged 8 to 16 years.

When only longitudinal data are considered, the findings are generally consistent, with boys exhibiting a large (>120%) increase in $\dot{V}O_{2max}$ between the ages of 8 and 16 years, with values

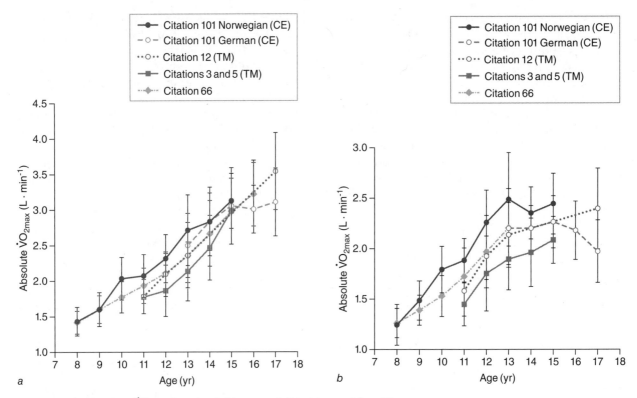

Figure 7.4 Absolute $\dot{V}O_{2max}$ data for *(a)* boys and *(b)* girls aged 8 to 17 years.

continuing to rise during adolescence, resulting in a doubling of $\dot{V}O_{2max}$ between 11 and 17 or 18 years of age (10, 52, 66, 101, 116). Girls also show a progressive increase in $\dot{V}O_{2max}$ between the ages of 8 and 13 years, with $\dot{V}O_{2max}$ leveling off at around 14 years of age. Change during adolescence is less consistent in girls, with some studies showing an increase, some reporting a leveling off, and others showing a decline in $\dot{V}O_{2max}$ between ages 14 and 16 years (10, 52, 101).

Direct comparison of absolute $\dot{V}O_{2max}$ values between healthy children and those with a chronic disease is not possible. Clinical studies generally report values for mixed ages and sex, often including young adults within study cohorts. This is because many of these disorders are not commonplace, because of (un)willingness to be tested and because of contraindications to testing. Examples of absolute $\dot{V}O_{2max}$ across a range of cardiac, pulmonary, and musculoskeletal disorders are provided in table 7.2 for reference.

As the child ages, he or she also grows. Body size is highly correlated with $\dot{V}O_{2max}$ ($r \sim 0.70$), and $\dot{V}O_{2max}$ is usually scaled to body mass to account for the changes in size. Conventionally

scaling $\dot{V}O_{2max}$ to body mass is achieved simply by dividing absolute $\dot{V}O_{2max}$ (ml · min^{-1}) by body mass (kg) to produce a ratio-standard value in ml · kg^{-1} · min^{-1}. Figure 7.5 provides examples of mass-related $\dot{V}O_{2max}$ in healthy 8- to 17-year-olds. A different pattern of development in $\dot{V}O_{2max}$ is apparent compared to absolute $\dot{V}O_{2max}$. Whereas absolute $\dot{V}O_{2max}$ increases with age in boys, mass-related $\dot{V}O_{2max}$ remains unchanged in most studies of boys aged 8 to 17 years, with values of about 48 to 55 ml · kg^{-1} · min^{-1} (figure 7.5*a*). One exception is a cycle ergometer study with German boys, where a gradual decline in mass-related $\dot{V}O_{2max}$ was apparent in boys with increasing age (101). In girls, a progressive decline from about age 11 years in mass-related $\dot{V}O_{2max}$ was evident, from about 50 to 35 ml · kg^{-1} · min^{-1} (figure 7.5*b*).

Mass-related $\dot{V}O_{2max}$ is more commonly reported in clinical studies than absolute $\dot{V}O_{2max}$. Data in table 7.2 illustrate that those with respiratory conditions normally show larger values of $\dot{V}O_{2max}$ (32-44 ml · kg^{-1} · min^{-1}) than those with cardiac conditions (20-34 ml · kg^{-1} · min^{-1}). This is in accord with current thinking that it is most likely reduc-

Table 7.2 Maximal Oxygen Uptake in Children With a Variety of Cardiac, Pulmonary, and Musculoskeletal Disorders

Reference	Sex	Age (yr)	Disorder	TM or CE	$\dot{V}O_{2max}$ (l·min^{-1})	$\dot{V}O_{2max}$ (ml·kg^{-1}·min^{-1})
DISORDERS THAT MAY AFFECT CARDIAC OUTPUT						
Opocher et al. 2005 (77)	F/M	8.7	Fontan	TM	0.90 ± 0.33	27.0 ± 5.0
Singh et al. 2007 (102)	F/M	9.6	Fontan	CE	-	26.3 ± 9.6
Blank et al. 2012 (27)	F/M	11.5	Isolated congenital complete atrio-ventricular block (paced)	CE	1.31 ± 0.5	34.4 ± 9.5
Buys et al. 2012 (32)	F/M	16.4	Transposition of the great arteries (Senning procedure)	TM	-	33.6 ± 6.7
Friedberg et al. 2013 (44)	F/M	11.9	Tetralogy of Fallot	-	-	30.6 ± 6.6
Guimarães et al. 2008 (47)	F/M	8.6	Heart failure (dilated cardiomyopathy)	TM	-	19.6 ± 5.8
Giardini et al. 2011 (45)	F/M	13.5	Heart failure (dilated cardiomyopathy)	CE	-	28.6 ± 10.0
DISORDERS THAT MAY AFFECT A-VO$_2$ DIFF						
de Groot et al. 2009 (36)	F/M	10.3	Spina bifida	TM	1.23 ± 0.6	34.1 ± 8.0
Pérez et al. 2013 (80)	F	10	Cystic fibrosis	TM	-	31.9 ± 6.9
	M	10	Cystic fibrosis	TM	-	38.7 ± 6.7
Pianosi et al. 2005 (81)	F	13	Cystic fibrosis	CE	-	37.2
	M	10	Cystic fibrosis	CE	-	41.2
Werkman et al. 2001 (110)	F	13.8	Cystic fibrosis	CE	1.6 ± 0.6	36.5 ± 6.3
	M	13.7	Cystic fibrosis	CE	1.9 ± 0.5	44.2 ± 8.8
Werkman et al. 2011 (111)	F/M	14.6	Cystic fibrosis	CE	1.9 ± 0.6	38.9 ± 7.4
Stevens et al. 2009 (104)	F/M	12.7	Pulmonary disorders: cystic fibrosis, noncystic fibrosis bronchiectasis, ciliary dyskinesia	CE	-	35.0 ± 8.0
Madsen et al. 2013 (60)	F/M	14.8	Primary ciliary dyskinesia	CE	-	37.9
DISORDERS THAT MAY AFFECT PERIPHERAL MUSCLE FUNCTION						
Brehm et al. 2013 (31)	F/M	10.5	Cerebral palsy GMFCS I and II	CE	-	39.3 ± 8.2
Balemans et al. 2013 (22)	F/M	9.9	Cerebral palsy GMFCS I	CE	-	35.5 ± 1.2
Balemans et al. 2013 (22)	F/M	10.3	Cerebral palsy GMFCS II	CE	-	33.9 ± 1.6
Balemans et al. 2013 (22)	F/M	10.0	Cerebral palsy GMFCS III	CE	-	29.3 ± 2.5

CE = cycle ergometer test. F = females. GMFCS = Gross Motor Function Classification System, which grades severity (I = mild, II = moderate, III = severe). M = males. $\dot{V}O_{2max}$ = maximal oxygen uptake. TM = treadmill test.

tions in SV that have the most pronounced effect on $\dot{V}O_{2max}$.

As children grow they also mature, but individual biological clocks run at different rates, and this affects both physiological function and $\dot{V}O_{2max}$. Few studies have investigated $\dot{V}O_{2max}$ in relation to maturation, perhaps because the assessment of maturation is challenging. The use of an anthropometric-derived marker of age to peak height velocity, which only requires the measurement of standing and seated height, weight, and date of birth, provides an easy-to-use indicator of physical maturity and is encouraged where possible (67).

When analyzed using ratio scaling it has been reported that maturation does not exert an additional effect on $\dot{V}O_{2max}$ over and above that due

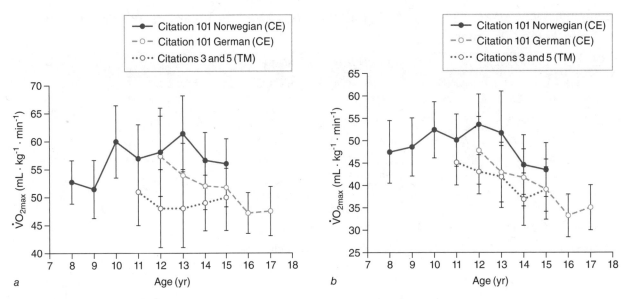

Figure 7.5   Mass-related $\dot{V}O_{2max}$ for *(a)* boys and *(b)* girls from ages 8 to 17 years.

to body mass, but the true relationship between $\dot{V}O_{2max}$ and maturation has been obscured by inappropriate scaling (108). Ratio scaling "overscales," benefiting lean children and penalizing obese children, and several studies have illustrated how ratio scaling leads to a different interpretation of the change in $\dot{V}O_{2max}$ with growth and maturation when compared to alternative methods of controlling for body size such as allometry (log-linear analysis of covariance with mass as the covariate).

When examining the development of $\dot{V}O_{2max}$ by age and growth, allometric scaling presents the same developmental pattern with growth as absolute $\dot{V}O_{2max}$ with age. That is, with body mass appropriately controlled for, $\dot{V}O_{2max}$ increases in boys from childhood through adolescence and into young adulthood, whereas in girls it increases until about age 14 years, when a leveling off is observed (108).

In addition to age and growth, maturation independently influences $\dot{V}O_{2max}$. Research by Armstrong and colleagues (9) demonstrated in a sample of 200 12-year-old children that while no changes in $\dot{V}O_{2max}$ are apparent with stage of maturation when using a ratio-scaled $\dot{V}O_{2max}$ in ml $\cdot$ kg^{-1} $\cdot$ min^{-1}, allometrically scaled $\dot{V}O_{2max}$ increases with increasing maturation, resulting in 12% and 14% differences between prepubertal and late pubertal girls and boys, respectively. These findings were replicated in a longitudinal study by the same group (12), and the same incremental effects of maturation were apparent on $\dot{V}O_{2max}$

independent of chronological age and body mass when multilevel regression modeling was applied. Intriguingly, when adiposity (skinfold thickness) was included in the analyses, maturation remained a significant covariate, but the magnitude of the effect was reduced, indicating a strong relationship between maturation and lean body mass.

The approach chosen to normalize $\dot{V}O_{2max}$ for body size exerts considerable influence on the degree of difference in $\dot{V}O_{2max}$ between obese and lean children and adolescents. When expressed as an absolute value, $\dot{V}O_{2max}$ is greater in the obese child or adolescent. The total body mass ratio standard most likely creates too big a difference in favor of the lighter child by overscaling for total mass and without consideration of the large difference in fat mass. The most appropriate, and body fat independent, method of scaling $\dot{V}O_{2max}$ in obese youngsters is probably by scaling to lean body mass (38). When scaled to lean body mass, $\dot{V}O_{2max}$ is much less impaired in the obese child (~10% lower) compared to healthy weight children of the same sex and comparable in age and maturation (48, 59).

These data illustrate just how influential body size and composition are upon our interpretation of $\dot{V}O_{2max}$ with ensuing age, growth, and maturation. Lean body mass appears to be the predominant influence on $\dot{V}O_{2max}$ through adolescence, but both chronological age and stage of maturation are additional explanatory variables, independent of body size and fatness. Lean body mass has not conventionally been used because of the complexity of assessment, but with the advent of easier-to-use methods, such as bioelectrical impedance, air-displacement plethysmography, and DXA, scaling $\dot{V}O_{2max}$ to lean body mass should become more commonplace.

It is clear that differing methods for accounting for size alter our interpretation of $\dot{V}O_{2max}$ during growth and maturation in both boys and girls. The real challenge for the clinician is to decide how best to express and therefore interpret $\dot{V}O_{2max}$ in the clinical context. Rowland (100) argues that when $\dot{V}O_{2max}$ is related to endurance performance, the ratio standard ($ml \cdot kg^{-1} \cdot min^{-1}$) is an acceptable scaling technique. This argument is supported by data that show running speed (a performance measure) relates to $\dot{V}O_{2max}$ when $\dot{V}O_{2max}$ is scaled to body mass of near unity, that is, the ratio standard (71). Ultimately the goal of much clinical exercise testing is to understand whether the child with a health problem can perform the same amount of endurance exercise as a healthy child. Therefore, although it clouds physiological understanding of true aerobic fitness until alternative scaling techniques become commonplace, it is probably prudent for the clinician to use the ratio standard.

Sex

$\dot{V}O_{2max}$ is consistently higher in boys than in girls, even prior to puberty, and by late adolescence this difference can be as much as 40%. Prepubertal girls and boys have similar muscle mass and hemoglobin concentration, yet $\dot{V}O_{2max}$ is still higher in boys than in girls. This difference has been as pronounced as 12% and widens through adolescence (14). Nearly all clinical studies pool data on boys and girls. There are two exceptions in those presented in table 7.2, which are two studies of children and adolescents with cystic fibrosis. Both report higher $\dot{V}O_{2max}$ values for boys than for girls, with a difference of about 17%, but the wide age range is worth noting. Sex differences in $\dot{V}O_{2max}$ have generally been attributed to differences in habitual physical activity, body composition, and blood hemoglobin concentration.

Habitual physical activity in the child characteristically comprises substantial periods of sedentary time (>70% of a waking day) interjected with short bouts of movement (3-17 s) (26, 60). These bouts of activity are neither long enough nor intense enough to enhance $\dot{V}O_{2max}$ and there is, at best, a weak relationship between physical activity and $\dot{V}O_{2max}$ (19). Therefore habitual

physical activity is unlikely to contribute to sex differences in $\dot{V}O_{2max}$.

There is a small but detectable difference in muscle mass throughout childhood, with lean mass about 10% greater in boys than in girls even prior to puberty (107). In contrast there is a dramatic disparity in muscle and fat mass between the sexes during puberty, with large increases in muscle mass in boys and large increases in fat mass in girls through adolescence. By mid-adolescence nearly 55% of total mass in boys is muscle, with only about 12% to 14% fat (17). Muscle accounts for about 40% to 45% of total mass in girls, but body fat increases to about 25% (63). The dramatic pubertal changes in muscle and fat mass contribute to the widening of the sex difference in $\dot{V}O_{2max}$ with age. Boys' greater muscle mass not only facilitates oxygen utilization during exercise but also enhances venous return to the heart via an augmented peripheral muscle pump, therefore boosting SV (33). This may explain why $\dot{V}O_{2max}$ remains higher in boys than in girls even when adiposity, body mass, stature, and age are all accounted for.

There is a marked increase in hemoglobin concentration in boys during adolescence compared to girls, whose values plateau (35). The potentially greater oxygen-carrying capacity in boys might be expected to account for the sex difference in $\dot{V}O_{2max}$, but when hemoglobin concentration was added to a multilevel regression model of $\dot{V}O_{2max}$, a nonsignificant parameter estimate was obtained with 11- to 17-year-olds (12).

Normal Values

Consideration of whether a $\dot{V}O_{2max}$ value in a patient is normal requires a normal comparison group. The ideal comparison group would be a representative sample from the same population as the patient group, with data on girls and boys across a range of age groups, ideally with an assessment of maturation. These values should be obtained using the same equipment and protocols used in the clinical population. Unfortunately, data regarding normative values for $\dot{V}O_{2max}$ by these criteria do not exist. For this reason, creating normal values from one's own laboratory experience is preferred rather than relying on those published in the literature.

Table 7.3 provides a guide for average mass-related $\dot{V}O_{2max}$ for both treadmill and cycle ergometer exercise in healthy children from mean values

Is Maximal Oxygen Uptake Related to Habitual Physical Activity?

The relationship between $\dot{V}O_{2max}$ and habitual physical activity (HPA) in young people has been explored for over 40 years. Early studies estimated HPA from heart rate (HR) monitoring, and large cross-sectional studies of children and adolescents consistently reported no significant relationships between $\dot{V}O_{2max}$ and either moderate or vigorous physical activity (2, 6). A 3 yr longitudinal study of over 200 11- to 13-year-olds used multilevel modeling to control for age, maturation, and sex and reported that not only was $\dot{V}O_{2max}$ not a significant explanatory variable for either moderate or vigorous physical activity, but over the 3 yr period the percentage of time in both moderate and vigorous physical activity decreased, whereas $\dot{V}O_{2max}$ in both absolute terms and with body mass controlled increased (10).

The advent of smaller and more sophisticated motion sensors stimulated a resurgence of interest in the level and various dimensions of physical activity and its relationship with $\dot{V}O_{2max}$. In some studies with large sample sizes, statistically significant relationships were reported between physical activity estimated using accelerometers and $\dot{V}O_{2max}$, but the correlations were generally weak and, at best, accounted for a very small percentage (~9%) of the variance in $\dot{V}O_{2max}$ (37, 41). Increases of 2% to 5% in $\dot{V}O_{2max}$ and 30% in HPA over a 15 year period (ages 13-27) were observed in the longitudinal Amsterdam Growth and Health Study, and it was concluded that, "no clear relationship can be proved between physical activity and $\dot{V}O_{2max}$" (53).

The lack of relationship between $\dot{V}O_{2max}$ and HPA is almost certainly because of the genetic component of $\dot{V}O_{2max}$ (~50%) and because under normal conditions children seldom experience the duration or intensity of exercise required to enhance $\dot{V}O_{2max}$. The magnitude of change in $\dot{V}O_{2max}$ in children is modest (~10%), even after the most rigorous of training interventions (intensity >85% $_{max}$) (16). In contrast, recent innovations in motion sensing techniques have revealed that the child's day characteristically comprises substantial periods of sedentary time (>70% of a waking day) interjected with short bouts of movement (3-17 s) mostly of low intensity, with some moderate but seldom vigorous activity (1, 64). Physical activity has a substantial therapeutic benefit for children who suffer from chronic disease, but a meaningful relationship with $\dot{V}O_{2max}$ remains to be proven.

published over the past 30 years. It should be noted these are intended as a guide only and are not intended to be used as normative values. Mean $\dot{V}O_{2max}$ values for children with cardiac, pulmonary, and musculoskeletal disorders are rarely over 40 ml $\cdot$ kg^{-1} $\cdot$ min^{-1} (table 7.3). In both cystic fibrosis and heart failure the prognostic value of $\dot{V}O_{2max}$ has been demonstrated. In a study of 109 cystic fibrosis patients (72), survival was predicted by $\dot{V}O_{2max}$, with 83% survival rates in the fittest patients (those with ≥82 % predicted $\dot{V}O_{2max}$) and only 22% survival in those with the lowest $\dot{V}O_{2max}$ (≤58% predicted $\dot{V}O_{2max}$). More recently, these findings were replicated, with $\dot{V}O_{2max}$ values below 32 ml $\cdot$ kg^{-1} $\cdot$ min^{-1} predicting mortality in a group of cystic fibrosis patients (81). $\dot{V}O_{2max}$ has also predicted length of hospitalization in cystic fibrosis patients (80).

In adults with heart failure, $\dot{V}O_{2max}$ is a key predictor of mortality and is used to identify heart transplant candidates. One-year mortality is predicted by a $\dot{V}O_{2max}$ of 14 ml $\cdot$ kg^{-1} $\cdot$ min^{-1}, and therefore only patients with a $\dot{V}O_{2max}$ above 14 ml $\cdot$ kg^{-1} $\cdot$ min^{-1} are suitable candidates for a heart transplant (70). The prognostic value of $\dot{V}O_{2max}$ in children with heart failure is not as distinct. Guimarães et al. (47) did not find that $\dot{V}O_{2max}$ differentiated survivors from nonsurvivors in a group of children with dilated cardiomyopathy, all of whom were waiting for a heart transplant. Exercise time, however, was predictive of survival, with an exercise tolerance of 19 min in survivors and only 13 min in nonsurvivors. In contrast, rate of death and clinical deterioration were predicted by $\dot{V}O_{2max}$ in a group of 82 youngsters with heart failure (45). A $\dot{V}O_{2max}$ ≤ 62% of the predicted value was predictive of nonsurvival. Giardini et al. (45) argue that the discrepancy between their findings and those of Guimarães et al. (47) most likely relates to the way in which $\dot{V}O_{2max}$ was expressed. Guimarães and colleagues (47) used absolute $\dot{V}O_{2max}$ to predict survival, and they did not take into account the variation in body size and age among the patients, whereas Giardini et al. (45) accounted for size and age by expressing the data as a percentage of the age- and sex-predicted value.

Dissimilarity in testing approaches in the clinical setting, heterogeneous groups of differing ages and sexes, and varying data-processing techniques make it hard to compare $\dot{V}O_{2max}$ values across studies. A lack of good normative data is also a challenge for comparing clinical groups with healthy children. It would be prudent for clinical exercise facilities to build up solid data banks of clinical and healthy $\dot{V}O_{2max}$ values. This will require voluntary exercise testing in healthy

Table 7.3   Average Maximum Oxygen Uptake Values Reported for Healthy Boys and Girls by Ergometer

Sex	Age (yr)	Treadmill (ml $\cdot$ kg^{-1} $\cdot$ min^{-1})	Cycle ergometer (ml $\cdot$ kg^{-1} $\cdot$ min^{-1})
Boys	8–18	45–55	40–50
Girls	≤13	40–50	35–45
Girls	>13	35–45	30–40

children, which is considered a limitation because it is believed that only those who are interested in exercise and therefore have higher levels of aerobic fitness volunteer. Conversely, as Booth and Lees (30) argue, it is indeed the physically active who should be the control group given the considerable evidence that inactivity and sedentariness are leading causes of disease.

Conclusion

Measuring $\dot{V}O_{2max}$ in the clinical setting greatly improves both diagnostic and prognostic information available from an exercise test. To maximize the usefulness of the measurement, clinicians should choose equipment and design protocols with the child in mind and develop their own normative values. Care should be taken when interpreting $\dot{V}O_{2max}$ values because of the changes that occur with age, maturation, and growth. $\dot{V}O_{2max}$ also shows sex-divergent development, with absolute $\dot{V}O_{2max}$ increasing throughout childhood in boys and girls but values continuing to increase into young adulthood in boys while they level off by mid-adolescence in girls. When $\dot{V}O_{2max}$ is expressed as a ratio with body mass, values in boys remain stable between 8 and 18 years of age, while in girls a gradual decline in $\dot{V}O_{2max}$ is seen. With these marked differences in development, it is imperative that data are not pooled for boys and girls. Improving technology is making noninvasive assessment of the components of $\dot{V}O_{2max}$ possible during exercise. Future exercise testing in children may see an integrative approach that considers noninvasive assessment of $\dot{V}O_2$, oxygen delivery, extraction, and utilization in synchrony, extending our understanding of the processes that regulate $\dot{V}O_{2max}$ in health and disease.

Other Measures of Aerobic Fitness

Robert P. Garofano, EdD

An assessment of aerobic fitness by measurement of maximal oxygen uptake ($\dot{V}O_{2max}$) is important in quantifying exercise potentials or limitations of exercise capacity in children and adolescents. As outlined in chapter 7 and elsewhere in this book, knowledge of a youngster's $\dot{V}O_{2max}$ is useful in interpreting reported symptoms of chest pain, shortness of breath, or unexplained fatigue. $\dot{V}O_{2max}$ provides insights into whether children with chronic illness can safely participate in the activities of their choice (54). In those with heart disease, level of aerobic fitness is critical to well-being (36) and identification of $\dot{V}O_{2max}$ allows health caregivers to make decisions about the patient's participation both in everyday activities such as stair climbing and in physical education and sports. $\dot{V}O_{2max}$ can be a critical element, too, in clinical decision making, such as timing for surgery and heart transplantation (18, 22, 30, 64).

While $\dot{V}O_{2max}$ by treadmill or cycle testing is considered the gold-standard measure of physiological aerobic fitness, in some cases it is not feasible to ask a subject to provide a maximal exercise effort on a cycle or treadmill. Usually this involves patients with significant exercise limitations due to either cardiac or noncardiac disease, but "maximal" exercise is often very difficult for patients who are obese, have particularly low fitness levels, or are incapable of comprehending testing instructions. The lack of a metabolic cart to measure gas exchange variables also precludes the assessment of $\dot{V}O_{2max}$, even when a true exhaustive effort can be provided.

In such situations, a number of surrogate measurements are available that can provide an estimate of $\dot{V}O_{2max}$. Some of these methods use submaximal heart rate response as an indicator of aerobic fitness level. Others seek to define aerobic fitness by submaximal gas exchange measures (ventilatory anaerobic threshold, oxygen uptake efficiency slope). The highest measured workload on a cycle ergometer or endurance time on a treadmill protocol can also provide insight into the level of aerobic potential (72, 75). This chapter will review these alternative testing approaches, focusing on their feasibility and accuracy. It is important to recognize that all of these testing protocols and measures are different and not necessarily interchangeable. The protocol chosen depends on the clinical question being addressed, the resources available in the laboratory to answer this question, and the limitations of the testing protocol chosen.

Peak Workload

$\dot{V}O_2$ is a measure of the amount of aerobically derived energy needed to perform work (94). Not surprisingly, then, a strong relationship exists between peak work achieved on an exercise test and $\dot{V}O_{2max}$ (4).

Work in watts on a cycle ergometer has a linear relationship with both heart rate and $\dot{V}O_2$ (4, 6). Because of less influence of body weight on the external workload, testing with a cycle ergometer has been considered more useful than the treadmill when comparing work in children of different ages and sizes. Several studies have demonstrated very good correlation between measured and estimated $\dot{V}O_2$ in normal subjects using a cycle ergometer (4, 35, 56, 82) in adults and in younger subjects (32, 35, 57). The results have been impressive. $\dot{V}O_2$ can be predicted with very good accuracy from equations derived from peak work on a cycle ergometer in subjects as young as age 8 (table 8.1). Regression equations have produced statistical relationships as high as $r = .90$ and $r = .95$ with a standard error of $1.62 \text{ ml} \cdot \text{kg}^{-1} \cdot \text{min}^{-1}$ of $\dot{V}O_2$, respectively (33, 49). When applying scaling methods, body weight in kg, or body weight to the two-thirds power, good correlation still exists, $r = .86–.89$ (3). The same study demonstrated a $3 \text{ ml} \cdot \text{kg}^{-1} \cdot \text{min}^{-1}$ (8%) difference between measured and calculated peak $\dot{V}O_2$ from cycle ergometry in 247 children. It should be noted that peak $\dot{V}O_2$ is

Table 8.1 $\dot{V}O_2$ Reference Equations

Reference	Age (yr)	Sex	Equation
Dencker et al. 2007 (34)	8–11	Female	$\dot{V}O_{2peak}$ (ml/min) = 240 + (9.99)(max work in W)
		Male	$\dot{V}O_{2peak}$ (ml/min) = 115 + [(10.225)(max work in W)] + [(4.95)(body weight in kg)]
Hansen et al. 1989 (50)	9–11	Female and male	$\dot{V}O_{2max}$ (ml/min) = (12)(max work in W) + (5)(body weight in kg)
Jung et al. 2001 (56)	13–18	Female	$\dot{V}O_{2max}$ (ml/min) = (9.39)(max work in W) + (7.7)(body weight in kg) – (5.88)(age) + 136.7
		Male	$\dot{V}O_{2max}$ (ml/min) = (10.51)(max work in W) + (6.35)(body weight in kg) – (10.49)(age) + 519.3

an expression of aerobic capacity, while peak work involves both aerobic and anaerobic systems.

Total exercise time on the treadmill using the Bruce protocol has been evaluated for children. Early work of Cumming et al. (27) developed percentiles for total treadmill time using the Bruce protocol in children aged 4 to 18. He grouped the children by age and gender. The more time on the treadmill, the higher the percentile, and therefore the higher level of fitness. He also reported heart rate and blood pressure response, but they did not factor into the percentile rating. The authors did correlate treadmill time to peak $\dot{V}O_2$ in a separate set of subjects, which demonstrated good correlation ($r = .85$) but not as high as in studies using the cycle ergometer. Percentile aerobic fitness rating for the Bruce protocol has been used in many exercise laboratories for many years and is still popular today.

More recent studies have demonstrated similar treadmill times (89, 100) using the Bruce protocol. These studies report similar time values and also have added a 5th percentile and a 95th percentile not described in the original paper of Cumming et al. $\dot{V}O_2$ was not measured in these studies.

Formulas exist to estimate the oxygen cost of treadmill exercise in adults. Testing manuals have been published by the American College of Sports Medicine (ACSM) to calculate the $\dot{V}O_2$ in ml · kg^{-1} · min^{-1} for a specific speed and grade for many years (4). With these equations, the oxygen cost for the highest workload achieved for a particular subject can be calculated. If no gas exchange measurements are available, this method could be useful, particularly in a mature adolescent subject. The calculated oxygen cost can be compared to predicted values or to a chart of various activities

to evaluate whether the child has an appropriate functional capacity to participate in that activity (1, 5, 78). As an important caveat, however, such equations are not expected to be accurate for smaller children. In these subjects the calculated oxygen consumption for walking or running may not truly reflect the cost of locomotion, since these children have smaller stride lengths and greater stride frequency when performing the same work as mature subjects. The lower level of exercise economy in young children will increase the oxygen cost of a given workload compared to adults (4). Therefore, the calculated oxygen cost of a certain stage of the Bruce or other protocol may underestimate the measured $\dot{V}O_2$ for younger children (4, 99). Some prediction equations have been created with the subject holding the handrails during exercise, while others have not, and this will change the oxygen cost of the exercise. The $\dot{V}O_2$ formulas in the ACSM manual assume hands off handrails.

The relationship between the amount of work performed and the estimated $\dot{V}O_2$ has been explored for use in children with various diseases (congenital heart disease, cystic fibrosis, juvenile rheumatoid arthritis). Using the cycle ergometer, $\dot{V}O_2$ could be accurately predicted in adolescents with cystic fibrosis (97). In a group of 60 subjects with cystic fibrosis with mild to severe lung disease (37% to 147% of predicted for forced expiratory volume in 1 s [FEV$_1$]), the $\dot{V}O_2$ was predicted from work in watts with an $r^2 = .91$. That correlation is consistent with the values found in normal subjects. For patients with idiopathic juvenile arthritis that prediction of $\dot{V}O_2$ from peak work in watts was equally as impressive as the other studies, with an $r^2 = .94$ (32).

The predicted values for oxygen consumption from workload estimates during exercise have not generally been accurate in children or adults with congenital heart disease (36, 86). Mean reported values for percent predicted peak $\dot{V}O_2$ have ranged from 42% to 89% depending on the specific lesion. The poor correlation observed in children with congenital heart disease is probably due to the complex nature of blood flow and abnormal cardiac anatomy. In addition, many subjects with congenital heart disease have chronotropic incompetence due to their disease, surgeries, or medications (17, 24). Heart rate will not reach predicted values, and may have a steeper slope, as a function of $\dot{V}O_2$.

When investigators are seeking to quantify the fitness level of large populations (such as classes or schools), predicting $\dot{V}O_2$ from the highest workload achieved on a cycle ergometer may be the best option. Adding another measure, such as heart rate, could enhance the evaluation of effort. Testing then requires a cycle ergometer, a heart rate monitoring device, and experienced personnel to conduct the exercise test. Treadmill endurance time would be another option, but treadmills are more difficult to transport and require a source of electricity.

Ventilatory Anaerobic Threshold

The ventilatory anaerobic threshold (VAT) is an important noninvasive indicator of effort and measurement of the level of fitness or exercise capacity. Determination of VAT is not dependent on a maximal work test, but it provides insights into level of aerobic fitness as a predictor of $\dot{V}O_{2max}$.

The VAT is the measured $\dot{V}O_2$ at the point during a progressive test where exercise cannot be continued from solely aerobic metabolism (95). Normal-functioning chemoreceptors are very sensitive to increases in CO_2 as the body tries to maintain homeostasis with the rise in PCO_2 with exercise. Increased CO_2 production resulting from an increase in the production of lactic acid and a fall in bicarbonate concentrations increases ventilatory drive. Early research has demonstrated that the higher the amount of lactic acid at the time of fatigue, the higher the exercise capacity (93, 96). Others have agreed with this finding in

studies of the rise in blood lactate in response to heavy bouts of exercise in trained and untrained individuals (93, 94). Wasserman et al. (96), using arterial blood samples, described this point of sudden rise in lactate concentration during progressive exercise as the anaerobic threshold and related it to a specific and reproducible oxygen consumption. To avoid the necessity of invasive measurements of arterial or venous blood, the ventilatory anaerobic threshold was introduced, based on measures of gas exchange in expired air at the mouth. The history and refinement of the techniques to determine the VAT have been described in detail in previous published papers (14, 44, 63, 83, 94).

There are two major noninvasive methods for measuring the VAT. One is the ventilatory equivalent method, which defines the VAT as the point where ventilation increases disproportionately to the oxygen consumption, increasing the $V_E/\dot{V}O_2$, in the absence of a similar rise in the $V_E/\dot{V}CO_2$ (43, 93) (figure 8.1).

The second option is the V-slope method, or more currently, the modified V-slope method (43). Here the VAT is the point where the slope of the increase in $\dot{V}CO_2$ is steeper than the slope of the increase in $\dot{V}O_2$ as work increases (figure 8.2). The modification referred to here is simply the addition of averaging the breath-by-breath data, most commonly at 10, 15, or 20 s intervals. Both of these methods have been accepted for use for the measurement of the VAT in adults and children.

The V-slope method has been used more often with children than the ventilatory equivalent model. Children will have a more erratic breathing pattern during progressive exercise than adults (46, 76, 99), and it is believed that the more irregular the breathing pattern, the less accurate the ventilator equivalent method becomes (40, 46). One study of 22 healthy children reported detection of the VAT by the V-slope method in all subjects (40). In that same group there was between 16% and 18% failure rate of detecting VAT from the ventilatory equivalent method. Other authors have reported undetectable VAT points in 2.5% to 20% of normal subjects (51, 94) (table 8.2).

The cycle ergometer has advantages over the treadmill for detection of the VAT, since this point is observed best with equal and consistent increases in work over the entire exercise study (63). Ramp protocols on the cycle ergometer are

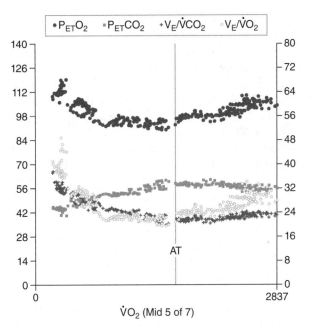

Figure 8.1 Ventilatory anaerobic threshold in one subject using the ventilatory efficiency method. $P_{ET}O_2$ reflects the end-tidal partial pressure of O_2, and $P_{ET}CO_2$ is the end-tidal partial pressure of CO_2. The solid vertical line in the center indicates the anaerobic threshold.

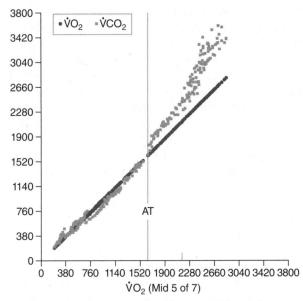

Figure 8.2 Ventilatory anaerobic threshold in one subject using the V-slope method. The solid vertical line in the center indicates the ventilatory anaerobic threshold.

Table 8.2 Six-Minute Walk Responses in Various Populations

Reference	Age (yr)	Status	Distance in meters (F/M)	Borg	S_pO_2	Heart rate pre to post	Blood pressure pre to post
Li et al. 2005 (59)	13–15	Normal subjects	F 637 ± 39 M 691 ± 66	No	No	Yes: 88–148 ± 19 bpm	No
Geiger et al. 2007 (45)	3–18	Normal subjects	F 656–661 M 667–728	No	No	No	No
Lammers et al. 2008 (58)	4–11	Normal subjects	M 470 ± 59	No	Yes: 97%–99%	Yes: 102 ± 19–136 ± 12 bpm	No
D'Silva et al. 2011 (29)	7–12	Normal subjects	F 548 ± 45 M 671 ± 86	Yes: NR	Yes: 96%–97%	Yes: 83 ± 2–104 ± 3 bpm	Yes: 109 ± 2–122 ± 2 mmHg
Barboza de Andrade et al. 2013 (11)	6–16	Asthma	F 425 ± 93 M 435 ± 137	No	No	NR	No
Ulrich et al. 2013 (88)	5–17	Normal subjects	F 608 ± 55 M 626± 65	No	Yes: no change	Yes: reported by age group	Yes: MAP reported by age group
Cahalin et al. 1996 (21)	41–57	Heart failure NYHC 3	M 310 ± 100	No	No	No	No
Lucas et al. 1999 (61)	Adults	Heart failure	All 393 ± 104	No	No	No	No
Ross et al. 2010 (79)	41–67	Mix HF, COPD, PH	F 347 ± 45 M 361 ± 136	No	No	No	No
Casanova et al. 2011 (23)	40–80	Normal subjects	F 555 ± 81 M 585 ± 96	Yes: NR	Yes: 97–95	No	No

bpm = beats per minute. COPD = chronic obstructive pulmonary disease. F = females. HF = heart failure. M = males. MAP = mean arterial pressure. NYHC 3 = New York Heart Class 3. NR = not recorded. PH = pulmonary hypertension.

Six-Minute Walk Test

The six-minute walk test measures the greatest distance one can walk in 6 minutes. The subject is instructed to walk as fast and as long as he or she can, but running is not permitted. According to American Thoracic Society (ATS) standards (7), subjects can either receive encouragement from testing staff or not. The latter approach removes some intra-administrator variability because test examiners can vary in their level of enthusiasm. Besides distance, the six-minute walk test also usually includes a scoring system for shortness of breath or other symptoms of fatigue (such as the 10- or 20-point Borg or 10-point Caler scale [12, 18, 39, 42]). The addition of other measurements such as oxygen saturation, heart rate, and blood pressure has been explored to enhance the quality of the test.

Table 8.2 provides data from several studies providing normal responses in children. The regression models in all of these studies indicated that age, height, and gender explained approximately 50% of the differences in walking distance. As the child matures, walking distances will more than likely increase, maybe not linearly each year but at certain times. Most influential is probably the time of peak height velocity, the maximum rate of growth, or the point where a child grows the quickest over a period of time. We cannot accurately predict walking distances in children based on the existing research. Changes in age and height can dramatically change the subject's walking distance. Instead of using regression equations, using normal ranges by age in years might be a better way to group normal walking distance results.

There are some studies that conclude that the 6-min walk distance correlates well with peak oxygen consumption, and some predict peak $\dot{V}O_2$ from the six-minute walk results (37). It is logical to assume that the farther one can walk in 6 min, the greater peak $\dot{V}O_2$. The studies predicting $\dot{V}O_2$ from the accumulated distance from the six-minute walk test are not abundant, are mostly performed in adults, and offer different conclusions (60). At this time the use of the six-minute walk test to predict peak oxygen consumption in children is not advised. However, changes in walk time on repeated testing (with recognition of the expected positive effect of increasing age) may prove useful in demonstrating changes in clinical condition in patients with heart or lung disease (67) (see chapters 3, 11, and 19).

best for this; the computer can increase the work rate very consistently. The treadmill will have larger changes in work rate because it is body weight dependent and because changes in speed and grade will affect children differently. One may walk at a specific mph or incline level while another needs to run; this changes the oxygen cost of the work. Also, with weight-dependent exercise, starting at too high a workload (e.g., 2.0 mph and 10% incline) for an obese child could cause early fatigue, and the work intensity of VAT will not be achieved. Changes in various treadmill protocols are not consistent between stages.

The $\dot{V}O_2$ at VAT is a highly reproducible measure that provides insight into aerobic exercise capacity in children. Strong associations exist between $\dot{V}O_2$ at VAT and peak $\dot{V}O_2$, providing estimates of aerobic fitness from submaximal data. Even if the percent of peak $\dot{V}O_2$ may vary among children of different ages, it is a clear representation of aerobic fitness.

Different ranges are observed when reviewing published reports of $\dot{V}O_2$ at VAT as a percent of the peak $\dot{V}O_2$ in children (table 8.3). Published studies have reported values from 45% of the peak $\dot{V}O_2$ to 75% of the peak $\dot{V}O_2$ (93). This discrepancy may be due in part to the fact that the percent of the $\dot{V}O_2$ where the VAT occurs falls with age (26). The $\dot{V}O_2$ at VAT has been reported in prepubertal children to be between 71% and 75% of peak $\dot{V}O_2$, and 45% to 61% of peak $\dot{V}O_2$ at ages that assume the completion of puberty (74, 77, 94). This age difference occurs because the ability to produce lactic acid during exercise depends on sexual maturity as the metabolic machinery evolves during puberty (83). Younger children may rely more on oxidative fat metabolism during exercise than adults or children who are past puberty. Therefore the age of the subject can make it more difficult to use VAT to determine $\dot{V}O_2$, certainly in the younger subjects.

Ventilation becomes more efficient as children become older. There is a reduction in the $V_E/\dot{V}CO_2$ ratio as well as anatomical dead space to tidal volume ratio (V_D/V_T) as children age (20). Lung volume increases as height increases regardless of age. The increase in total lung volume, and specifically peak exercise tidal volume, are much greater than the increase in the size of the mouth, bronchus, and trachea. Larger lungs have more alveoli and more gas exchange surface area to eliminate more CO_2 per breath. Giardini et al. (46) have demonstrated the decrease in $V_E/\dot{V}CO_2$ as children progressed to young adulthood at the VAT and also at the peak workload.

Table 8.3 Values for Ventilatory Anaerobic Threshold

Reference	Age (yr)	Protocol	Method	Peak HR (bpm)	Peak $\dot{V}O_2$ (ml $\cdot$ kg^{-1} $\cdot$ min^{-1})	$\dot{V}O_2$ at VAT (ml $\cdot$ kg^{-1} $\cdot$ min^{-1})	VAT % of peak $\dot{V}O_2$	Failure to measure VAT
NORMAL VALUES FOR VENTILATORY ANAEROBIC THRESHOLD								
Cooper et al. 1984 (26)	6–17	Cycle	$V_E/\dot{V}O_2$	NR	YF 38 YM 42 OF 34 OM 50	YF 23 YM 26 OF 19 OM 27	YF 61 YM 64 OF 58 OM 55	5%
Reybrouck et al. 1985 (77)	5–18	Treadmill	$V_E/\dot{V}O_2$	NR	F 41.9 M 51.3	F 26.8 M 30.9	F 64 M 60	2.5%
Washington et al. 1988 (94)	7–12	Cycle	$V_E/\dot{V}O_2$	191-196	41–47	30–35	71–75	18%
Ohuchi et al. 1996 (74)	8–21	Treadmill	V-slope	NR	46.4	20.7	44.6	9%
Fawkner et al. 2002 (40)	11–12	Cycle	V-slope	NR	1.65 L/min	0.87–0.9 L/min	55	0
Groen et al. 2010 (48)	10–15	Cycle	$V_E/\dot{V}O_2$	193	49.3	31.4	64	NR
Ten Harkel et al. 2011 (85)	8–18	Cycle	V-slope	F 186 M 184	F 42 M 47	F 26 M 28	F 62 M 61	NR
Mucci et al. 2013 (69)	9–11	Cycle	V-slope	186	39.4	22.1	56	0
Müller et al. 2013 (70)	11–14	Cycle	V-slope	187	42.4	21.5	51	0
VALUES FOR SUBJECTS WITH CHD								
Ohuchi et al. 1996 (74)	8–21	Treadmill	V-slope	NR	29.1	18.7	67.5	12.5%
Groen et al. 2010 (48)	11–17	Cycle	$V_E/\dot{V}O_2$	164	33.7	27.5	81.7	NR
Müller et al. 2013 (70)	11–14	Cycle	V-slope	175	35.5	20.7	59	0
VALUES FOR SUBJECTS WITH LUNG DISEASE								
Groen et al. 2010 (48)	14–16	Cycle	$V_E/\dot{V}O_2$	188	42.5	25.4	59.7	NR

bpm = heart rate in beats per minute. CHD = congestive heart disease. F = females. M = males. NR = not recorded. OF = older females. OM = older males. YF = young females. YM = young males.

The VAT is sensitive enough to detect improvements in fitness (43, 69). Some studies have reported an increase in the $\dot{V}O_2$ at VAT with training in prepubertal children (69). When peak $\dot{V}O_2$ is not achieved, either by effort or by design, it is possible to detect changes in fitness by performing submaximal studies and measuring the $\dot{V}O_2$ at the VAT before and after training (69) or other intervention, such as bed rest from surgery.

The VAT is also useful in children with special conditions. It has provided a good indicator of fitness in children with cystic fibrosis and those born premature, but the results are not as consistent for children with congenital heart disease. Children with cystic fibrosis demonstrated a $\dot{V}O_2$ at VAT similar to that of healthy children (97). Children who were born premature also demonstrated similar normal findings (55).

In children with heart disease, the reported findings have varied, from more normal values such as 59% (70) and 67.5% (74) to values of 81.7% of peak $\dot{V}O_2$ (48). Such values did not appear to be dependent on level of peak $\dot{V}O_2$. Patients with some of the less complex congenital cardiac lesions have peak $\dot{V}O_2$ values closer to values in normal children. More complex heart disease will affect oxygen delivery—for example, the inability of the heart rate to increase with exercise—or problems with stroke volume—for example, when the right ventricle is the systemic ventricle. This will greatly affect peak $\dot{V}O_2$ in these individuals. The study that reported the highest $\dot{V}O_2$ at VAT in heart disease patients also reported $\dot{V}O_2$ values at VAT that were very similar to the normal values (see table 8.3). These patients also had lower peak heart rates, which could be related to effort or chronotropic incompetence.

Submaximal Testing Protocols

A number of different testing approaches have been developed whereby $\dot{V}O_{2max}$ can be estimated by values of heart rate or $\dot{V}O_2$ obtained during submaximal exercise. While these can be applied to subjects who fail to provide a true exhaustive effort (i.e., consistent with criteria for a maximal test), a certain sacrifice of accuracy in assessing aerobic fitness must be accepted. Such submaximal tests may be applicable when evaluating large groups of subjects in epidemiologic studies or when equipment to determine gas exchange variables is not available.

Physical Work Capacity

The goal of submaximal testing is to produce a significant level of exercise—although not peak—to gather physiological data by which one can draw educated conclusions about physiological reserve or fitness. The Åstrand-Rhyming nomogram was one of the first developed to predict $\dot{V}O_2$ from a submaximal work test on a cycle ergometer (4). The Åstrand-Rhyming nomogram predicts maximum oxygen consumption from the linear relationship of heart rate response to several increasing workloads (4, 25). Work is increased on a cycle ergometer until the subject achieves a predetermined heart rate. Most commonly this is 170 bpm in healthy adults. The higher the amount of work needed to achieve that heart rate, physi-

cal work capacity at a heart rate of 170 bpm, or PWC 170, the higher the $\dot{V}O_{2max}$ for that subject. Original work by Åstrand et al. used six-minute stages, achieving steady state in each, defined by a plateau in heart rate before changing to the next workload (6).

In children a shorter stage length has been used, 2 and 3 min long, lowering total test time and enhancing compliance (66). Rowland (80) reported that children can reach a heart rate plateau in 2 min. To better understand stage length and to predict peak $\dot{V}O_2$ from PWC 170, investigators compared three different stage lengths in 11- to 16-year-old children (15). They found that each protocol produced different predicted peak $\dot{V}O_2$ values. It was observed that PWC 170 was weakly to moderately correlated to measured peak $\dot{V}O_2$, and these authors concluded that it was not considered a strong surrogate for peak $\dot{V}O_2$. Interestingly, the 2 min stage protocol correlated best to peak $\dot{V}O_2$ ($r = .51$). For children between the ages of 11 and 16, the shorter protocol is acceptable, providing a shorter testing session so that more subjects can be tested.

Heyman et al. studied the utility of submaximal exercise testing on a cycle ergometer in healthy 9- to 14-year-old boys (52).They used 3 min stages at each workload to get to a near steady-state metabolism and to avoid leg muscle fatigue that could limit how much work was performed. This was followed up by a second experiment with a different group of subjects, using the testing protocol formulated from the results of the first experiment. They produced a prediction equation for peak watts using height. This increased the workloads by a slightly larger amount in taller subjects for the second group. The second group also performed a second test without $\dot{V}O_2$ measurements. With this approach, improvement was observed in the correlation between measured and calculated $\dot{V}O_2$, from $r = .67$ (boys) to .93 (girls) to $r = .84$ (boys) to .96 (girls). The correlation when the measured peak $\dot{V}O_2$ was predicted from the calculated peak $\dot{V}O_2$ for all subjects was $r = .97$ for L/min and $r = .98$ when corrected for body weight in kilograms. Also in this study, work in watts predicted peak $\dot{V}O_2$ better than heart rate, $r = .99$ versus $r = .94$, respectively.

In a study of 14 boys with type 1 diabetes, good correlation was found between PWC 170 and peak $\dot{V}O_2$, $r = .81$, using a 2 min cycle ergometer protocol (56). PWC 170 has also been used to evaluate differences in calculated peak $\dot{V}O_2$ between normal subjects and patients with idiopathic scoliosis (32). Other special populations of children may not be able to achieve a peak heart rate of 170 bpm, limiting the usefulness of the submaximal test (54, 57, 81). It seems clear that the best predictor of peak $\dot{V}O_2$ is from peak work on the cycle ergometer.

Smaller children may not fit the available standard cycle ergometers, and modifications are needed for the seat height and crank arm length. Some labs will have two cycle ergometers to accommodate all sizes of children.

Submaximal $\dot{V}O_2$

Two studies measured $\dot{V}O_2$ in a group of obese children and reported a very good correlation, $r^2 = .996$ and $r^2 = .75$, respectively, between measured peak $\dot{V}O_2$ and predicted peak $\dot{V}O_2$ calculated from submaximal heart rate and $\dot{V}O_2$ during treadmill walking (18, 71). The protocol selected for the HALO (18) project was not a standard treadmill protocol; instead, the protocol was derived from studies of obese subjects in which the subjects selected their own walking speed. The self-paced walking speed was maintained or increased slightly early in the protocol for comfort, and incline was increased every 4 min until 85% of age-predicted maximum heart rate. These authors extrapolated the submaximal heart rate and $\dot{V}O_2$ responses to the age-predicted maximal heart rates. The regression value they reported is very high for predicted $\dot{V}O_2$ for treadmill exercise.

The second study proposed using heart rate and speed from a 4 min submaximal walking protocol (70). After two minutes of rest following a 4 min walk protocol, the subjects performed a more traditional progressive treadmill test to volitional fatigue. The experimenters validated their prediction equation in a second group of subjects and demonstrated good results in predicting $\dot{V}O_{2max}$ ($r = .89$). For the HALO project (18), two issues need to be pointed out. First, each subject achieved 85% of predicted heart rate, and the measured $\dot{V}O_2$ during the submaximal exercise study was 93% of peak. Thus, this is very close to a peak exercise study, accounting for the very high correlation with $\dot{V}O_{2max}$. Second, the authors used the formula 220 – age to calculate the maximum predicted heart rates for their subjects. Other researchers (62, 81, 92) disagree with this age-predicted maximum calculation in children, certainly younger children. Not every subject can achieve 100% of the age-predicted maximum heart rate at peak exercise,

and therefore a single target number should not be used (62). Peak heart rate has been shown to be lower on a cycle ergometer (4), and therefore peak heart rate ranges should be adjusted to account for the testing protocol used.

Oxygen Uptake Efficiency Slope

The oxygen uptake efficiency slope (OUES) was originally developed as a submaximal, effort-independent measurement of the efficiency of the body in extracting oxygen from ambient air and using this oxygen to perform work (8). With parameters of minute ventilation on the x-axis and oxygen consumption on the y-axis, a plot between the two provides the oxygen consumption at certain ventilation parameters during exercise. OUES is the slope of this relationship with values of ventilation and oxygen uptake, made linear by expressing values by log transformation (10). It has been reported that the OUES is protocol independent because similar values have been obtained on two different treadmill protocols (9, 90). The differences between ergometers, cycle or treadmill, has not been explored for the same population.

The OUES has been described as the $\dot{V}O_2$ per tenfold increase in ventilation (2), referring to the base 10 log. The steeper the slope, the more efficient the subject is at using the oxygen for the amount of air taken in during exercise. There is, in fact, a well-established relationship between V_E and $\dot{V}O_2$ during progressive exercise (4, 85): The slope of V_E is known to increase at a higher rate than $\dot{V}O_2$ at and after the VAT (94). OUES is influenced by lactate accumulation at higher workloads and dead space ventilation. Both of these responses can alter ventilation or effective pulmonary ventilation.

Other factors can influence the OUES. Increased $\dot{V}CO_2$ production with an early VAT (as often seen in patients with heart disease) will increase ventilation. Increased dead space ventilation, as with lung disease, will reduce the amount of air available to provide oxygen to the lung, resulting in alveolar hypoventilation. The consequent derangement in ventilation:perfusion balance can be detected in resting and exercise oxygen saturation (17).

In children OUES is affected by changes due to growth and maturation (2). Increases in body height and mass, as well as muscle mass, specifically in boys, increase both V_E and $\dot{V}O_2$. Normal changes with puberty will affect OUES, and as many studies point out, caution should be used when interpreting these results and when comparing normal subjects to children with various conditions (2, 38, 65).

Adult studies with healthy subjects and subjects with heart disease have shown good correlations of OUES to $\dot{V}O_{2peak}$. Since it was originally developed for a submaximal purpose, investigators have divided the results of OUES into percentages of the total exercise time or peak $\dot{V}O_2$. Most of the studies have performed one peak work test and then shortened the data points to some portion of the peak $\dot{V}O_2$—OUES 50% or OUES 75%, for example. The objective of such investigations has been to compare the OUES slopes to determine if the submaximal portion of the study would have provided the same results as the whole data set. Several studies, in fact, have reported that the submaximal data set will produce an OUES slope that was not statistically different from the OUES slope from all the data points. More often the data sets that are a higher percentage of the total have performed better (86, 87). These studies have reported very good correlations between different OUES levels of 75%, 80%, and 90% of peak $\dot{V}O_2$ and peak $\dot{V}O_2$ (r = .81-.94). The lowest data sets of OUES 25% and OUES 50% have been shown to be the least reliable. The correlation of OUES 25% was reported as r = .35 (31).

Some disagreement exists over authors' different use of techniques and ergometers in creating submaximal data sets to establish OUES. The grouping of data up to the VAT has been used, as have higher intensities where the respiratory exchange ratio (RER) is 1.0 or 1.1. There is also uncertainty whether to include resting data or to start with unloaded cycling or with the onset of work (2, 3). The breathing pattern may be erratic at rest and at very low workload conditions, becoming even more so as workload increases (31, 76). Others have reported that the last workloads increase ventilation significantly more than $\dot{V}O_2$ and lower the OUES slope (3, 38). Both of these factors could change the OUES 100% slope. The majority of the studies (15 of 17) included resting data (2, 17).

The VAT is often used as a reference point because of its strong relationship with peak $\dot{V}O_2$. Overall, a good correlation has been reported between $\dot{V}O_2$ at VAT and the OUES slope. Different methods of detecting the VAT can also influence results, while there is only one method to calculate the OUES (98).

There have been studies examining OUES in healthy children as well as in those with congenital heart disease, cystic fibrosis, and obesity. In general these have demonstrated a good correlation with peak $\dot{V}O_2$ (r = .76-.94). OUES 100% has correlated with peak $\dot{V}O_2$ and VAT in healthy children. There has also been good agreement between OUES at 75%, 80%, and 100% in normal subjects. However, the results in children with special conditions have not been as consistent as those for healthy subjects. OUES can be greatly influenced by conditions that affect oxygen delivery, changing $\dot{V}O_2$ as well as possibly increasing V_E. In certain heart disease patients where oxygen delivery is not affected greatly by their lesion, OUES is as reliable as with healthy children (17). Others (9) have reported good agreement between OUES 100% and OUES 90% but not OUES 75% in children with congenital heart disease. In that study, the mix of patients was extensive (seven different lesions) but did not include cyanotic heart disease.

Giardini et al. reported equally high correlations with OUES 100% and peak $\dot{V}O_2$, and OUES 50% and OUES 50% to 100% in patients with noncyanotic congenital heart disease (47). This was one of the few investigations that separated the data points between the first and second halves of the exercise study. This approach was based on the authors' belief that ventilation responses differ significantly in the latter half of an exercise test and OUES would therefore be affected. However, they demonstrated that in this group of patients the two OUES slopes did not differ significantly.

The most complex heart diseases studied in children are those with single ventricle or Fontan physiology. These patients have low peak $\dot{V}O_2$ and, in a parallel fashion, the lowest OUES slopes for 100%, 75%, and at the VAT (17). Similar to conditions of heart failure, poor oxygen delivery will drive ventilation, so these patients have very low OUES slopes by the increased ventilation and the low $\dot{V}O_2$. In a subset of patients who had received the Fontan surgical repair, those patients where OUES 75% or at VAT did not correlate with OUES 100% were those who demonstrated arterial oxygen desaturation at rest (17). Once again, a factor that will limit oxygen delivery will affect both of the variables in the OUES slope.

In children with lung diseases characterized by mild to moderate airflow obstruction (FEV_1 46%-107% of predicted), OUES and BSA values at 50% and 75% were significantly lower than those at OUES 100% of the peak $\dot{V}O_2$ (16). Body surface area was used to correct the data for differences in growth and maturation. In a comparison of healthy children to those with cystic fibrosis, only OUES and BSA 50% differed between the two groups. For the other comparisons of OUES and BSA 75% and 100%, OUES could not distinguish the cystic fibrosis group from the control group.

In a group of adult patients with cystic fibrosis, the authors investigated the reliability of OUES and other parameters (49). All cardiopulmonary exercise testing variables demonstrated very good reliability from test to retest (ICC values 0.86-0.94). They also found the OUES 70% and 80% did not differ from OUES 100%, and they correlated with peak $\dot{V}O_2$. The highest correlation existed between OUES 80% and both peak $\dot{V}O_2$ and OUES 100%. This group felt most comfortable using the OUES 80% to predict peak $\dot{V}O_2$ in patients with cystic fibrosis who had moderate lung disease. Since increased lung disease will decrease ventilation, this association may not be as useful in those with severe lung involvement.

OUES has been proposed as a submaximal measure of fitness in obese and overweight adolescents (53). Drinkard et al. performed exercise testing in overweight adolescents and controls (38). They calculated OUES at the VAT, 150% of the $\dot{V}O_2$ at VAT, and peak exercise. Although OUES correlated well with peak $\dot{V}O_2$, they found a magnitude bias, where OUES over-predicted peak $\dot{V}O_2$ from low-intensity data sets and under-predicted peak $\dot{V}O_2$ for higher intensity data sets in overweight subjects. The same results were found for the normal control group at OUES VAT and OUES 150%. In a similar study (65) involving only obese subjects aged 7 to 18, OUES 100% and OUES VAT both correlated well with peak $\dot{V}O_2$ (r^2 = .44-.91). When these authors compared their findings to other published data of a normal set of subjects, OUES was higher than normal with uncorrected data and lower than normal when OUES was corrected for body weight, BSA, or fat-free mass (63). This demonstrates that OUES is influenced by body size and level of sexual maturation of the subjects. Therefore OUES may be useful in normal children who cannot perform a peak effort but not ideal for predicting $\dot{V}O_2$. As with other submaximal exercise protocols to predict $\dot{V}O_2$, nothing is ideal or completely independent of the size of the child or how much work he or she performs. No current data are available on changes in OUES during puberty.

There are limited data suggesting the OUES can detect improvements in aerobic fitness from chronic training. Gademan et al. trained 34 heart failure patients and demonstrated a 14% improvement in peak $\dot{V}O_2$ (41, 91). They also found improvement of similar magnitude for the OUES slopes for 75%, 90%, and 100% of peak $\dot{V}O_2$. Others (19, 33, 68) have shown improvements in peak $\dot{V}O_2$ and OUES with 3 mo of supervised training. Although OUES at 90% and at a RER of 1.0 increased, other parameters showed improvements as well. The $\dot{V}O_2$ at VAT increased and the slope of $V_E/\dot{V}CO_2$ decreased. OUES did not seem to be superior to the other indicators in demonstrating improvements in fitness from training.

The consensus of the research data indicates that OUES is a useful submaximal measure of aerobic fitness that is effort independent. It has been shown to be reproducible and does not suffer the intra-observer or inter-observer discrepancies often encountered with VAT (90). The OUES is easy to calculate and is determined by many more data points than VAT, peak work, or peak heart rate. The consensus of the research data indicates that it is most useful at the higher percentages of the peak $\dot{V}O_2$ (that is, OUES 80% or OUES 90%). OUES below 45% to 50% of peak $\dot{V}O_2$ is less reliable and does not correlate as well with peak $\dot{V}O_2$ (73). There is conflicting evidence regarding whether OUES is equally useful in predicting peak $\dot{V}O_2$ in healthy subjects and in those with chronic heart or lung disease (13).

Conclusion

In summary, our quest is to determine children's functional capacity so they can engage in physical activity for all of the benefits exercise provides. We want to be safe and prudent in our recommendation for increasing activity, certainly when a child has a condition that may limit the level of exercise. Exercise testing is essential to achieve this goal. Simple measurement of workload on a cycle ergometer will provide information on how close we are to "normal." Peak workload on a cycle ergometer is a measure of very vigorous exercise. The calibration of the cycle is often very good, and the workloads are very consistent over fast and slow pedaling. It is therefore a very useful measure of functional capacity for many subjects of different sizes and levels of physical conditioning. Within-subject reliability is also very good, so we can compare the child's responses over time accurately. Work in watts measured on the cycle ergometer can be used in a calculation to predict the oxygen consumption, again giving us a measure of functional capacity.

Current research indicates that one can predict the peak $\dot{V}O_2$ from the $\dot{V}O_2$ at VAT with good accuracy in healthy children as well as in those with lung disease. Caution should be used when predicting peak $\dot{V}O_2$ from the $\dot{V}O_2$ at VAT in children with congenital heart disease. Metabolic measurements need to be available to calculate the VAT, which adds to the cost and resources needed to predict the $\dot{V}O_2$ from this method. The measurement of VAT becomes more useful when a submaximal effort has limited the exercise test. Age, comprehension, size, motivation, and protocol all factor into a submaximal response. The caveat of the method is when we assume that the $\dot{V}O_2$ at VAT is about 60% of the peak $\dot{V}O_2$. Physical conditioning and specifically a lack of physical fitness influence this method greatly, and this may lead to over- or underestimation of peak $\dot{V}O_2$. We know that physical deconditioning, either from being sedentary or as the result of a disease process, lowers the $\dot{V}O_2$ at the VAT.

OUES is a useful submaximal measure of aerobic fitness that is effort independent. This method also relies on the availability of metabolic measurements. It has been shown to be reproducible and does not suffer the intra-observer or inter-observer discrepancies often encountered with VAT. The OUES is easy to calculate and is determined by many more data points than VAT, peak work, or peak heart rate. The consensus of the research data indicates that it is most useful at the higher percentages of the peak $\dot{V}O_2$ (that is, cut points of OUES 80% and OUES 90%). OUES below 45% to 50% of peak $\dot{V}O_2$ is less reliable and does not correlate as well with peak $\dot{V}O_2$. There

is conflicting evidence regarding whether OUES is equally useful in predicting peak $\dot{V}O_2$ in healthy subjects and in those with chronic heart or lung disease. If a subject can achieve 80% or 90% of predicted peak $\dot{V}O_2$, many of us would consider that an adequate level of exertion. If the subject achieves 80% or 90% of peak $\dot{V}O_2$, we do not know if the OUES slope can be abnormal. My guess is probably not. The slope of the line that represents the V_E and $\dot{V}O_2$ relationship is strongly influenced by the onset of anaerobic metabolism. The V-slope method of VAT detection is based on the change in that slope. Therefore the choice of the cut points will influence the OUES slope. Another factor that will influence the OUES is the choice of ergometer. We know that the peak $\dot{V}O_2$ on the cycle ergometer is about 10% to 15% less than what is achieved on a treadmill. Therefore percent cut points will actually represent different amounts of $\dot{V}O_2$. This could be problematic if we did not have equations for both the treadmill and cycle ergometer.

The six-minute walk is a simple test that is easy to administer and has good reproducibility. There are two versions, encouraged and unencouraged. The unencouraged version may reduce variability among test administrators. Distance and dyspnea score are recorded. The words and phrases of the Borg scale may not be completely understood by young children. Other scales with pictures have been used to assess the child's perceived exertion. The six-minute walk test may be more useful in underserved areas where metabolic equipment is not available. Also, large groups of children can be tested at very little cost, and over time increases in distance walked and a lower rate of perceived exertion may relate to improved fitness. This topic has been covered in greater detail in chapter 3.

Cardiac Output Measurement Techniques

Darren E.R. Warburton, PhD, and Shannon S.D. Bredin, PhD

Cardiovascular (or cardiopulmonary) fitness is an important determinant of exercise performance and the capacity to carry out activities of daily living (3, 31, 32, 84). Cardiovascular fitness (traditionally defined) relates to the ability to transport and utilize oxygen during exercise and work, reflecting the collective efficiency of the lungs, heart, vascular system, and exercising muscles (84). As outlined throughout this text, leading exercise and clinical physiology laboratories will assess cardiovascular (aerobic) fitness during exercise tests that allow for the determination of peak or maximal aerobic power ($\dot{V}O_{2max}$, the maximal rate of oxygen consumption) (89). According to the Fick equation, $\dot{V}O_{2max}$ is the product of cardiac output ($\dot{Q}$) and arteriovenous oxygen difference (a-$\bar{v}O_2$ diff) (80). Cardiac output (L/min) is the product of heart rate (bpm) and stroke volume (ml/beat) (87). The a-$\bar{v}O_2$ diff refers to the difference in the oxygen content of arterial and venous blood and reflects the oxygen extraction at the tissue level (mg/dl) (80).

As discussed elsewhere in this text, the measurement of $\dot{V}O_{2max}$ in young people is generally more difficult to interpret (3, 89) than in adults owing to the variability of the physiological responses to maximal exercise seen during growth and maturation. Furthermore, the interpretation of aerobic fitness is confounded by the need to appropriately account for differences in body size (89, 90). However, the direct (or indirect) assessment of $\dot{V}O_{2max}$ in youth has important implications for optimal health- and performance-related physical fitness (84). Owing to the importance of $\dot{Q}$ for oxygen transport and optimal human performance, this chapter will briefly outline some of the traditional and novel means of assessing $\dot{Q}$ during exercise in young people. This includes a critical analysis of the strengths and weaknesses

of each technology with particular reference to the usability of each technology with children during exercise. Normal values expected for maximal $\dot{Q}$ in youth are presented in chapter 6.

Cardiac output is widely held to be a key (often primary) determinant of exercise performance in healthy people (84) and in those living with chronic medical conditions (81, 85). Cardiac output is related directly to oxygen transport, aerobic fitness, and optimal exercise performance (87). Thus, the accurate assessment of $\dot{Q}$ is important for the evaluation of the cardiovascular responses to exercise, physical training, and growth and maturation (87). As reviewed elsewhere, the measurement of $\dot{Q}$ under exercise conditions (especially maximal exercise conditions) is perhaps one of the most important yet difficult measures in exercise and clinical physiology (82, 83). A series of methodologies have been proposed and used to measure $\dot{Q}$ under resting and exercise conditions (82, 83).

Invasive Versus Noninvasive Techniques

Invasive and noninvasive measures of $\dot{Q}$ have been largely taken in adult populations; however, there is mounting evidence demonstrating the utility of many of these technologies in children and adolescents, particularly in clinical settings. This chapter will briefly review the major techniques used today and evaluate the evidence for their suitability in pediatric populations.

The criterion (often referred to as "gold standard") methods of measuring $\dot{Q}$ during resting and submaximal exercise are invasive procedures (such as the direct Fick and dye-dilution methods) that require venous or arterial cannulation (82, 87). Recently, experts in clinical and exercise physiology have argued that we must be cautious about considering the direct Fick and dye-dilution methods to be gold standard comparisons (87). Theoretically, the direct Fick is the optimal means of determining $\dot{Q}$ at rest and during exercise, and it has been the standard to which most other technologies were compared. In fact, both the direct Fick and dye-dilution methods have been shown to be reliable during steady-state conditions such as at rest and during submaximal exercise (82, 83). However, the validity of both measures during strenuous exercise has been challenged owing to the difficulty of achieving reliable measurements under non-steady-state conditions (a prerequisite for accurate measurement in the direct Fick) (82, 83). Moreover, the invasiveness and inherent risks associated with both technologies limit their use during strenuous to maximal exercise and in healthy individuals (82, 87). Most studies conducted with children using these methods have involved those who were already undergoing catheterization as part of their medical assessment (87). These techniques are very difficult to perform in traditional laboratory settings; they require sophisticated medical management and supervision (87). Thus, it is very difficult to mimic real-world settings (e.g., races or endurance events) with these technologies. Similar concerns have been raised about more invasive techniques such as thermodilution and lithium dilution, which evaluate the temporal changes in temperature and lithium chloride, respectively. Justifiable concern is also raised about the invasiveness of these technologies for use with healthy children. Rosenthal and Bush (65) stated that, "Catheterization is nowadays unacceptable in healthy children and the data collected under such a stress bear little resemblance to normal life."

A variety of noninvasive measures have been developed and validated for use at rest and during exercise in healthy populations owing to the inherent risks associated with invasive measures of $\dot{Q}$ (87). Commonly employed technologies include

- foreign gas rebreathing,
- Doppler echocardiography,
- impedance cardiography, and
- arterial pulse contour.

This chapter will outline the relative strengths and weaknesses of these approaches for the evaluation of $\dot{Q}$ in children. In particular, we provide insight into the reliability and validity of these measures for assessing $\dot{Q}$ during incremental to maximal exercise.

Direct Fick Method

The direct Fick method calculates $\dot{Q}$ via oxygen or carbon dioxide content from arterial and mixed venous blood in combination with $\dot{V}O_2$ or CO_2 elimination ($\dot{V}CO_2$) during steady-state conditions (87). This invasive methodology requires arterial blood sampling from a systemic arterial catheter (often the brachial artery) and a mixed venous blood sample from a pulmonary artery catheter (18, 82). Blood sampling in conjunction with expired air analysis of $\dot{V}O_2$ or $\dot{V}CO_2$ (82) allows for the calculation of $\dot{Q}$ via the direct Fick equation

$$\dot{Q} = \frac{\dot{V}O}{CaO_2 - C\bar{v}O_2}$$

where $\dot{Q}$ = cardiac output (L/min), $\dot{V}O_2$ = oxygen consumption (L/min), CaO_2 = arterial content of oxygen, and $C\bar{v}O_2$ = mixed venous content of oxygen.

The direct Fick method has consistently been shown to be reliable at rest and during submaximal steady-state exercise, often with coefficients of variation of 5% or less (34, 82). However, steady-state conditions are required for the attainment of accurate measures of $\dot{Q}$ (18, 82). This is seldom possible during vigorous or maximal exercise, thereby limiting greatly the use of the direct Fick assessment of $\dot{Q}$ at these exercise intensities (82). Furthermore, the direct Fick method requires specialized medical personnel and equipment (87). Without meticulous care, the reliability and validity of this procedure may be worse than other techniques (82). There are also several potentially fatal complications associated with ventricular catheterization (such as ventricular arrhythmias and fibrillation, and perforation of the pulmonary artery or right ventricle) (82, 87). However, this risk should not be overstated because these procedures are always conducted with trained health professionals, and the incidence of these complications is relatively low (87).

The direct Fick continues to be used with healthy individuals (largely adults) in clinical and limited research applications. However, owing to the aforementioned limitations, this technique is less often used with strenuous to maximal exercise conditions (especially with children and

How Should Cardiac Output Measures Be Adjusted for Body Differences?

Pediatric cardiologists and clinical exercise physiologists have increasingly recognized the importance of correcting measures of cardiovascular structure and function for differences in body size (13, 22, 78). It is well recognized that a larger body size will have a greater oxygen demand and therefore higher blood flow (13). Appropriate controls are needed (particularly in maturing children) to account for the influence of differences in body size on measures of cardiac structure and function. There is considerable debate about the optimal scaling method. The most frequently used scaling method involves the simple division of a cardiac measure (such as $\dot{Q}$) by a measure of body size (i.e., ratio [ratiometric] scaling) (13, 22). However, this technique does not account for the complex and often nonlinear relationship between cardiac function and body size (13). Also, as reviewed by Chantler and colleagues, traditional linear regression models may either over- or under-correct for the effects of body size, leading to potentially flawed conclusions (13). Moreover, a basic assumption of linear regression is constant variance (*homoscedasticity*) such that each response variable (e.g., $\dot{Q}$) has the same variance in error (standard deviation) regardless of the predictor (e.g., body mass). However, the literature indicates that $\dot{Q}$ can vary widely over the range of body mass (13, 22). Furthermore, several cardiovascular variables that have been scaled by ratiometric approaches remain related to body size after scaling, which indicates that appropriate normalization did not occur (i.e., the scaling procedures did not produce a body-size-independent cardiovascular measure) (22). Therefore, alternative scaling procedures have been widely advocated, particularly for use with children.

Allometric scaling involves the division of a cardiovascular measure (such as $\dot{Q}$ or stroke volume) by a body size measure raised to a scalar exponent (22). Simply stated, the allometric body size correction is described as $y = a(x)^b$, where y is the physiologic variable (i.e., cardiac output), x is body mass, a is the proportionality coefficient (scaling factor), and b is the allometric scaling exponent, derived from the linear equation $\log y = \log a + b(\log x)$ among participants in a particular group (54). The physiologic variable can then be "normalized" for body size in the individual subjects in this group as y/x^b. This approach has several advantages over traditional ratiometric scaling, one of which is that it removes the effects of body size on cardiovascular structure and function (22). For instance, Turley and colleagues, using data from the HERITAGE study, revealed that traditional ratio scaling, regardless of the body size variable (body surface area, weight, fat-free mass), did not appropriately normalize $\dot{Q}$ or stroke volume in individuals between the ages of 17 and 65 (77). Moreover, Rowland and colleagues revealed that allometric scaling of $\dot{V}O_{2max}$, $\dot{Q}_{max}$, and maximal stroke volume eliminated the effects of body size in premenarcheal girls, but ratiometric scaling approaches did not (66). Allometric scaling appears to be particularly effective at minimizing the effects of body size on cardiovascular measures in obese persons (22).

Several factors have been used as scaling variables to correct measures of cardiovascular structure and function. Body mass index, body surface area, and lean body mass have been commonly used as scaling variables (13). Cardiac measures (particularly cardiac dimensions, $\dot{Q}$, and stroke volume) are often scaled by the simple division of body surface area (22). This is a routine practice in adult cardiology and exercise physiology studies. However, it is a less common practice in pediatric settings. The scaling of cardiac measures via body surface area and body mass has been criticized widely (13, 22). As reviewed by Dewey et al. (22), anthropometric variables (such as body mass and body surface area) include tissues with a high metabolic potential (e.g., muscle) and tissues with a low metabolic potential (e.g., adipose tissue and extravascular fluid volumes). The influence of low-metabolic-potential tissues on body mass and body surface area measures is especially significant in obese patients and in those with high extracellular fluid volumes (e.g., patients with heart failure) (22). Therefore, scaling according to body surface area or body mass is not generally recommended. Instead, body size indices that reflect a tissue mass with high metabolic potential (such as fat-free mass or lean body mass) should be used to scale cardiac measures (22).

adolescents). Pediatric research using the direct Fick method is generally restricted to resting evaluations with children who are undergoing medical treatment that requires invasive catheterization. As previously mentioned, the risks associated with this technology are often deemed to be too high for use with healthy youth in exercise physiology settings. Moreover, the invasiveness of the procedure makes it very difficult to duplicate real-world exercise conditions, limiting greatly the ecological validity of the procedure (87). With the accuracy and reliability of noninvasive $\dot{Q}$ assessment technologies, the use of direct Fick to assess $\dot{Q}$ in healthy people is not advocated, particularly in young people (87).

Dye-Dilution Method

Dye dilution (also called indicator dilution) is another invasive $\dot{Q}$ assessment technique that has been compared extensively against the direct Fick method (62, 82). This technique has demonstrated relative accuracies similar to the direct Fick (62). In dye dilution, a bolus of dye (typically 1 ml indocyanine green) is injected into the venous circulation (at the pulmonary artery or close to the right atrium) and arterial blood is sampled continuously downstream (87). A dye concentration curve (over time) is created, allowing for the calculation of $\dot{Q}$ by dividing the volume of dye injected by the area under the curve (8, 20, 26). A larger $\dot{Q}$ will produce a dye-dilution curve that rises and falls more rapidly (82, 87).

Coefficients of variation range between 5% and 10% for the dye-dilution technology at rest and during submaximal exercise (82). Limited research has also demonstrated similar reliability during maximal exercise (26). Dye dilution has been used largely in adult populations (82) with pediatric usage being largely limited to critically ill children. Owing to methodological concerns (similar to those surrounding the direct Fick technique), the use of the invasive dye-dilution method is generally not recommended for determining $\dot{Q}$ in children (87).

Pulse dye densitometry is a modification of dye dilution that has been proposed to be more suitable for use in healthy persons (87). This less invasive procedure involves the injection of indicator dye (e.g., indocyanine green) via a central or peripheral venous line. Changes in the concentration of the dye are evaluated via pulse oximetry (i.e., spectrophotometry) (5, 43) on the finger or

nose. The oximetry probe detects changes in the light of specific wavelengths (2), providing information about the arterial concentration ratios of dye and hemoglobin (43, 87).

There continues to be debate about the validity and reliability of pulse dye densitometry, especially at low $\dot{Q}$ (5, 9, 36). Some studies (in adult patients) have demonstrated reasonable agreement with other invasive procedures such as thermodilution (35-37). However, several authors have cautioned about the use of this technology, even during resting conditions. For instance, Reekers and colleagues (63) compared indocyanine green dilution measured by pulse dye densitometry to simultaneous arterial blood indocyanine green concentration (dye dilution) and found wide limits of agreement for $\dot{Q}$ and central blood volume. The authors concluded that the reliability and clinical utility of pulse dye densitometry appears to be limited. Other studies have also noted that there is a high incidence of low or absent pulse dye densitometry signals (5).

Although limited in comparison to the adult literature, pulse dye densitometry has been increasingly used with youth in clinical settings (52) and with animal models (74). Similar to the adult literature, there have been reports of difficulty obtaining reliable pulse waveforms at rest in children (74). Moreover, there have been reports that obtaining valid and reliable measures of blood volume are problematic using this technology (1); however, others have demonstrated favorable comparisons with results using invasive blood volume measurements (52). Taguchi and colleagues (74) have proposed that a three-wavelength system may be more appropriate and reliable for use with human children.

Despite the potential of pulse dye densitometry, to our knowledge, no study has examined $\dot{Q}$ using this technology during incremental to maximal exercise (in adults or children) (87). One study compared pulse dye densitometry to another noninvasive technique (the Modelflow technique) during light stepping exercise (50). It remains unclear whether movement artifact during whole-body vigorous exercise will be a significant barrier to using this procedure (87). It should be noted that variations of this technology (making use of near infrared spectroscopy) have been used to measure relative changes in muscle perfusion from rest to maximal cycling exercise (33).

Owing to the lack of reliability often observed for absolute $\dot{Q}$ at rest and during exercise conditions, the invasiveness of the procedure, and its limited use with children, the use of this technology is not advocated for assessing $\dot{Q}$ with young, healthy children. Future advancements in the technology may enhance its reliability, validity, and usability in exercise settings. However, the invasiveness of the procedure will remain a significant barrier for use with children.

Thermodilution Method

The thermodilution technique is based on principles similar to those of the dye-dilution method with the exception that a bolus of cold fluid is injected (into the right atrium) instead of dye (82). Cardiac output is calculated by measuring the cooling of the blood using a temperature probe (thermistor) in the tip of the pulmonary artery catheter (87).

As reviewed by Warburton et al. (82) there are several limitations to the thermodilution technique that may markedly affect its reliability and validity. Key issues include unknown heat loss during the handling of the coolant before it enters the circulation, unknown coolant loss through the vessel wall, and physiological variations in the temperature of pulmonary blood (82, 87). Owing to issues like these, clinicians should be cautious when using this technique, particularly when attempting to validate other measures (82). For instance, several studies have reported a systematic overestimation of $\dot{Q}$ in comparison to other methods (82). However, studies in children have shown good agreement between the direct Fick and thermodilution methods ($r = .92$) with no systematic bias (95). Major limitations for use in children remain the invasiveness of the procedure and the need for sophisticated medical equipment and monitoring. Also, repeat $\dot{Q}$ measurements via thermodilution may lead to fluid overload in small subjects (95).

The transpulmonary thermodilution (or arterial thermodilution) method is a variation of traditional pulmonary artery thermodilution involving the injection of cold saline through a central venous catheter (often the superior vena cava and femoral vein) and the measurement of thermal dilution (via a thermistor-tipped catheter) in a central artery (such as the aorta or femoral artery) as opposed to the pulmonary artery (87). Using this technique, it is possible to gain information about preload (e.g., intrathoracic blood volume and global end-diastolic volume) and extravascular lung water (10, 11).

This technique reduces the risks associated with pulmonary catheters, but it remains invasive, requiring central venous and central artery catheterization (87). The cold saline injected into the right atrium is mixed with right ventricular blood, allowing for changes in temperature to provide an estimate of $\dot{Q}$ (49). Therefore, $\dot{Q}$ is calculated according to the temperature change across the cardiopulmonary system similar to traditional thermodilution (49). This procedure is attractive for critical care patients who already have central venous and arterial catheters (49). It also allows for the calculation of extravascular lung water and may be useful in the monitoring of intracardiac shunts (30). The transpulmonary indicator dilution method compared favorably to traditional thermodilution (i.e., the pulmonary artery catheter method) (51) and the direct Fick method (58, 76) in pediatric patients. Two recent studies from the same laboratory (10, 11) have demonstrated that the transpulmonary thermodilution technique was valid and reliable during exercise in adults. The feasibility, reliability, and validity of this technique in children during exercise remains to be determined.

Owing to the invasiveness of the procedures, questions surrounding the reliability and validity of the measure, limited exercise data, and the need for sophisticated testing personnel and equipment, the use of transpulmonary thermodilution cannot be advocated for the determination of $\dot{Q}$ in healthy young people during rest or exercise conditions (87). This procedure may be more suitable for children undergoing medical procedures that already involve central venous and arterial catheters.

Lithium Dilution Method

The lithium dilution technique was first described by Linton and colleagues (47) and involves the intravenous (via a peripheral or central vein) injection of isotonic lithium chloride solution (150 mM) into a peripheral (or central) vein (49, 87). An arterial lithium concentration-time curve is constructed by drawing arterial blood through a lithium sensor (40, 49, 87). The $\dot{Q}$ is calculated based on the lithium dosage and the area under the concentration-time curve (prior to recirculation) according to the formula

$$\dot{Q} = \frac{\text{Lithium dose} \cdot 60}{\text{Area} \cdot (1 - \text{PCV})}$$

where $\dot{Q}$ = cardiac output (L/min), area = the integral of the lithium dilution curve, and PCV = packed cell volume (calculated as hemoglobin concentration (g/dl)/34) to correct for the fact that lithium is only distributed in the plasma of blood (49).

The lithium dilution technique has been evaluated in children in a clinical setting, revealing a strong relationship with transpulmonary thermodilution (48). Moreover, this technique has been shown to provide findings similar to other techniques during resting conditions (42). There has been a slow but progressive increase in the use of the lithium dilution technique in exercise settings, including studies with healthy individuals (27), heart failure patients (41), and horses (25). These studies have demonstrated the possibility of measuring $\dot{Q}$ during light to maximal exercise conditions. A commercially available system that makes use of the lithium dilution technique is provided by LiDCO Ltd (www.lidco.com). This method (as discussed later) has been used in the intermittent calibration of $\dot{Q}$ assessed via continuous blood pressure. A particular strength of this approach is the potential to measure continuous (real-time, beat-to-beat) stroke volume and $\dot{Q}$ during exercise conditions (41).

Although the lithium dilution technique is often considered minimally invasive (i.e., there is no need for a central venous or pulmonary artery catheter), it remains too invasive for routine use at rest and during exercise in healthy children.

Foreign Gas Rebreathing Techniques

Foreign gas rebreathing techniques have been used extensively in clinical and exercise physiology settings with adults and children (82, 83, 86). Inert soluble gases can enter or leave the blood through the lungs, allowing for the calculation pulmonary blood flow (which is equivalent to $\dot{Q}$ during equilibrium). The $\dot{Q}$ is estimated by measuring the rate at which an inert, soluble gas enters and leaves the bloodstream via the lungs (24, 82, 87). The rate of disappearance of the inert, soluble gas is directly proportional to the flow of blood past the lungs, providing a valid and reliable measure of $\dot{Q}$ during resting and maximal exercise conditions (82). There are several foreign gas methods to assess $\dot{Q}$, including closed-circuit rebreathing (86), open-circuit rebreathing, and

single-breath constant exhalation methods (12). See previous reviews on these procedures for more details about their reliability and validity (82, 83).

N_2O Rebreathing Method

The nitrous oxide (N_2O) rebreathing method (as its names implies) involves the rebreathing and evaluation of the gas uptake of soluble N_2O. This technique can be used to examine $\dot{Q}$ in non-steady-state conditions, but it is affected by ventilatory abnormalities (82). The technique is generally more accurate and reliable during submaximal and maximal exercise than at rest owing to the slight hyperventilation that often occurs at rest (82, 87). Several investigations have examined the N_2O rebreathing method at rest and during exercise in adults (82). Increasing evidence (although still limited) has evaluated this procedure in children, including exercise conditions (93, 94, 96). In particular, a relatively new noninvasive device that uses N_2O rebreathing and a photo-acoustic gas analyzer (Innocor, Innovision, Odense, Denmark) has been used with children. This is a low-cost and portable alternative to traditional foreign gas rebreathing setups that require a mass spectrometer. As was predicted in 2008 (87), this device has increasingly been used with children under exercise conditions. The N_2O rebreathing procedure holds considerable promise for the evaluation of $\dot{Q}$ in young people (particularly during submaximal and maximal exercise) (87). The gas mixture is relatively easy to rebreathe and can be examined via fast-response analyzers (82, 87). The N_2O rebreathing technique is also quite reliable under strenuous and non-steady-state conditions, making it particularly attractive for use under maximal exercise conditions (82).

Acetylene Rebreathing Method

Several international exercise physiology and sport cardiology laboratories currently make use of the acetylene rebreathing method to assess cardiac output during exercise (87). As stated previously (82), "Many investigators consider it the most viable means of estimating $\dot{Q}$ during submaximal and/or maximal exercise" (p. 33). The acetylene rebreathing technique has been used in various research and commercial platforms, including open circuit, closed-circuit rebreathing, and single-breath constant exhalation procedures (82).

The closed-circuit acetylene procedure involves rebreathing a gas mixture containing 35% to 45% oxygen, 0.5% to 1.0% acetylene, 5% to 10% helium, and a balance of nitrogen (82). Many laboratories employ standard rebreathing frequencies and durations at rest (approximately 1 breath/1.5 s for 18 s) and during maximal exercise (approximately 1 breath/s for 10 s) (82). Expired concentrations of gas mixture are measured continuously, generally by a mass spectrometer interfaced with specialized software that calculates $\dot{Q}$ (82). A constant level of helium provides an objective indicator of the adequate mixing of the lung-bag rebreathing system (82). After the point of equilibration of acetylene, the rate of disappearance of acetylene is directly proportional to pulmonary blood flow, allowing for the estimation of $\dot{Q}$ (82). The mathematical formulae behind the technique has been described in detail elsewhere (82).

Acetylene rebreathing has well established reliability (5%-10%) under exercise conditions, and reliability improves with increasing exercise intensity (82). The technique has also been validated at rest and in exercise conditions against criterion methods, including the invasive direct Fick and dye-dilution methods (82).

The open-circuit acetylene technique involves the normal breathing of a mixture of two inert gases (soluble acetylene and insoluble argon) (29). In this technique, a wash-in period of 6 to 10 breaths of the inert gases is used with an attempt to maintain a normal breathing pattern (39, 83). Gledhill's laboratory was the first to use the open-circuit technique during incremental to maximal exercise in healthy adults (12). In the first study, Card and colleagues (12) demonstrated a strong relationship (r = .974) between the open- and closed-circuit acetylene systems during incremental to maximal exercise in healthy adult males. The authors acknowledged several advantages of this system, including ease of operation for the evaluator and participant, greater tolerance by participants, and the ability to breathe almost normally during the procedure (i.e., spontaneous breathing with no disturbance in the normal breathing pattern). This technique has been used increasingly in the literature, but its acceptance has been slower than anticipated by the original authors (12). The use of this technique has been hampered greatly by its current reliance on mass spectrometry and customized software. Nonetheless, the procedure has been shown to be reliable (7, 23) and valid (particularly for submaximal exercise conditions). The technique has compared favorably to a wide range of techniques, including the acetylene rebreathing closed-circuit technique, direct Fick, and thermodilution (4, 7, 29, 39). Several investiga-

tions have used this technique in healthy young and older adults during exercise (4, 7, 23, 39). This technique may be better for use with children owing to the use of spontaneous rather than forced breathing (53, 87).

Acetylene-based techniques have several advantages for assessing cardiac function during exercise (including maximal exercise) (87). In particular, the ability to accurately and reliably determine $\dot{Q}$ during strenuous and non-steady-state conditions is attractive for use in young people, adults, and clinical populations (82). Fast-response acetylene gas analyzers have been created, greatly reducing the costs of this methodology by removing the need for mass spectrometry (82, 87). It can be anticipated that laboratories will increasingly use this method to assess the $\dot{Q}$ of young people during exercise. Both the open- and closed-circuit acetylene systems appear to be appropriate for the determination of $\dot{Q}$ in young people.

Acetylene Single-Breath Constant Exhalation Procedure

The addition of rapid response infrared acetylene analyzers to commercially available metabolic carts (such as the Ergocard from Medisoft) represented a significant shift in what traditional exercise physiology laboratories could evaluate during exercise. Pulmonary blood flow and $\dot{Q}$ can be assessed using this system through the use of a single, prolonged, constant expiration of acetylene (87). Participants are required to breathe a gas mixture that contains 0.3% methane, 0.3% acetylene, 21% oxygen, 0.3% carbon monoxide, and the balance nitrogen (83). Exercising subjects are asked to inhale the gas mixture to total lung volume and hold their breath for approximately 2 s (allowing for acetylene absorption into tissues and gas distribution equilibration) (83, 97). Participants then exhale at a constant rate of 200 to 500 ml/s, and rapid response infrared sensors detect the changes in the respective gas concentrations, allowing for the determination of $\dot{Q}$ (83). Further details regarding the mathematical relationship between gas absorption and $\dot{Q}$ can be found elsewhere (83, 97).

Several investigations demonstrated the reliability and validity of the single-breath constant exhalation procedure (75). This technique has been shown to be reliable during incremental to maximal exercise conditions (23, 75). However, several investigators have noted the difficulty some participants have exhaling at a constant flow rate during strenuous (particularly maximal) exercise (23). In our experience, it has been difficult to conduct this procedure in children during strenuous exercise. Further research is warranted to examine the reliability, validity, and feasibility of this procedure for assessing $\dot{Q}$ in young people (87).

CO_2 Rebreathing Method

The CO_2 rebreathing (indirect Fick) method has been used extensively in clinical and exercise physiology settings (82). This technique involves the assessment of the CO_2 content of the blood using either the Collier (17) or Defares (21) methods. The $\dot{Q}$ is estimated from pulmonary capillary blood flow assessed using CO_2 as the indicator gas in the Fick equation (49, 87). There are several advantages to measuring CO_2; it is easy to measure, and arterial CO_2 content can be estimated from expired gas (49).

As reviewed previously (82), the Collier method requires that the participant rebreathe a gas mixture containing 10% to 20% CO_2 until an equilibrium is observed. At the equilibrium plateau, there is no further exchange of CO_2 between the alveolar-capillary membrane and the rebreathing bag. It is assumed that the CO_2 at this equilibration point represents the partial pressure of mixed venous CO_2 (P_vCO_2). The partial pressure of mixed arterial CO_2 (P_aCO_2) can either be assessed from the blood (including an arterial blood sample, a capillary blood sample from the fingertip or ear lobe, or an arterialized venous blood sample) or estimated noninvasively from end-tidal CO_2 ($P_{et}CO_2$) immediately prior to the start of rebreathing. Correction factors are sometimes applied to P_aCO_2 to correct for alveolar-arterial differences and to P_vCO_2 to correct for gas-blood differences (82).

In the Defares method, participants rebreathe from a bag containing a low concentration of CO_2 (i.e., 0%-5%). During the rebreathe procedure, the CO_2 increases in an exponential manner toward the P_vCO_2 (82). The exponential slope is used to determine the point of equilibrium using the end-tidal CO_2 values allowing for the determination of P_vCO_2 (82). Unlike the Collier method, no correction is applied in the Defares method for gas-to-blood pCO_2 differences owing to the fact that this procedure does not require the attainment of an equilibrium pCO_2 (87).

In both methods, P_vCO_2 or P_aCO_2 are used with an appropriate CO_2 dissociation curve to estimate CO_2 content. The strengths and limitations of these two methods have been outlined in detail elsewhere (82). Briefly, the CO_2 rebreathe method provides accurate and reliable estimations of $\dot{Q}$ during submaximal exercise. A distinct advantage of this procedure is the simple need for fast-response CO_2 analyzers; sophisticated and costly quadrupole mass spectrometers are not required. The key limitations of the procedure include the fact that individual changes in pH, venous oxygen saturation, and temperature cannot be controlled for when using a standard dissociation curve for CO_2 (eliminating the need for blood sampling). Most researchers using this technology argue that these changes have a negligible effect on the determination of $\dot{Q}$; however, this assumption may not be correct during exercise (82). Also, the accuracy and reliability of the procedure are somewhat limited at rest. A major limitation of the technology is the need for steady-state conditions for the valid measurement of $\dot{Q}$ (82). Therefore, most researchers who use the CO_2 rebreathe method only examine $\dot{Q}$ at exercise intensities below 85% of maximum. However, some have successfully used the Defares method during maximal exercise in adults (28). The Collier is extremely difficult to use during high-intensity exercise owing to the high concentration of CO_2 (82).

The CO_2 rebreathe technology has been used extensively in adults and children (87). Research conducted in young people (similar to adult literature) has revealed several strengths and limitations of the CO_2 rebreathing procedure, including greater accuracy during exercise than at rest (6, 96), $\dot{Q}$ varying depending on the method used to estimate P_vCO_2 and dead space (56, 57, 70), higher day-to-day variability in children than in adults (57), differences related to body surface area (e.g., less accuracy in children with small body surface area and small tidal volumes) (45), and inaccuracy with certain patient populations, such as those with poor hemodynamic stability, severe pulmonary disease, or intrapulmonary shunt (49, 59). Research has indicated that studies measuring $\dot{Q}$ in children using the equilibrium method should apply a downstream correction factor to improve the validity of the procedure (38).

Collectively, the CO_2 rebreathe method has been accepted for use during submaximal exercise in adults and young people. It is currently not advocated for use during maximal exercise conditions. Also, special measures appear to be required for valid and reliable measures in children. If these precautions are not taken, the reliability and validity of the procedure appears to be lower in young people than in adults (87). With the rebreathing technique, the assumption is made that pulmonary blood flow is equal to systemic blood flow. In certain patients this is not true. For instance, in patients with atrial septal defects, ventricular septal defects, patent arterial ducts, and with other left to right shunts, this technique will overestimate the cardiac output because pulmonary blood flow in these patients is greater than systemic blood flow. Conversely, in patients with right to left shunts, such as those with fenestrated Fontans, gas rebreathing techniques will underestimate systemic blood flow because systemic blood flow is greater than pulmonary blood flow.

Doppler Echocardiography

Doppler echocardiography has been used widely with young people, particularly at rest and during submaximal exercise. This technology measures stroke volume by measuring the velocity of blood in the aorta, pulmonary artery, or mitral valve and the diameter of the blood vessel through which blood is flowing (87). In the Doppler method, an ultrasound wave is transmitted through blood flowing through a vessel (often the aorta) (83). The suprasternal notch is the most common location for the Doppler probe during exercise; this allows the ultrasound wave to pass through the blood flowing through the ascending aorta (83).

The movement of red blood cells causes a shift in the frequency of the reflected ultrasound waves (i.e., a Doppler frequency shift), which yields a measure of blood velocity according to the formula

$$V = \frac{\Delta f \times c}{2f \times \cos\theta}$$

where Δf is the shift (change) in frequency (Hz), f is the frequency of the original ultrasound wave (Hz), c is the velocity of sound in tissue (1540 m/s) and θ is the angle between the blood flow and the ultrasound signal (83).

The stroke volume (ml/beat) is calculated as the product of the velocity time integral and the cross-sectional area of the blood vessel (67, 83). The $\dot{Q}$ is calculated as a product of stroke volume and heart rate.

A key advantage of Doppler echocardiography is its ability to provide noninvasive, beat-to-beat measurements of cardiac function at rest, during submaximal exercise, and during maximal exercise (with appropriate care). However, the equipment is quite expensive and requires a high level of training to use (clinical and national certification in sonography), and movement artifact has a significant effect on the reliability and validity of the methodology (particularly during strenuous exercise) (83). The strengths and weaknesses of this technology have been reviewed extensively (67, 83).

The validity of Doppler echocardiography in resting and exercise conditions has been well established (when conducted by highly trained sonographers and using appropriate controls) (83). Several investigations have also reported good reliability of $\dot{Q}$ derived via Doppler echocardiography during exercise in children (55, 68, 69, 79). Most laboratories that use Doppler echocardiography do so during light- to moderate-intensity exercise, and they apply stringent controls to minimize the effects of movement artifact. This includes using highly trained clinical sonographers with specialized training in exercise echocardiography and cycle or supine ergometry. For laboratories using two-dimensional (with Doppler) echocardiography, these controls in addition to the evaluation of cardiac function in the left lateral decubitus position are often employed (83).

In a systematic evaluation of the literature (83) we revealed that Doppler echocardiography provides reasonable estimates of resting $\dot{Q}$ when compared to established invasive and noninvasive techniques. Doppler echocardiography also was able to provide reasonable estimates (on a beat-by-beat basis) of temporal changes in $\dot{Q}$ from rest to submaximal exercise. Moreover, there was good reliability for the measure at rest and during submaximal exercise (with coefficients of variation of 10%-15%). However, there was strong evidence that Doppler echocardiography underestimates absolute values of $\dot{Q}$ (by approximately 15%-20%) during exercise in comparison to other noninvasive and invasive measures. It is certainly more difficult to obtain maximal measures of $\dot{Q}$ using Doppler echocardiography (16, 71), although the procedure appears to be more valid than two-dimensional echocardiography (16). Various limitations of Doppler echocardiography are identified in the literature along with potential explanations for the errors seen at higher intensities of exercise. These include the effects of elevated breathing frequencies, movement artifact, and the difficulty of keeping the transducer in the same location (in addition to several other technical issues that are discussed later in this section). However, we emphasized the important information that can be derived from Doppler echocardiography, and we also highlighted how most other techniques (including the criterion methods) are limited during near-maximal exercise conditions.

Chew and colleagues (15) conducted a 20 yr review of the literature related to Doppler assessments of $\dot{Q}$ in critically ill children. The authors reported that the precision of Doppler measurements of $\dot{Q}$ approximate 30% in comparison to the direct Fick, dye-dilution, or thermodilution methods. The inter- and intra-observer repeatability was similar to other techniques (range 2%-22%), and the bias was generally less than 10% (despite considerable variation in studies). The authors concluded that Doppler echocardiography was acceptable for tracking changes from baseline in children but less accurate and reliable than other methods for reporting absolute values.

A more recent comprehensive review by Rowland and Obert (67) evaluated the reliability of Doppler echocardiography during exercise, highlighting the strengths and weaknesses of the procedure. The evidence indicated good construct and concurrent validity and similar reliability to other measures of $\dot{Q}$ during exercise. The authors (established experts in Doppler echocardiography) also highlighted how this technology is particularly suited for use for young children. The authors emphasized that the technology provides limited risk and minimal discomfort to the participant and can be used during incremental exercise. The authors also acknowledged a series of limitations of the procedure, including the reliance on some form of stabilization of the participant during exercise, leading most researchers to use cycle ergometry when using Doppler echocardiography. Other factors that affect the reliability and validity of the procedure include the potential for aortic cross-sectional area change during exercise, transducer angulation, turbulence and alteration of a flat velocity profile in the aorta with increasing $\dot{Q}$, and uncertainty about the location of the appropriate measurement site for the aortic outflow area (67).

In summary, Doppler echocardiography has considerable advantages for assessing $\dot{Q}$ in young people. It can provide a noninvasive determination of beat-by-beat changes in $\dot{Q}$ during exercise (including non-steady-state conditions) with a reliability and validity that are similar to many other invasive and noninvasive devices. Questions remain about its absolute accuracy during maximal exercise and its applicability to exercise modalities other than cycle ergometry.

Impedance Cardiography

Impedance cardiography (transthoracic electric bioimpedance) evaluates $\dot{Q}$ by passing a small (4 mA), high-frequency (100 kHz) alternating current through the chest (83). Recording electrodes are placed at specific locations on the upper body to allow for the monitoring of changes in electrical impedance. There are many configurations and recommendations for the optimal measurement of changes in thoracic impedance. However, the traditional setup often uses four pairs of electrodes in addition to standard electrocardiography recordings (49). Many laboratories will use eight disposable electrodes (four pairs of dual sensors consisting of transmitting and sensing electrodes) placed on the base of the neck and lower chest (at the level of the sternal-xiphoid process junction) to transmit the electrical current and record impedance changes in the thoracic cavity (49). Band electrodes are also commonly used in impedance cardiography.

Impedance cardiography measures changes in electrical resistance (impedance) that occur in the thoracic cavity as blood volume increases and decreases during systole and diastole, respectively (49, 87). Changes in thoracic impedance during systole are thought to reflect changes in stroke volume. The continuous assessment of impedance allows for the evaluation of beat-by-beat changes in cardiac function. Several formulas have been used to calculate stroke volume from changes in thoracic impedance, and these are described in more detail elsewhere (83).

Considerable debate exists regarding the accuracy and reliability of impedance cardiography, particularly under exercise conditions (83). The theoretical constructs behind the technology have also been widely criticized (83). Early work revealed low to moderate correlations between impedance cardiography and other technologies. However, more recent studies revealed good reliability and accuracy in comparison to other techniques (83).

Impedance cardiography was often seen to face great challenges under exercise conditions (particularly strenuous and maximal exercise). To overcome these challenges, many early investigators would use exercise pauses or breath-holding procedures, which greatly limited the ecological validity of the technology (particularly during exercise) (83).

In recent years, the technology has benefited from many advancements, particularly related to its utility during exercise in healthy and clinical populations. A growing body of research has demonstrated the ability of impedance cardiography to assess the temporal exercise-related changes in $\dot{Q}$ (44, 60). Owing to its ease of administration and noninvasiveness, impedance cardiography has been used extensively with children, providing important insight into cardiovascular dynamics on a beat-by-beat basis and in an operator-independent manner. Of importance, impedance cardiography provides minimal intrusion to young participants (i.e., simply requires the placement of electrodes on the neck and chest).

There are several commercially available systems that use impedance cardiography with advanced filtering techniques. The PhysioFlow has been used in laboratory (14, 64) and ambulatory settings, including prolonged exercise trials, with success. According to the manufacturer, this methodology does not rely on the assessment of baseline impedance (which is affected by hydration status, blood resistivity, and distance between electrodes). This device is also well tolerated and reliable when used with children, including during maximal exercise (88).

In summary, despite widespread criticism of its theoretical construct, recent research has shown that impedance cardiography can track temporal changes in $\dot{Q}$ during exercise conditions. The technique has many advantages for use with children, including being noninvasive, portable, versatile, cost-effective, suitable for measuring beat-by-beat changes in cardiac and hemodynamic function, and usable in settings where there is less technical training and experience (83).

Arterial Pulse Contour Method

The arterial pulse contour method permits a continuous, beat-to-beat evaluation of $\dot{Q}$ from arterial pressure waveform tracings. Available systems generally provide an estimate of the aortic pressure waveform from a peripheral artery such as the brachial artery (invasively via catheter or noninvasively via tonometry)

or noninvasively from the fingertip (49). The stroke volume can be assessed using various models of the systemic circulation as described elsewhere (49). Stroke volume (and therefore $\dot{Q}$) can be estimated from the diastolic or systolic portions of the arterial pressure waveform (49). The pulse contour method provides measures of relative changes in $\dot{Q}$, and a calibration against another valid measure of $\dot{Q}$ is often advocated for the estimation of absolute values (72). The methodology (and its various modifications) are described in detail elsewhere (46, 61, 92). Despite criticism, arterial pulse contour methodologies have been shown to provide adequate agreement with other techniques in adults (19, 61) and clinical populations (92).

The Modelflow software developed by Wesseling and coworkers (91) has been incorporated into the Finapres blood pressure monitoring system. Therefore, several exercise physiology laboratories from around the world could measure cardiac function noninvasively via arterial blood pressure waveforms derived from finger plethysmography in combination with the Modelflow software. The Modelflow system has been used increasingly in the literature and tested against other techniques (73, 91). However, questions remain about its ability to accurately measure absolute values of stroke volume and $\dot{Q}$, particularly during exercise. Calibration against other techniques has been advocated for the accurate assessment of absolute stroke volume and $\dot{Q}$, limiting the potential utility of the Modelflow system for children.

Pulse contour analysis techniques have generally been conducted in adults and clinical populations. The ease of use and noninvasive nature of the procedure may make it attractive for use in young people. However, it is important to highlight that it is extremely difficult to obtain a reliable pulse contour during movement and exercise. Movement artifact is a serious problem when attempting to utilize this technology.

In summary, the arterial pulse contour analysis method shows promise for noninvasive and continuous measurement of $\dot{Q}$ under resting conditions; however, its ability to track $\dot{Q}$ during exercise is currently limited. Moreover, calibration using other techniques is advocated if accurate measures of $\dot{Q}$ are required. It is currently not advisable to use this technology for exercise conditions in children owing to its limited reliability and accuracy during exercise and the availability of other noninvasive procedures that can assess cardiac function on a beat-by-beat basis.

Conclusion

There are many invasive and noninvasive techniques for assessing $\dot{Q}$ in young people. Most of these techniques were developed for adults in clinical situations, and therefore they are often not optimal for use with young people. Recently, a series of techniques have been developed that allow for the assessment of $\dot{Q}$ at rest and during exercise. Many of these techniques are noninvasive and suitable for use with young people. Each technique has its relative strengths and weaknesses. It is ultimately up to individual clinicians or researchers to determine the optimal means of assessing $\dot{Q}$ for their clients.

Assessing Myocardial Function

Thomas W. Rowland, MD

Impairment of contractile or relaxation properties of the ventricular myocardium is a common end point in the pathogenesis of cardiac disease. Whether as the result of chronic volume overload, pressure work, ischemia, or metabolic disorders, deterioration of myocardial functional capacity is often a final common result of cardiac disorders. The health status of the heart muscle serves, then, as one of the strongest predictors of morbidity and mortality in patients with cardiac disease. Therefore, insights into the systolic and diastolic functional status of the myocardium both at rest and during exercise are important to clinicians seeking to establish a prognosis or determine the timing of surgical and medical interventions.

Increases in inotropic (systolic) and lusitropic (diastolic) ventricular function are critical to augmenting cardiac output during progressive exercise. As described in chapter 4, a failure to enhance contractility (and, in a parallel fashion, myocardial relaxation) as workload increases will diminish stroke volume response, limiting maximal cardiac output, $\dot{V}O_{2max}$, and endurance.

Given these considerations, we should expect the assessment of systolic and diastolic myocardial responses to exercise to serve as a useful tool for establishing the severity and prognosis of cardiac abnormalities. Adult cardiologists have a long history of measuring heart function by clinical exercise testing, particularly with radionuclide angiography, given the frequent issues of congestive heart failure in older patients. Early on, concerns about myocardial functional status were limited for pediatric patients with congenital heart disease, whose pathophysiology typically involved hypoxemia, pulmonary hypertension, and critical obstructions. More recently, however, dramatic advances in pediatric cardiac surgery have permitted survival and extension of life in many complex cases of congenital heart disease,

creating new clinical problems that include progressive deterioration in myocardial function. This trend has created an increasing need to more completely assess myocardial function both at rest and during exercise in pediatric patients.

The ideal technique for assessing myocardial functional responses to exercise would be

- noninvasive and feasible, being easy to perform and measure,
- accurate and reliable,
- capable of estimating myocardial performance *during* exercise,
- free of radiation risk,
- capable of providing clinically predictive information, and
- not requiring a "maximal" exercise effort.

Unfortunately, at present, despite considerable progress in measurement technologies, no such technique has been clearly identified. This chapter will provide a state-of-the art review of current methodological approaches to estimating myocardial function with exercise. We will give special attention to newer echocardiographic techniques whose clinical utility is still in the early stages of development but which bear promise for applicability in the future.

In this chapter, inotropic functional capacity, or myocardial contractility reserve, is defined as the increase observed in the velocity and force of heart muscle contraction in the course of a progressive exercise test. This differs from the standard physiological definition, which indicates the extent and speed of myocardial contraction when loading conditions (preload, afterload) are kept stable. As indicated in chapter 4, increases in the force and velocity of myocardial contraction can be influenced by an increase in precontraction fiber length, a rise in heart rate, a reduction in afterload, sympathetic stimulation, and action of circulating catecholamines. As all of these influences are operant in a standard, progressive exercise test performed in the upright position, it is not possible to easily identify the effects of individual factors on myocardial functional responses (29). Consequently, the definition used in this chapter is empirically derived and encompasses the combined effects of these factors.

Systolic Time Intervals

Measurement of the duration of the components of the systolic portion of the cardiac cycle (systolic time intervals, or STIs) was one of the first methods for noninvasively estimating myocardial contractility. At rest, the left ventricular ejection time (LVET) increases with contractile responses to augmented preload, while the pre-ejection period (PEP, between myocardial activation and blood ejection) shortens. Thus, a close inverse relationship can be demonstrated between the ratio of PEP to LVET and the ventricular ejection fraction.

Using STIs determined by simultaneous measurement of electrocardiogram, carotid pulse trace, and phonocardiogram, estimates of inotropic function were used clinically to assess the severity of various forms of congenital and acquired heart disease in youth. Difficulties arose when applying STI measures during exercise because of artifact created by body motion. Nonetheless, normal values for STIs during submaximal cycling were reported by Vavra et al., who found similar shortening of PEP and LVET in boys and men (35).

The STI technique was cumbersome to perform and required high-fidelity equipment and close attention to detail (16). Moreover, the pathophysiological underpinnings of STI findings were complicated, and results varied according to type of heart disease. Consequently, this approach to assessing myocardial function was supplanted by other techniques, particularly radionuclide angiography and echocardiography. As will be noted later in this chapter, however, measurement of systolic ejection time (in relation to stroke volume) by Doppler echocardiographic methods continues to be useful in estimating cardiac contractility with exercise.

Radionuclide Exercise Testing

Measurement of radiation counts over the heart by first-pass or gated blood-pool methods after injection of radioactive tracers provides infor-mation about ventricular systolic and diastolic volumes during supine or upright exercise. This time-honored technique has been commonly used to estimate exercise-related changes in ventricular ejection fraction as a marker of global myocardial functional reserve in adult patients, particularly those with coronary artery disease. Findings have correlated nicely with indices of myocardial contractility recorded during cardiac catheterization.

Myocardial radionuclide imaging has not been as enthusiastically embraced in the assessment of pediatric patients. The radiation dose in a single study is less than that normally received by a patient undergoing standard cardiac catheterization, but this concern limits both the number of subjects examined and the number of serial examinations performed (10). The method is also constrained by expense, the need for technical expertise, and, in children, the absence of data obtained from healthy control subjects.

DeSouza et al. reported a rise in average ejection fraction from 63% at rest to 81% at peak supine exercise using gated equilibrium nuclear angiograms in 25 subjects aged 8 to 18 with familial hypercholesterolemia but no clinical evidence of heart disease (9). Parrish et al. found an average rise in left ventricular ejection fraction of 14% in 32 children aged 5 to 19 without significant heart disease (22). The magnitude of increase in ejection fraction with exhaustive exercise in these reports of "normal" children is similar to that described in healthy adult populations (2), in whom a rise of less than 10% has been identified as abnormal (10).

Nuclear stress tests have been used to evaluate ventricular function in pediatric patients. These studies have usually been conducted in the setting of long-term follow-up after surgery for complex congenital heart disease, such as atrial switch for transposition of the great arteries (21), cyanotic abnormalities with external conduits (33), and tricuspid valve atresia (1). In these studies, abnormal ejection fraction responses to exercise have typically been observed in about half of the cases, with many patients demonstrating a decline in ejection fraction.

Pattern of Stroke Volume Response

As described in chapter 4, during progressive exercise in the upright position, stroke volume initially rises by approximately 25% at low intensities and then remains relatively stable to the point of exhaustion. This pattern is similar in otherwise healthy persons with high and low levels of aerobic fitness, the curve being shifted superiorly in the former. During the course of such an exercise test, inotropic and lusitropic functions rise as a means of maintaining the same stroke volume as the ejection period shortens.

In patients with myocardial dysfunction, the ability to augment contractile force is insufficient to maintain stroke volume as workload increases, and stroke volume can be expected to decline with increasing exercise intensity (8, 28) (figure 10.1). Therefore, the assessment of the *pattern* of change in stroke volume during exercise by any of the methods outlined in the previous chapter can be useful in identifying patients with myocardial dysfunction.

By itself, a low stroke volume value at peak exercise does not necessarily demonstrate depressed myocardial function, since a reduced peak stroke volume at exhaustion is characteristic of both a reduction in aerobic fitness (low $\dot{V}O_{2max}$) in a healthy child and impaired myocardial function with heart disease. However, the latter is more likely in the person with either a marked depression of maximal stroke volume or a progressive reduction in maximal stroke volume on serial testing over time. Expected normal values for maximal stroke index (related to body surface area) are 50 to 60 ml/m^2 in boys and 45 to 55 ml/m^2 in girls with the Doppler ultrasound or carbon dioxide rebreathing techniques (27). Peak values are reported to be approximately 10% lower when using the thoracic bioimpedance method.

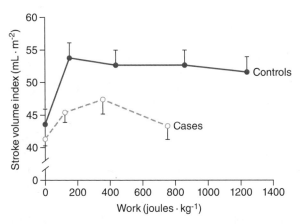

Figure 10.1 Pattern of stroke volume response to progressive upright exercise in healthy adolescent subjects (controls) and young patients with myocardial dysfunction (cases).

The relationship between oxygen uptake and work rate also provides an indirect measure of stroke volume response. During a progressive cycle exercise test, $\dot{V}O_2$ should rise linearly with increasing work (watts); the normal slope of $\dot{V}O_2$ versus work is 9.0 to 11.0 ml/(min · W). A depressed slope or a decline in trajectory of this relationship as work intensity rises is compatible with a depression in myocardial functional reserve (37).

Oxygen Pulse

Since the equipment capable of estimating stroke volume or cardiac output is not available in all exercise testing laboratories, clinicians have often relied on the *oxygen pulse* (absolute oxygen uptake divided by heart rate) as an indirect means of assessing stroke volume responses to exercise and, by extension, myocardial contractile health. According to the Fick equation,

$$\dot{V}O_2 = \text{stroke volume} \times \text{heart rate} \times (C_{aO2} - C_{vO2})$$

where the last two terms indicate arterial and mixed venous oxygen content, respectively. By rearrangement of this equation, oxygen pulse reflects the product of stroke volume and arterial venous oxygen content difference. Values are therefore influenced by individual differences not only in stroke volume but also in the augmentation of oxygen extraction at the muscle level as exercise work intensifies.

The measurement of oxygen pulse during an exercise test can provide insights into the change in stroke volume during progressive exercise—and thus myocardial function—by the assessment of

- the pattern of oxygen pulse as exercise progresses, and
- the values of oxygen pulse at peak exercise.

As indicated previously, stroke volume normally rises early in a progressive test and then levels off at low work rates. Concomitantly, arterial venous oxygen difference, dictated by the rate of oxygen extraction at the level of the skeletal muscle, increases steadily, usually in a curvilinear fashion. In a normal person, oxygen pulse is expected to rise progressively as the product of the two.

In subjects with normal myocardial functional responses to exercise, oxygen pulse should progressively rise as exercise intensity increases. This can be graphed by the software that accompanies many commercial metabolic assessment systems.

While this graph is usually curvilinear, a flattening or decline in oxygen pulse plotted against workload during progressive exercise suggests a dampened stroke volume response, reflecting limitations in myocardial contractile reserve (19).

A linear rise is normally observed in the relationship of heart rate to $\dot{V}O_2$ per kg body mass during a progressive exercise test, with an expected slope of 2.5 to 4.0 beats per ml/(kg · min) (6). When plotted against $\dot{V}O_2$, a *rise* in heart rate trajectory (i.e., increased slope) with exercise is consistent with a diminished stroke volume and inotropic response (7).

Published reports indicate that oxygen pulse at maximal exercise provides a reasonable—but not precise—estimate of peak stroke volume. Among 44 trained soccer players and active nonathletic adolescents, Unnithan and Rowland (34) found a correlation coefficient of r = .73 (p < .05) at peak exercise between oxygen pulse and stroke volume (estimated by Doppler echocardiography) (figure 10.2). This mimics values described in adults during moderate-intensity exercise by Norris et al. (20).

Washington et al. (36) provided normal data for oxygen pulse at maximal exercise related to body surface area during cycle testing in healthy children and adolescents (table 10.1). It should be reemphasized that since oxygen pulse is the product of the difference between stroke volume and arterial venous oxygen, factors influencing the latter (particularly anemia) will affect the validity of oxygen pulse as a predictor of stroke volume.

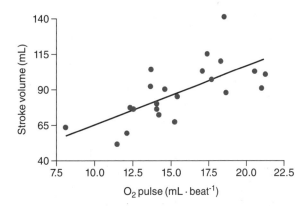

Figure 10.2 Relationship between stroke volume and oxygen pulse ($\dot{V}O_2$/heart rate) at peak exercise in adolescent soccer players and nonathletic subjects (r = .73).

Reprinted, by permission, from V. Unnithan and T. Rowland, 2015, "Use of oxygen pulse in predicting Doppler-derived maximal stroke volume in adolescents," *Pediatric Exercise Science* 27: 12-418.

Likewise, an impaired chronotropic response to exercise may also affect the reliability of the oxygen pulse to estimate stroke volume.

Doppler Echocardiographic Techniques During Exercise

In the course of the historical evolution of clinical echocardiography, a number of techniques have developed to effectively assess not only cardiac anatomy, dimensions, and blood flow but myocardial systolic and diastolic function as well. Some, particularly those involving measurements of changes in ventricular size (such as ejection fraction and shortening fraction), are often challenging to obtain during exercise. Others, particularly Doppler measures of transmitral filling velocity and myocardial longitudinal function (tissue Doppler imaging), have been found to be both feasible and reliable in maximal exercise studies in healthy youngsters (3, 30). These approaches require exercise on a cycle ergometer (for patient stability) and a step (typically 3 min) rather than a ramp protocol (for submaximal measurements during steady state). Recent reports indicate the applicability of such approaches in the clinical setting; however, their utility and predictive value remain to be addressed.

Systolic Function

A number of markers of systolic function can be obtained by Doppler echocardiographic techniques during a progressive cycle exercise test. These include left ventricular systolic ejection rate and longitudinal myocardial velocity during ventricular contraction.

Systolic Ejection Rates

Obtaining curves of velocity over time in aortic outflow by Doppler echocardiography from the suprasternal notch during exercise permits a calculation of both stroke volume and systolic ejection time. Calculation of the volume ejected over time provides an indirect measure of ventricular pressure and serves as an index of changes in myocardial force and velocity (4). The left ventricular ejection time normally shortens from approximately 0.25 s to 0.18 s in the course of a progressive test.

Calculated values of systolic ejection rate need to be adjusted for left ventricular size when making

Table 10.1 Average Values for Oxygen Pulse at Maximal Cycle Exercise Related to Body Surface Area in 70 Girls and 81 Boys Aged 7 to 13

Body surface area (m^2)	Oxygen pulse (ml/beat)
0.90	5.4
1.00	6.2
1.10	6.9
1.20	7.9
1.30	8.8
1.40	9.8
1.50	11.0
1.60	11.9
1.70	12.7

Data from Washington et al. 1988.

group or inter-individual comparisons or when making measurements serially in the same patient over time. It is most convenient to assess the normalcy of responses in ejection rate by simply calculating a maximum-to-resting ratio. Among healthy youths this value has been reported to be 1.5 to 2.0 (29).

DeSouza et al. compared systolic ejection rates during maximal semisupine exercise testing of 27 patients (aged 14.4 ± 3.2 yr) treated with high-dose anthracyclines (>260 mg/m^2) for childhood cancer to normal control subjects (8). Among the patients, ejection rate (adjusted for body surface area) rose from 131 ml/(m$^{-2} \cdot$ s^{-1}) at rest to 194 ml/(m$^{-2} \cdot$ s^{-1}) at peak exercise (+48%), while in the healthy children values increased from 165 ml/(m$^{-2} \cdot$ s^{-1}) to 272 ml/(m$^{-2} \cdot$ s^{-1}) (+65%).

Tissue Doppler Imaging

Estimation of longitudinal myocardial velocity during ventricular contraction can be achieved by tissue Doppler imaging (TDI-S), which differentiates high-amplitude, low-frequency ultrasound signals of myocardial velocities from those of blood flow. In an apical four-chamber view, TDI-S waves are recorded at the lateral or septal aspect of the mitral valve annulus, indicating the integral of systolic velocity from base to apex. TDI-S can be obtained either as peak values by pulse wave acquisition or as mean values by color flow mapping.

TDI-S as a measure of contractile function correlates closely with ventricular ejection fraction (26) but can be affected by transducer beam angu-lation, translational motion of the heart, and tethering effects (where motion of one segment may influence that of adjacent myocardium). Studies in healthy subjects have demonstrated the feasibility of obtaining TDI-S at maximal exercise intensities and have shown acceptable levels of test-retest reliability, with coefficients of variation of 5.1% (3) to 7.4% (30). Investigations in healthy youths indicate that lateral TDI-S typically increases by a factor of 2.2 to 2.6 during a maximal exercise test (29).

Diastolic Function

Echocardiographic techniques have also provided information about diastolic functional responses to exercise. Not surprisingly, these lusitropic characteristics parallel those of systolic adaptations to an acute exercise challenge. Echocardiographic assessment during an exercise test may then provide useful information about impairment of diastolic function associated with cardiac abnormalities. Diastolic function measures include longitudinal myocardial relaxation during ventricular diastole and mitral inflow velocity.

Tissue Doppler Imaging

In a manner similar to assessing myocardial longitudinal velocity during systole, tissue Doppler imaging provides the rate of relaxation of longitudinal myocardial fibers during ventricular diastole. This velocity, termed TDI-E', normally increases approximately twofold from rest to maximal exercise. It reflects not only the relaxation properties of the myocardium but also the effect of suction created by ventricular systolic contraction.

Transmitral Pulse Wave Doppler

The Doppler E wave marking the velocity of transmitral valve ventricular filling in early diastole reflects the pressure gradient between the "upstream" factors (left atrial volume, distensibility) and those "downstream" (left ventricular relaxation properties, ventricular suction). During a progressive exercise test, as measured by pulse wave Doppler at the tips of the mitral valve leaflets in an apical four-chamber view, the peak E velocity in a healthy subject approximately doubles, which translates into a 4.5-fold increase in left atrial-ventricular pressure gradient. This rise in transmitral gradient permits a stable volume of left ventricular filling (equal to the stroke volume) and constant left ventricular preload as the diastolic filling period falls. The late diastolic A wave reflecting atrial contraction has not been considered in Doppler exercise studies because E and A waves fuse at low exercise intensities.

That this increase in gradient during exercise in a healthy subject is a manifestation of enhanced ventricular diastolic function (i.e., a decline in downstream pressure) rather than increased left atrial pressure is indicated by the stability or slight decline in the ratio E/E' observed as the exercise test progresses. Although controversial, E/E' is generally considered a marker of left ventricular filling pressure. In patients with limitations of diastolic function, it can be expected that responses of both mitral E and TDI-E' velocities with exercise will be dampened, while E/E' will rise.

Clinical Experience

Doppler echocardiographic techniques in clinical populations have been described in both adult and pediatric patients. The feasibility of this approach has generally been documented during supine or semisupine cycle exercise at low-to-moderate workloads. While clinical information has been derived from such studies, the utility of Doppler measures of responses to exercise in predicting disease outcomes, particularly compared to echocardiographic findings at rest, remains to be clarified.

Adult Studies

Moon et al. demonstrated that responses of TDI-E' and TDI-S to low-grade exercise were approximately double in hypertrophic cardiomyopathy patients with low levels of fibrosis compared to those with greater evidence of myocardial scarring (detected by late gadolinium enhancement) (18). In other studies of patients with hypertrophic cardiomyopathy, change in TDI-E' was found to be depressed during exercise compared to controls and also to correlate with peak $\dot{V}O_2$ (11, 32). In older patients with heart failure, E/E' at peak exercise has been reported to predict all-cause mortality (31), and TDI-S can predict $\dot{V}O_{2max}$ (17). Kang et al. demonstrated that changes in TDI-S velocities with exercise correlated with the degree of mitral valve regurgitation in patients with nonischemic cardiomyopathies (14).

Pediatric Studies

Poerner et al. reported that patients after atrial switch (Senning or Mustard procedure) for transposition of the great arteries had lower TDI-E' values at rest and during exercise than healthy subjects (24). The study by Harada et al. indicated that following surgery for tetralogy of Fallot, patients had depressed TDI-S and TDI-E' responses of the right ventricle when exercising to a heart rate of 135 bpm (12). Chen et al. found lower TDI-S and TDI-E' values at rest in young patients following arterial switch for transposition of the great arteries, but the magnitude of response to exercise was similar (5).

Stress Echocardiography

Stress echocardiography generally means cardiac ultrasound studies performed immediately after an exercise challenge. Because this technique is used after exercise, the cardiac ultrasound measurements are not affected by artifact created by body motion. The exercise may be performed on a treadmill or cycle ergometer. On the other hand, since hemodynamic variables change rapidly after cessation of maximal exercise, recovery echocardiogram studies are hampered by the lack of a convenient means of standardizing measurements in inter-individual or group comparisons or in a single patient on repeated studies. In reported usage of this technique in the literature, recordings are typically made within 30 to 60 s postexercise with a heart rate >80% of that at peak exercise (23). A number of different exercise protocols have been used, including findings after exercise to a particular percentage of $\dot{V}O_{2max}$, symptom-limited exercise, and exercise at fixed submaximal workloads (25).

Technologies on the Horizon

Newer technologies have recently been developed that detect changes in myocardial function at rest, and these have been applied as well during exercise. While such approaches are conceptually sound, certain methodologic challenges (motion artifact, tachycardia, frame rates) must be overcome before they can become feasible in the clinical exercise testing laboratory setting.

Ventricular strain and strain rate. These ultrasound techniques provide information about the rate of myocardial displacement. They are independent of insonation angle and tethering effect, and they can provide insights into myocardial function in longitudinal and radial dimensions.

Ventricular twist. During ventricular systole, the contraction of spiral fibers arranged helically in the ventricular wall causes a wringing-out action to occur clockwise at the cardiac base (viewed from below). Measurement of the extent and rate of this twist (and untwist in diastole) by echocardiographic methodologies may provide perhaps the most realistic assessment of myocardial function. Normal values have been described at rest and at low levels of exercise in healthy populations.

Three-dimensional echocardiography. Three-dimensional ultrasound images of the heart offer the potential for a more accurate measurement of chamber volumes, shape, and contractility. Acquisition of three-dimensional images is hampered during exercise by tachycardia and poor signal quality.

Magnetic resonance imaging. Techniques are being developed to use magnetic resonance imaging to assess ventricular volumes during exercise. These techniques avoid the influence of respiratory variation and increases in heart rate.

Stress echocardiography was initially developed to assess abnormalities in ventricular segmental wall motion as a means of identifying and quantifying coronary artery disease in adults. Many clinicians have favored using a pharmacologic challenge to stimulate myocardial perfusion demands (such as the dobutamine stress test) instead of exercise, which may be poorly tolerated by some patients.

Both of these techniques have been used in pediatric patients to address clinical questions about the adequacy of myocardial perfusion. Stress echocardiography appears to be particularly useful in the identification of patients with significant coronary artery involvement following Kawasaki disease. Studies such as that by Hijazi et al. have indicated that stress echocardiography can indicate significant coronary artery involvement in patients with Kawasaki disease by identifying segmental wall function abnormalities in the face of negative ECG findings on standard exercise stress tests (13). Similar issues regarding ischemia-related myocardial dysfunction can be addressed by stress echocardiography in young patients with congenital coronary artery anomalies and post-transplant graft atherosclerosis (23).

The use of postexercise echocardiography has been extended to assess ventricular function in patients with myocardial and valvular disease in both adult and pediatric populations (25) and by the use of myocardial perfusion imaging techniques to replace echocardiographic assessment (15). Among patients with congenital heart disease, most of these reports indicate deviations from normal in myocardial function. However, as noted by Robbers-Visser et al., "outcomes of these studies cannot be easily compared [because] there is a wide variety in the age and diagnosis of the patients, the type of measurements with echocardiography, and the type of stressor. In general, the clinical relevance of these abnormal findings with stress-testing is unknown" (25, p. 555).

Conclusion

The contractile and relaxation properties of the myocardium serve as key predictors of morbidity and mortality from heart disease at all ages. It is intuitively attractive to think that challenging myocardial functional capacity with exercise should give us insights into myocardial dysfunction that is not evident in the resting state. To this end, many diagnostic methods to assess inotropic and lusitropic ventricular function have been applied during exercise. However, no optimal means has yet been identified to accurately, noninvasively, and safely assess myocardial function during exercise with clear predictive value for children who have heart disease. However, early experience with new ultrasound techniques has provided evidence that clinically valuable markers of myocardial function with exercise may be developed.

Pulmonary Function

Patricia A. Nixon, PhD

The pulmonary system plays an integral role at rest and during exercise for transporting gases to and from the alveoli and for the exchange of gases between the alveoli and pulmonary capillaries, ultimately preserving blood acid-base balance. In healthy children, the pulmonary system is generally not considered a limiting factor for exercise capacity; however, in children with cardiopulmonary or other disorders, the pulmonary system may not function sufficiently to meet the demands of exercise, thus contributing to impaired exercise tolerance and limited exercise capacity.

There are a number of objectives for assessing pulmonary function as part of the exercise evaluation of the child, including those with and without disease. Measurement of expired gas volumes and concentrations is essential for determining aerobic fitness as reflected in peak oxygen uptake ($\dot{V}O_{2peak}$). Exercise stress testing may provoke or accentuate cardiopulmonary abnormalities that may not be evident at rest and may help to explain exercise intolerance or limitations to exercise. Repeated testing is valuable for assessing changes in these responses that may occur with disease progression or intervention (e.g., pharmacologic, surgical, exercise). Exercise testing can also be used to diagnose and evaluate exercise-induced problems such as bronchoconstriction, arterial oxyhemoglobin desaturation, and chest pain of noncardiac origin. Test results will also provide the basis for prescribing exercise that is safe for the child with the goal of improving exercise tolerance and aerobic fitness, as well as boosting the confidence of the child and concerned parent.

Protocols

The objective of the exercise test will determine the most appropriate testing mode as well as the protocol for obtaining the desired information,

as shown in table 11.1. The most commonly used protocols involve progressive increases in work with the end points of either volitional exhaustion, the appearance of signs or symptoms, or a predetermined submaximal end point (e.g., 85% of age-predicted peak heart rate). The increase in work rate can be incremental, with the rate increasing at the beginning of each stage (e.g., each minute if 1 min stages), or ramp-like with the work rate increasing linearly. It is generally recommended that the work increase at a rate that makes the test last about 8 to 12 min (14). This time frame is sufficient for examining responses to exercise but short enough to avoid undue fatigue, or perhaps boredom, in the child.

Choice of the appropriate testing protocol and interpretation of the test results require some understanding of the physiology of the pulmonary system and the measurement of pulmonary function at rest and during exercise. The pulmonary system provides

1. the mechanical pump and conduits for gas flow to and from the alveoli, and
2. the respiratory membrane, including the alveolar and capillary walls, across which diffusion of oxygen and carbon dioxide occurs.

Both work together to help maintain acid-base balance of the blood.

Pulmonary Function at Rest and During Exercise

The mechanical aspect of ventilation ($\dot{V}_E$—expressed as volume expired per minute) is determined by the tidal volume (V_T the volume of air expired each breath), the frequency of inspiration/expiration per minute (f_b), and the patency of airways and their resistance to airflow. Lung volumes

Table 11.1   Testing Mode, Protocol, and Measurements According to Objectives of Pulmonary Exercise Testing

Objective	Mode	Protocol	Measurements and observations
Aerobic fitness	Treadmill or cycle ergometer	Progressive incremental or ramp to exhaustion	$\dot{V}O_{2peak}$; VAT; OUES
Exercise tolerance	Treadmill; cycle ergometer; walking	Progressive incremental or ramp; 6-min walk test	$\dot{V}O_{2peak}$; $\dot{V}_E/\dot{V}CO_2$; $P_{ET}CO_2$, S_pO_2 Exercise tidal F-V loops with resting maximal F-V loop Symptoms (e.g., dyspnea, fatigue, chest pain) and signs (e.g., oxyhemoglobin desaturation, cyanosis, coughing)
Exercise-induced bronchoconstriction	Treadmill (preferred) or cycle ergometer	6–8 min at intensity = 80%–90% of peak heart rate or 40%–60% of MVV with spirometry pre- and 1, 3, 5, 10, 15, 20 min postexercise	Heart rate; FVC; PEF; FEV_1; FEF_{25-75}; S_pO_2; symptoms and signs
Exercise-induced hypoxemia	Treadmill; cycle ergometer; walking	Progressive incremental or ramp; submaximal steady state; 6-min walk test	S_pO_2 in lieu of arterial measurements of S_aO_2 and P_aO_2

FEV_1 = forced expiratory volume in 1 s. FEF_{25-75} = forced expiratory flow between 25% and 75% of FVC. FVC = forced vital capacity. MVV = maximal voluntary ventilation. OUES = oxygen uptake efficiency slope. P_aO_2 = arterial oxygen pressure. PEF = peak expiratory flow rate. $P_{ET}CO_2$ = end-tidal carbon dioxide pressure. S_aO_2 = % oxyhemoglobin saturation in arterial blood. S_pO_2 = % oxyhemoglobin saturation from pulse oximetry. VAT = ventilatory anaerobic threshold. $\dot{V}_E/\dot{V}CO_2$ = ratio of minute ventilation to volume of carbon dioxide produced. $\dot{V}O_{2peak}$ = peak oxygen uptake.

and capacities associated with breathing are depicted in figure 11.1a. The tracing produced by spirometry can only assess dynamic lung volumes, that is, those directly manipulated by inspiration and expiration. The determination of functional residual capacity (FRC) and the calculations of residual volume (RV) and total lung capacity (TLC) require additional measurement via plethysmography or nitrogen washout or helium-dilution methods (108).

Parts a and b of figure 11.1 differ by the variables depicted on the x- and y-axes. In figure 11.1a, the tracing shows lung volumes and capacities (y-axis) associated with breathing with time on the x-axis. Figure 11.1b depicts the more commonly measured maximal flow-volume (F-V) loop of a healthy child obtained via standard spirometric techniques (66, 82). The F-V loop provides information on dynamic lung volumes (x-axis) and expiratory flow rates (y-axis), the latter reflecting airway patency, or conversely, airway obstruction. Both tracings are obtained by having the child breathe normally to provide resting tidal volumes followed by a forced maximal inspiration, and then a forced maximal expiration. These are the measures commonly examined from the F-V loop.

1. The forced vital capacity (FVC), the maximal volume expired forcefully after a maximal inspiration.

2. The peak expiratory flow rate (PEF), which occurs early in the forced expiration and is somewhat effort dependent.

3. The forced expiratory flow rates at 25%, 50%, and 75% of the FVC with the forced expiratory flow rate between 25% and 75% of FVC (FEF_{25-75}) considered to be a measure of smaller airway function.

4. The forced expiratory volume in the first second (FEV_1), which provides a measure of larger airway patency.

5. The ratio of FEV_1 to FVC (FEV_1/FVC), which is examined to determine if reduced flow reflects actual obstruction or is merely a consequence of overall reduced lung volumes.

As previously mentioned, several measures noted in the figures cannot be determined from spirometry and require additional testing. These include RV, which is the residual volume that remains in the lungs after a maximal expiration,

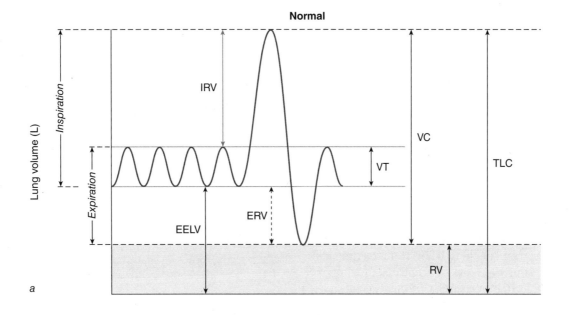

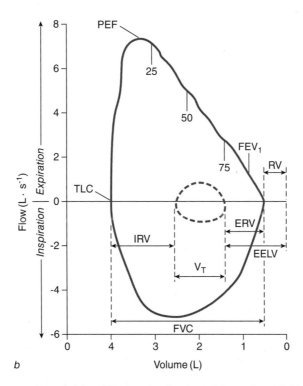

Figure 11.1 *(a)* Lung volumes and capacities. The tracing begins with resting tidal breathing, followed by a maximal inspiration and then maximal expiration, and returning to resting breathing. RV, TLC, and EELV are depicted but require other methods to measure. V_T = tidal volume. IRV = inspiratory reserve volume. ERV = expiratory reserve volume. EELV = end-expiratory lung volume. VC = vital capacity. RV = residual volume. TLC = total lung capacity. *(b)* Resting maximal flow-volume (F-V) loop obtained via standard spirometry using the same breathing maneuver. In addition to the volumes and capacities noted in *a*, flow rates are noted. FVC = forced vital capacity. PEF = peak expiratory flow. FEV_1 = forced expiratory volume in 1 s. The numbers 25, 50, and 75 correspond to forced expiratory flow rates at 25%, 50%, and 75% of FVC.

and the end-expiratory lung volume (EELV), which is the volume that remains in the lungs at the end of expiration and would equal RV if the expiration were maximal. The RV along with the FVC comprise the TLC.

The changes that occur in response to exercise are presented in figure 11.2. To meet the increasing ventilatory demands, the V_T increases by tapping into the inspiratory and expiratory reserves, reaching approximately 45% to 60% of VC (5, 36) but staying well within the F-V loop obtained from the maximal inspiratory/expiratory maneuver at rest as noted in figure 11.2b (53). In figure 11.2b, it can also be seen that the EELV is lower with exercise than it is at rest. This reduction optimizes the inspiratory muscle length and reduces the elastic load and thus the work of breathing, during exercise (53). At higher exercise intensities, increases in $\dot{V}_E$ are met primarily by increases in breathing frequency (f_b) (not depicted in the figures).

The $\dot{V}_E$ at peak exercise increases with age and is generally higher in boys than in girls of the same age and in aerobically trained compared to untrained children (5, 7, 75). The higher $\dot{V}_E$ in older children can be partially attributed to the higher V_T associated with a larger body size (78, 97). Conversely, the f_b declines with increasing age. In response to progressive exercise, adults meet the increasing ventilatory demands at lower intensities with greater reliance on increasing V_T up to about 60% of VC and plateauing around 60% of $\dot{V}O_{2max}$. At higher exercise intensities, higher $\dot{V}_E$ requirements are met by increasing f_b. A study of younger prepubescent children reports a similar pattern of response to progressive exercise, although the V_T at peak exercise only reached about 40% of VC (5). The shallower, faster breathing of younger children may reflect compensation for their smaller, less compliant lungs. In contrast, some evidence suggests that younger children exhibit an alternate pattern by increasing V_T over f_b to meet higher ventilatory demands (13).

One additional variable that is commonly measured with resting pulmonary function testing to compare with minute ventilation during exercise is maximal voluntary ventilation (MVV). For this maneuver (performed at rest), the child is asked to breathe deeply but as quickly as possible to try to ventilate as much air as possible during a 12 to 15 s period. The results are then extrapolated to 1 min and considered a measure of mechanical ventila-

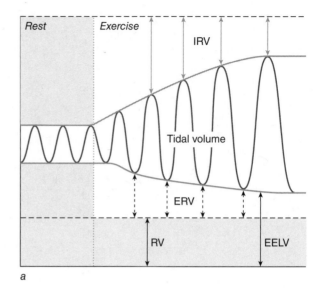

a

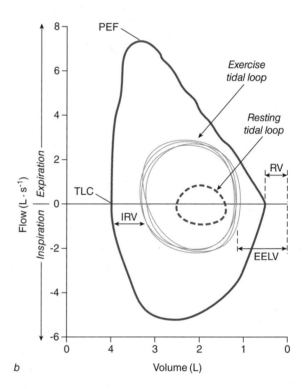

b

Figure 11.2 Changes in lung volumes during exercise *(a)* in an exercise tidal F-V loop of a healthy 14-year-old female. Tidal loops obtained near maximal exercise are superimposed on her resting maximal F-V loop. In both *a* and *b*, the V_T increases by encroaching on the IRV and ERV. *(b)* Note that the exercise tidal loop does not touch on the expiratory portion of the resting F-V loop. Also note the decrease in EELV toward residual volume that reduces the elastic load and consequently, the work of breathing.

tory capacity (L/min). The measure is compared with the maximal $\dot{V}_E$ achieved during exercise expressed as a ratio ($\dot{V}_E$/MVV). In healthy children and adults, the ratio generally lies between 60% and 70%, suggesting that 30% to 40% of breathing reserve remains, and ventilatory capacity is not a limiting factor to exercise (36, 111).

A maximal $\dot{V}_E$ that approaches MVV is thought to suggest possible ventilatory limitation to exercise. A $\dot{V}_E$/MVV less than 60% may suggest a submaximal effort, but it can also reflect other sources of exercise limitation. In general, the ratio has some limitations and should be interpreted with caution. First, the $\dot{V}_E$ during exercise is an involuntary response, whereas the MVV is a voluntary maneuver, and the difference is evident when $\dot{V}_E$ at peak exercise exceeds the MVV obtained at rest (i.e., $\dot{V}_E$/MVV > 100%). When the MVV maneuver is examined in the context of the F-V loop, tidal breathing occurs at very high lung volumes (high EELV) approaching TLC, greatly increasing the elastic load and work of breathing (53). In contrast, during maximal exercise, tidal breathing occurs on a more optimal portion of the F-V loop with reduced EELV, elastic load, and work of breathing. Exercise also induces bronchodilation in healthy individuals, which would likely increase the MVV. Consequently, an MVV obtained postexercise might provide a better comparison with $\dot{V}_E$ obtained at peak exercise. Furthermore, reliable efforts to achieve MVV may be difficult to obtain in children, and for this reason some investigators suggest that MVV can be estimated by multiplying the FEV_1 by 35 (32, 35). However, a recent study of children with cystic fibrosis (CF) found that MVV was more accurately predicted by the formula MVV = 27.7 FEV_1 + 8.8 (PredFEV_1), and the researchers concluded that actual measurement of MVV was preferable (102).

Possible mechanical ventilatory limitations during exercise may be better reflected in the examination of the exercise tidal flow-volume loop plotted within the resting F-V loop obtained with a maximal inspiration/expiration maneuver (51). As noted in figure 11.2*b*, during maximal exercise, the tidal volume increases but does not encroach on the resting maximal F-V loop, and the end-expiratory lung volume (EELV) is lower, reducing with elastic load and work of breathing.

Examination of the exercise tidal F-V loops can be useful for identifying ventilatory limitations to

exercise in various pediatric diseases. As shown in figure 11.3, in the child with an expiratory flow limitation (as indicated by the reduced expiratory flow and concave appearance of the expiratory portion of the resting F-V loop), the exercise tidal loop encroaches on the resting maximal F-V loop. In some cases, the exercise tidal loop exceeds the resting F-V loop, most likely reflecting bronchodilation with exercise that is not present on the pre-exercise voluntary resting F-V loop (indicating that comparison with a postexercise F-V loop should be examined as well). It can also be noted in the figure that the tidal F-V loop is shifted to the left, resulting in a higher EELV. This shift will improve flow at higher lung volumes, but it will

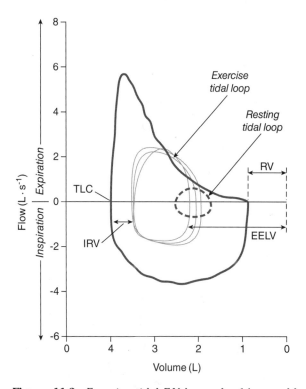

Figure 11.3 Exercise tidal F-V loop of a 14-year-old male obtained near maximal exercise are superimposed on his resting maximal F-V loop. Note the concavity of the expiratory portion on the resting F-V loop, which is consistent with airway obstruction. During exercise, the tidal loop encroaches on the expiratory portion of the resting F-V loop, suggesting expiratory flow limitation. Also note the shift in tidal breathing toward TLC and the increase in EELV consistent with dynamic hyperinflation, which improves airflow at higher lung volumes but increases the elastic load and thus the work of breathing.

also increase the elastic load and work of breathing and reduce the inspiratory flow reserve (53). In contrast, patients with restrictive changes in pulmonary function (including those with congestive heart failure or congenital heart disease) have been shown to breathe at extremely low lung volumes during exercise (despite normal $\dot{V}_E$/MVV values), which may contribute to dyspnea and exercise intolerance (53, 84). Because most commercial exercise testing carts include the measurement of resting and exercise tidal F-V loops, these measures should provide better assessment of possible mechanical ventilatory limitations to exercise than the traditionally measured $\dot{V}_E$/MVV.

Coupling of $\dot{V}_E$ and $\dot{V}CO_2$

The increase in ventilation during exercise occurs mainly in response to central and peripheral chemoreceptor stimulation by increased levels of CO_2 in the blood. This is produced metabolically by exercising muscle cells as well as by the buffering of lactic acid by bicarbonate and the eventual fall in blood pH at higher exercise intensities. Ventilation ($\dot{V}_E$) during exercise is influenced by the volume of CO_2 produced ($\dot{V}CO_2$), the partial pressure of CO_2 in the arterial blood (P_aCO_2), and the ratio of dead space volume to tidal volume (V_D/V_T), and it can be determined by the following equation (94):

$$\dot{V}_E(BTPS)=\frac{0.863 \times \dot{V}CO_2(STPD)}{PaCO_2 \times (1-\frac{V_D}{V_T})}$$

As shown in figure 11.4, there is a positive linear slope between $\dot{V}_E$ and the volume of CO_2 expired ($\dot{V}CO_2$) throughout exercise, except near maximal exercise where the slope steepens slightly, reflecting the respiratory compensation (RC) point at which $\dot{V}_E$ was further stimulated by the fall in pH. The increase in ventilation helps to regulate the P_aCO_2 in the blood and maintain acid-base homeostasis by expelling CO_2. The $\dot{V}_E$ will also be affected by the portion of each breath that does not participate in gas exchange, that is, the dead space volume (V_D). The V_D consists of the inhaled air that doesn't reach the alveoli, as well as the air that reaches the alveoli but does not participate in gas exchange with the pulmonary capillaries. In healthy children and adults, about 30% of inspired V_T is comprised of dead space (V_D) (69). During exercise, the V_D/V_T (measured from both finger capillary blood and end-tidal gas composition) has been shown to decrease to approximately 0.20 to

0.25 in both children and adults, promoting better alveolar ventilation ($\dot{V}_A$) and gas exchange with perfused capillaries ($\dot{Q}$) (99). A more recent study reported estimates of V_D/V_T (based on expired gases) as low as 0.11 and 0.13 at peak exercise in 6- to 17-year-old girls and boys, respectively, and no correlation with age (69). Children with airway obstruction (e.g., cystic fibrosis) may exhibit greater dead space breathing during exercise due to mismatching between ventilated alveoli and perfused capillaries ($\dot{V}_A/\dot{Q}$). Consequently, $\dot{V}_E$ will have to increase to dispose of CO_2 to maintain acid-base balance and to provide adequate O_2 to the working muscles.

Children tend to have higher $\dot{V}_E$ for their body size and $\dot{V}CO_2$ produced (and thus $\dot{V}O_2$ consumed) for a given level of work compared to adults (20, 88, 97). Because V_D/V_T does not appear to be higher in children than in adults during exercise (69), the higher $\dot{V}_E$ likely reflects a lower P_aCO_2 set point (77) for stimulating ventilation. Measurement of P_aCO_2 from arterial blood gas can be painful and is generally not warranted in children. Examination of the measurement of CO_2 tension in expired air at the end of expiration ($P_{ET}CO_2$) may provide some information about P_aCO_2, bearing in mind the physiological assumptions (18) and understanding that the difference between P_aCO_2 and $P_{ET}CO_2$ will vary with V_T, f_b, and CO_2 output (54). With progressive exercise, the $P_{ET}CO_2$ typically increases above resting levels, corresponding to

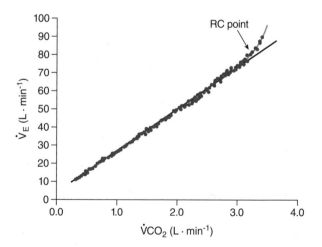

Figure 11.4 Plot of $\dot{V}_E$ versus $\dot{V}CO_2$ obtained during progressive exercise test to exhaustion in a 14-year-old male. Note the linear increase in $\dot{V}_E$ relative to increasing $\dot{V}CO_2$ until the respiratory compensation (RC) point, where $\dot{V}_E$ increases further in response to decreasing pH.

increased $\dot{V}CO_2$ production and its delivery to the lungs. With more intense levels of exercise, the buffering of lactic acid increases the f_b and alveolar ventilation to maintain acid-base balance, and the $P_{ET}CO_2$ subsequently falls. Failure to decrease or even continued increase in $P_{ET}CO_2$ suggests hypoventilation (increased P_aCO_2), which may be seen in children with cystic fibrosis and has been associated with worse prognosis (49, 74). It should be noted that children may exhibit a low $P_{ET}CO_2$ prior to or at the start of testing, reflecting anxiety-induced hyperventilation. As the test proceeds, metabolic factors will generally override the psychogenic influences and dictate the ventilatory response.

Ventilatory Efficiency Slope

The higher $\dot{V}_E$ relative to $\dot{V}CO_2$ produced results in a steeper slope for the change in $\dot{V}_E$ versus the change in $\dot{V}CO_2$. This is called the ventilator efficiency slope, or $\dot{V}_E/\dot{V}CO_2$ slope. The slope has been shown to decline with age (figure 11.5) from values of 33.0 ± 4.5 and 33.3 ± 5.2 for 8- to 10-year-old girls and boys, respectively, to 27.8 ± 3.6 and 26.8 ± 3.9 in 16- to 18-year-old girls and boys, respectively (21), with values stabilizing in adult women (27.8 ±

2.6) but declining further in adult men (24.3 ± 2.2) (70). The inverse association with age is further supported by a study of 175 healthy Dutch children aged 8 to 18 by Ten Harkel et al. (104), who determined the regression equation for predicting $\dot{V}_E/\dot{V}CO_2$ slope of (0.64 × age) + 38 for both girls and boys. For these studies the slope was determined using data from the entire test (including the nonlinear increase between the RC point and peak exercise), making the value somewhat effort dependent. Restricting calculation of the $\dot{V}_E/\dot{V}CO_2$ slope to data from the linear portion up to the RC point does not require, nor is it dependent on, a maximal effort, but it yields lower values. In a study of 243 10- to 17-year-olds (128 males) comparing the two slopes, Giardini et al. (33) found a substantially lower mean $\dot{V}_E/\dot{V}CO_2$ slope of 24.5 ± 3.0 based on data up to the RC point versus 28.3 ± 4.0 when data above the RC point were included. When using the submaximal data, a slope < 28 has been reported to be normal for children based on the linear portion up to the RC point (93).

The higher slope in children compared to adults is consistent with a lower set point for regulating P_aCO_2, as well as lower CO_2 stores or reduced lactic acid production associated with the less developed anaerobic capacity in younger children (20). The $\dot{V}_E/\dot{V}CO_2$ slope has been shown to be steeper in children with expiratory flow limitations—in compensation for elevated dead space—and in those with pulmonary blood flow maldistribution (e.g., congenital heart defects and pulmonary hypertension) (12, 83). The slope has also been shown to correlate with markers of disease severity (e.g., pulmonary artery pressure and pulmonary vascular resistance) as well as with worse prognoses in children with pulmonary hypertension (89) and adults with congestive heart failure (34, 59).

Pulmonary Gas Exchange

In order for gas exchange to take place, there must be adequate alveolar ventilation to perfusion ($\dot{V}_A/\dot{Q}$) matching; that is, ventilated alveoli must be matched with pulmonary capillaries perfused with blood. Hypoventilation results when perfused capillaries are aligned with underventilated alveoli (resulting in an increase in P_aCO_2) as may occur with airway obstruction. Conversely, mismatching of ventilated alveoli with underperfused capillaries will result in hypoxemia (a drop in P_aO_2), which may occur with reduced cardiac output and pulmonary arterial pressure or elevated pulmonary vascular resistance.

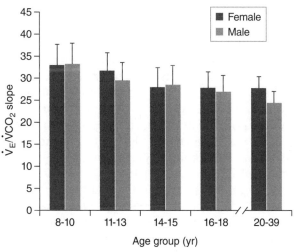

Figure 11.5 Bar graph depicting the decline in the slope of $\dot{V}_E$ versus $\dot{V}CO_2$ by age in both females and males. Data for children and adolescent age groups are taken from those reported by Cooper et al. (21) for 169 8- to 18-year-olds (60.4% Caucasian, 17.8% Asian, 19.5% Hispanic, 1.8% African American). The data for 20- to 39-year-olds was taken from results reported by Neder et al. (70) on 20 females and 20 males (race or ethnicity not reported). Testing mode was cycle ergometry for both studies.

The efficiency of pulmonary gas exchange can be estimated from the difference between the alveolar and arterial PO_2 ($\Delta P_{A\text{-}a}O_2$). In healthy people, improvements in $\dot{V}_A/\dot{Q}$ matching with exercise will increase the $\Delta P_{A\text{-}a}O_2$ to values as high as 20 to 30 mmHg. As the P_aO_2 should not change with exercise, a $\Delta P_{A\text{-}a}O_2 > 30$ or 35 mmHg is considered potentially abnormal and values > 50 or 55 mmHg indicate impaired gas exchange (3, 26). Substantial decreases in P_aO_2 may occur in persons with pulmonary vascular disease and in those with more severe obstructive lung disease. Although studies reporting $\Delta P_{A\text{-}a}O_2$ with exercise in children are less common (perhaps due to the need to obtain arterial blood gas measurements, which can be painful), one study (46) reported greater increases in $P_{A\text{-}a}O_2$ in a group of children with more severe (22.5 mmHg) compared to less severe (15 mmHg) exercise-induced asthma during submaximal exercise, although the differences were not significant. In contrast, the P_aO_2 values during exercise were significantly lower in the children with more severe versus less severe exercise-induced asthma (84 vs. 92 mmHg, respectively). Because P_aO_2 provides the driving pressure for saturation of hemoglobin with oxygen (S_aO_2) at the lung, some information about S_aO_2 and potential hypoxemia can be obtained noninvasively by pulse oximetry with the measurement commonly referred to as S_pO_2. The S_pO_2 is measured by a sensor attached to the finger, ear, or forehead. The sensor transmits and receives light wavelengths that are absorbed differently by oxygenated and deoxygenated hemoglobin. Depending on the measurement site, accuracy is affected by poor circulation, increased levels of carboxy- and met-hemoglobin, dark skin pigment, ambient light, movement, fingernail polish, and pierced ear holes (64, 73). Studies report both over- and underestimation of true arterial S_aO_2 values during exercise, with greater discrepancies when actual S_aO_2 values are less than 90% (64, 79).

In healthy people, the S_pO_2 should be near 100% at rest, and it should remain at this level during exercise. A fall in $S_pO_2 \geq 4\%$ or values $\leq 88\%$ during exercise suggest hypoxemia and are deemed clinically significant (3). Although the cutoff of $S_pO_2 \leq 88\%$ is sometimes used as a criterion for termination of the exercise test, substantially lower values have been observed with no reports of untoward effects. In general, hypoxemia and increasing metabolic acidosis will limit the child's ability to continue exercise. In children with congenital heart disease, reduced S_pO_2 levels are not uncommon

at rest and decrease significantly during exercise (38). Children with more severe airway obstruction (e.g., CF, asthma) may also exhibit decreases in S_pO_2 with exercise but to a lesser degree (47, 74). In both children and adults, evidence suggests that greater levels of oxyhemoglobin desaturation may be observed with submaximal exercise (such as a six-minute walk test) than with maximal exercise, perhaps due to greater reliance on aerobic metabolism submaximally than on greater anaerobiosis and higher minute ventilation with maximal exercise (15, 71).

Assuming adequate $\dot{V}_A/\dot{Q}$ matching, the diffusion of gases between alveoli and capillaries is dependent on the gas partial pressure gradient between the two as well as the red blood cell transit time, the solubility of the gas in the liquid and tissues, the amount and quality of hemoglobin, and the permeability and thickness of the alveolar-capillary membrane (50). The diffusion capacity of the lungs is commonly assessed at rest by measuring the diffusion of carbon monoxide (DL_{CO}) using the single-breath technique (60). In children, the resting DL_{CO} is generally lower in females and is directly associated with age, height, and total lung capacity (55). It has been shown to be predictive of oxyhemoglobin desaturation during exercise in persons with obstructive lung disease (58). Because the measurement of DL_{CO} generally requires controlled breathing maneuvers and breath holding, its measurement during exercise is limited particularly in children, with some studies opting to compare measurements obtained pre- and postexercise. Anderson and Godfrey found DL_{CO} to be linearly related to $\dot{V}O_2$ obtained with different exercise intensities, with peak values being approximately three times greater than resting values (4), generally attributed to increases in pulmonary capillary perfusion (48). In a study comparing pre- and postexercise measurement in children, postexercise DL_{CO}, although initially higher, was shown to fall below pre-exercise levels, commensurate with decreases in pulmonary blood flow (40). The same study reported greater decreases in children with CF than in healthy controls, despite similar pre-exercise values, supporting the usefulness of measuring it with exercise stress (31).

$\dot{V}_E$, $\dot{V}O_2$, VAT, OUES

Although CO_2 is the primary stimulus for $\dot{V}_E$, examination of the relationship between $\dot{V}_E$ and $\dot{V}O_2$ provides information about the ventilatory anaerobic threshold (VAT) as well as the efficiency of ventilation relative to $\dot{V}O_2$ as reflected in the oxygen uptake efficiency slope (OUES). During progressive exercise, $\dot{V}_E$ increases linearly with increasing $\dot{V}O_2$ up to a point or threshold where a disproportionate increase in $\dot{V}_E$ is observed relative to $\dot{V}O_2$. The VAT is believed to reflect that point where aerobic metabolism requires supplementation with anaerobic metabolism to meet the increasing metabolic demands of exercise (110). The disproportionate increase in $\dot{V}_E$ occurs in response to increased CO_2 being produced metabolically (via glycolysis) as well as from the buffering of lactic acid via bicarbonate. Consequently, it provides an estimate of lactic acid threshold using noninvasive methods, and it reflects the child's ability to meet exercise demands aerobically. Thus it correlates with $\dot{V}O_{2peak}$ (7, 45) and can be used to reflect aerobic fitness when a maximal effort is not obtained or is deemed unsafe. It has also been shown to have good test-retest reliability and is sensitive to changes in fitness associated with aerobic exercise training (45, 61).

In addition to the plot of $\dot{V}_E$ versus $\dot{V}O_2$, the VAT can be determined by plotting other variables, such as $\dot{V}CO_2$ and $\dot{V}O_2$ (V-slope method) (103) or $\dot{V}_E/\dot{V}CO_2$ and $\dot{V}_E/\dot{V}O_2$ versus time (8) as described in chapter 8. The VAT of healthy children ranges between 58% and 83% of $\dot{V}O_{2peak}$ and generally decreases with age, is higher in boys than in girls, and is lower in children with chronic disease (91, 92, 109). It is best determined from a progressive exercise test during which work rate increases in a ramp-like fashion, enabling easier detection of changes in linear associations between the variables plotted in the graph. Computerized metabolic carts will calculate the VAT, but visual verification of the threshold is often warranted. In addition, the VAT may be difficult to detect in children who have very low fitness levels associated with disease and in those who exhibit hypoventilation at higher work rates.

In contrast to VAT, the determination of the OUES is based on the linear relation of $\dot{V}O_2$ versus the log of $\dot{V}_E$ throughout exercise. It does not require a maximal effort and is reported to be a submaximal index of cardiorespiratory functional reserve that reflects both pulmonary dead space and metabolic acidosis (7). It has been shown to correlate significantly with $\dot{V}O_{2max}$ ($r = .94$), $\dot{V}O_{2peak}$ ($r = .88-.92$), and VAT (2, 7, 63). It is generally higher in boys than in girls, increases with age, and is correlated with height and body surface area ($r = .835$ and .861, respectively); and it has been shown to

be significantly lower in children with congenital heart disease (11, 63). Consequently, OUES values are sometimes reported relative to body surface area or lean body mass or weight to reduce variability and enable comparisons of children of different sizes. The methods for determining OUES vary considerably; some studies calculate the slope using all data points up to $\dot{V}O_{2peak}$, whereas others include data up through 50%, 75% or 90% of exercise duration, making it difficult to compare studies and identify what is considered normal.

Assessment of Exercise-Induced Bronchoconstriction

An exercise challenge test is useful for identifying exercise-induced bronchoconstriction (EIB) in the child who reports breathlessness or exhibits a wheeze or cough during or following exercise. The stimulus for EIB is thought to be the high levels of ventilation and subsequent airway cooling and drying that stimulate the release of inflammatory mediators such as histamine and cysteinyl leukotrienes, leading to constriction of the airways (39).

The goal of testing is to increase minute ventilation rapidly, which can be achieved by increasing the work rate to a level of 80% to 90% of maximal predicted heart rate (220 − age) or 40% to 60% of MVV within the first 2 to 3 min of exercise and then sustaining exercise at this high intensity for an additional 4 to 6 min (22). Testing should be conducted on a motor-driven treadmill (preferable) or a cycle ergometer with nose clips in a cool, dry environment (inspired air temperature <25 °C and <50% relative humidity) to promote airway cooling and drying (22). The degree of bronchoconstriction and the sensitivity of the test are greater in much colder air (−20 °C) (16). Ideally, heart rate should be determined by an electrocardiogram, which can also rule out any arrhythmia or ischemic changes. Arterial oxyhemoglobin saturation (S_aO_2) should be monitored noninvasively using a pulse oximeter. Spirometry should be performed pre-exercise and as soon as possible after exercise terminates, preferably 1 but no later than 3 min postexercise, with repeated measurements at 5, 10, 15, and 20 min postexercise. Exercise-induced bronchoconstriction is determined from the percent change in FEV_1 (L) from pre-exercise baseline values. A 10% decrease in FEV_1 is considered to be abnormal, and a 15% decrease is commonly used as a diagnostic criterion for EIB (22). The mathematical calculations for percent change should be supported by observable changes (e.g., concavity) in the expiratory flow-volume loop. The time to reach the lowest FEV_1 value (the *nadir*) postexercise has been shown to correlate with age, with children under age 10 reaching the nadir as quickly as 3 min postexercise (107). Consequently, the early postexercise measurements are important in the pediatric population.

Assessment of Exercise-Induced Arterial Hypoxemia

Exercise-induced arterial hypoxemia (EIAH) has also been reported in healthy prepubescent children as estimated by pulse oximetry. Nourry et al. (76) found that 7 of 24 children exhibited exercise-induced arterial hypoxemia, defined as a 4% or greater decrease in S_pO_2 from rest to maximal exercise. Four children were considered to exhibit mild hypoxemia (S_pO_2 between 93% and 95%) and three exhibited moderate hypoxemia (S_pO_2 of 88%-93%). In contrast, Laursen et al. (57) defined EIAH as a $S_pO2 \leq 91\%$ and reported no EIAH in 19 healthy prepubescent girls. However, several girls exhibited 2% to 4% decreases from resting values. Studies of young adults (highly trained males and females of various fitness levels) provide evidence of EIAH as supported by decreases in P_aO_2 and S_aO_2 determined from arterial blood gas measurements, which may be associated with inadequate alveolar ventilation, increased $\dot{V}_A/\dot{Q}$ inequality, diffusion limitation associated with high pulmonary blood flow and short red blood cell transit times (24, 76), or perhaps compromised morphology of the lung's alveolar capillary interface (42). Although similar explanations for EIAH are plausible in children, verification of its occurrence via direct arterial measurements are needed in view of the potential errors associated with the noninvasive estimates from pulse oximetry.

Asthma

Asthma is the most common chronic disease in children, affecting nearly 7 million (9.3% of) U.S. children (17). It is characterized by chronic inflammation of the airways, which contributes to airway hyperreactivity, and by episodic or in more severe cases, persistent airway obstruction and respiratory symptoms (27). It is believed that the majority of children with asthma exhibit EIB, which may affect exercise tolerance as well as participation in physical activity (112). As reviewed by Welsh et al. (114), studies examining aerobic fitness in children provide mixed results, with some reporting comparable fitness and others reporting reduced fitness in asthmatic children compared to their nonasthmatic peers. The inconsistencies may reflect differences in measurement of fitness (e.g., laboratory vs. field tests), differences in levels of severity of the children's asthma, the relatively small sample, selection bias in choice of nonasthmatic children for comparison, use of bronchodilator therapy prior to testing, and perhaps inappropriate statistical analyses (114). Some studies suggest that fitness is comparable as long as levels of physical activity are similar (30, 98). Pharmacologic advances such as inhaled corticosteroid therapy have reduced the number of hospitalizations and emergency department visits, suggesting that the severity of disease has lessened (1). Consequently, the results of studies conducted prior to these advances may not accurately reflect the current impact of the disease on exercise tolerance.

In the past two decades, surprisingly few studies of children report ventilatory parameters assessed during exercise testing. Santuz et al. (98) reported lower $\dot{V}_E$ at peak exercise in sedentary and regularly active children when compared to healthy controls (matched on age, height, weight, and habitual level of physical activity), but no difference in $\dot{V}_E$ was observed in asthmatic children who participated in organized or competitive sports compared to their peers. During submaximal exercise, the asthmatic and nonasthmatic children had a similar $\dot{V}_E$, but it was achieved by a lower f_b and a higher V_T in the asthmatic children, which would lessen dead space breathing and improve alveolar ventilation. In contrast, Moraes et al. (67) reported no differences among newly diagnosed adolescents with mild intermittent asthma (n = 20), mild persistent asthma (n = 13), and their healthy peers (n = 36) for $\dot{V}O_2$, $\dot{V}_E$, $\dot{V}_E/\dot{V}O_2$, or ventilatory reserve at maximal exercise. Berntsen et al. (9) likewise found no difference in $\dot{V}O_2$ or $\dot{V}_E$ at peak exercise in a group of 86 13-year-old adolescents with asthma compared to 76 same-aged peers without asthma. Resting lung function was slightly lower in the asthmatic children but generally within normal limits. In a multiple regression model that included skinfold thickness, resting pulmonary function, and physical activity, analysis indicated that skinfold thickness and vigorous physical activity participation (in asthmatics only) were the primary independent predictors of $\dot{V}O_{2peak}$ in the study.

Expiratory flow limitations after exercise are well documented in children with asthma, and a recent study reported postexercise inspiratory flow limitation as well (25). However, no study has reported measurement of exercise tidal loops to assess ventilatory mechanics in children with asthma during exercise. Consequently, the results of one study of adults are presented (52). Exercise tolerance and pulmonary mechanics during exercise were compared between eight adults with asthma and six adults without asthma (52). Asthmatic adults had similar larger airway function but reduced smaller airway function at rest, and $\dot{V}O_{2peak}$ was slightly lower (but not significantly) in asthmatic compared to control subjects (104% vs. 130% of predicted). At maximal exercise, the groups attained similar $\dot{V}_E$, but the asthmatic subjects had higher mean values for $\dot{V}_E/\dot{V}CO_2$ and lower $P_{ET}CO_2$ than the control subjects. Superimposing the exercise tidal loop on the resting maximal F-V loops produced similarities to figure 11.3, revealing increases in EELV with the expiratory flow of the tidal loop encroaching on the resting F-V loop during exercise, indicative of expiratory flow limitation. Interestingly, the authors noted variability in EELV and in the degree of exercise flow limitation during interval exercise (intensity switching from 60% to 40% of $\dot{V}O_{2peak}$), and they suggested that the variability reflected changes in bronchomotor tone (i.e., bronchodilation vs. bronchoconstriction) at different exercise intensities.

Recent studies reporting measurements of arterial blood gases are rare in children with asthma. One recent study (62) collected expired gases and earlobe capillary blood samples in eight adolescents with mild intermittent asthma (MIA), eight with mild persistent asthma (MPA), and 12 nonasthmatic control subjects to examine gas exchange during exercise and to estimate P_aO_2 before and after maximal exercise stress. Before exercise, the children with MPA had significantly lower P_aO_2 (75.1 ± 6.6 mmHg) than those with MIA (81.7 ± 6.7 mmHg) and control (83.3 ± 4.9 mmHg) subjects, suggesting hypoxemia. Postexercise P_aO_2 did not differ among groups, but this was attributed to a decrease in the control subjects because the P_aO_2 of the children with MPA did not change significantly. Older studies (28, 37) report gas exchange abnormalities (elevated $P_{A-a}O_2$ difference) at rest that improved during exercise, which the authors attributed to better $\dot{V}_A/\dot{Q}$ matching. A more recent study (43) of habitually active 18- to 45-year-old men and women with mild-to-moderate asthma examined arterial blood gases during exercise. Eight of 21 subjects exhibited oxyhemoglobin desaturation (S_aO_2 ≤94%) during prolonged

exercise and were classified as the Lo-S_aO_2 group, and those whose S_aO_2 remained above 94% were classified as the Hi-S_aO_2 group. Despite similar fitness levels, the P_aCO_2 of the Hi-S_aO_2 group fell from rest to 34.0 ± 2.7 mmHg at exhaustion, whereas that of the Lo-S_aO_2 group fell during the first minute of exercise but then rose progressively, reaching 39.8 ± 4 mmHg at exhaustion, suggesting hypoventilation and impaired gas exchange.

Given the prevalence of asthma in children, as well as reports of reduced fitness, more studies examining ventilatory mechanics and gas exchange parameters during exercise are warranted in children with asthma to determine their potential contributions to reduced fitness and exercise intolerance.

Cystic Fibrosis

Cystic fibrosis is an inherited life-shortening disease associated with defective epithelial ion transport that results in thick mucus that blocks the airways and pancreatic ducts and ultimately leads to impaired pulmonary function and undernutrition (96). Both of these factors contribute to exercise intolerance that occurs with disease progression. The measurement of FEV_1 obtained from spirometry at rest has been shown to correlate with $\dot{V}O_{2peak}$ ($r = .5$) (74), and those with more severe airway obstruction (FEV_1 <50% predicted) have significantly lower $\dot{V}O_2$ peak than those with milder (FEV_1 >65% predicted) or moderate (FEV_1 = 50%-64% predicted) (80). Even those with mild-to-moderate airflow obstruction have been shown to have significantly reduced fitness when compared to healthy children ($\dot{V}O_{2peak}$ of 49.9 ± 7.9 vs. 40.9 ± 7.8 ml · kg^{-1} · min^{-1}, respectively) (11). Despite this lower fitness level, the $\dot{V}_E$ at peak exercise in children with CF is similar to that of healthy persons (11, 80), most likely reflecting the need for children with CF to increase $\dot{V}_E$ to compensate for elevated dead space as indicated by an elevated V_D to V_T ratio (11), and dead space ventilation may be worsened by utilizing a greater f_b (105). Ventilatory inefficiency may also be evident in an elevated $\dot{V}_E/\dot{V}CO_2$ slope (68). Changes in ventilatory mechanics during exercise have also been reported (90, 105). The exercise tidal volume loop imposed on the resting maximal F-V loop is similar to that shown in figure 11.3 for those with mild airway obstruction. However, with increasing disease severity, tidal expiratory flow commonly exceeds the maximal expiratory F-V loop obtained at rest, and the tidal breathing shifts toward total lung

Standardized Guidelines and Normal Reference Values

For the measurement of resting pulmonary function, the American Thoracic Society and the European Respiratory Society have combined forces to recommend standardized testing guidelines and interpretive strategies (10, 65, 66, 82, 95, 108). In previous years, pediatric reference values for pulmonary function were mostly limited to Caucasian populations (86, 100), but in more recent years, data gathering has been more inclusive of minority populations (41, 87, 101). Guidelines for exercise testing have also been published by these societies (3, 6, 22, 26, 29) as well as by the American Heart Association (81), but standardization of methods has not been instituted like it has with pulmonary function testing. "Normal" values for various exercise parameters may be found in several publications (2, 5, 21, 33, 63, 69, 77, 78, 104), but these values are specific to the samples studied and the testing methods used. Representative, population-based reference values are urgently needed and should be established through standardization of testing methods and development of a national registry. In the meantime, each exercise laboratory needs to establish its own "normal" values by testing a representative sample of healthy boys and girls of varying ages, races, and ethnicities.

capacity, increasing EELV and limiting inspiratory reserve (90, 105).

Changes in gas exchange also become evident with worsening pulmonary obstruction. Oxyhemoglobin desaturation may occur during exercise but is generally limited to those with an FEV_1 <50% of predicted (47). Patients with an FEV_1 <60% of predicted have been found more likely to retain CO_2 (as reflected in the failure of $P_{ET}CO_2$ to decrease significantly with metabolic acidosis), indicating hypoventilation and $\dot{V}_A/\dot{Q}$ abnormalities, or perhaps, altered sensitivity to the CO_2 stimulus (19, 23). Carbon dioxide retention during exercise (independent of resting FEV_1) has been associated with greater decline in FEV_1 over 3 yr (49). In a study of 108 children and adults with CF (72), a $P_{ET}CO_2$ >41 (vs. ≤36) mmHg or an S_aO_2 at peak exercise ≤93% (vs. >93%) mmHg was each associated with more than double the risk of mortality 8 yr later. However when adjusted for other variables, including FEV_1, $\dot{V}O_{2peak}$, and the presence of *Burkholderia cepacia*, $\dot{V}O_{2peak}$ proved to be the only significant independent predictor of mortality; those with lower aerobic fitness ($\dot{V}O_{2peak}$ ≤58% of predicted) had 3.2 times the risk of dying in the subsequent 8 yr compared to those with higher aerobic fitness ($\dot{V}O_{2peak}$ ≥82% of predicted).

Exercise intolerance may also be associated with less skeletal muscle mass associated with undernutrition. However, Moser et al. (68) showed that $\dot{V}O_{2peak}$ remained lower in CF subjects than in healthy controls even when expressed relative to muscle cross-sectional area (CSA). Muscle CSA was only slightly smaller in the CF subjects, suggesting qualitative differences in skeletal muscle affecting exercise tolerance. Pinet et al. (85) reported lower quadriceps CSA and strength in CF subjects compared to healthy controls, but they also reported similar diaphragm mass and abdominal muscle thickness, suggesting that peripheral muscle wasting may occur at the expense of maintaining respiratory muscle mass and function. Other studies suggest that changes in muscle metabolism and O_2 uptake kinetics may affect fitness and exercise tolerance in persons with CF (44, 56, 106, 113).

Collectively, these studies emphasize the value of exercise testing for identifying abnormal responses that may contribute to exercise intolerance as well as determining prognoses for children with disease.

Conclusion

Exercise stress testing provides information about aerobic fitness and exercise tolerance in the healthy child as well as in those with chronic disease. It is useful for identifying maladaptive responses of the pulmonary system to exercise stress that are not apparent at rest, and it helps to elucidate their contribution to exercise intolerance. It can be used to examine changes that occur with disease progression or treatment or growth, and ultimately, it may have prognostic value in some pediatric populations with chronic disease.

EXERTION-BASED APPLICATIONS

Indications for exercise testing in youth vary widely. Each clinical question requires insight into how information gained by exercise testing can best serve the clinician in establishing cardiopulmonary diagnosis, prognosis, and response to treatment. Part III deals with those clinical situations that commonly benefit from the insights gained during exercise testing.

Congenital and Acquired Heart Disease

Michael G. McBride, PhD, and Stephen M. Paridon, MD

Decades of improved diagnostic and surgical techniques and perioperative care have dramatically improved short- and long-term outcomes in children with congenital heart disease. Even with complex congenital heart disease, most children are now surviving into adulthood. As a consequence of these improved results, much of the care for this population has become focused on long-term cardiac function and quality of life. This includes the physical activities associated with daily living, as well as recreational activity and competitive sports. Fortunately, most cardiac defects are minor and have little or no impact on cardiovascular performance at rest or during exercise (69). There are, however, subsets of anatomically and physiologically complex congenital heart defects that result in significantly abnormal hemodynamics.

Formal exercise testing is an integral part of the evaluation of children with confirmed or suspected congenital cardiovascular disease. Most of the diagnostic testing available to the pediatric cardiologist is performed in the resting state and at times in the sedated state. However, children and adolescents are seldom sitting or lying quietly during their waking hours. Therefore, exercise testing provides a diagnostic tool for the pediatric cardiologist that more closely mimics the activity of children and adolescents outside the clinical environment. Exercise testing provides a unique opportunity to evaluate the cardiovascular system under stress and to assess its integrated function with the body's other organ systems. Testing may be used to assess acute changes in the patient's condition, and serial testing can be used to observe longitudinal changes. For cardiac patients, evaluation in the exercise physiology laboratory can be used for multiple purposes, including the assessment of overall function, determining the causes of exercise limitations or elicited symptoms, safety stratification for sports or recreational activities, rehabilitation planning, and evaluating surgical and medical therapies, including subclinical findings such as myocardial ischemia. Furthermore, cardiopulmonary exercise testing also allows the clinician to assess the interaction of the cardiovascular system with the musculoskeletal and pulmonary systems in a state that is more likely to resemble normal daily activity. This type of testing is particularly useful in assessing the relative contributions of various organ systems to exercise performance in patients with congenital heart defects.

Basic cardiopulmonary function during exercise is discussed in the first section of this book. We will not repeat this except to describe the unique response to exercise that may be associated with individual classes of cardiac defects. This chapter focuses on six different categories of congenital and acquired heart disease:

1. Simple two-ventricle defects
2. Obstructive lesions
3. Complex two-ventricle defects
4. Single-ventricle physiology
5. Primary arrhythmias and channelopathies
6. Acquired heart disease and cardiomyopathies

Within each category we describe the anatomical features of each defect, various factors that affect exercise performance, and appropriate testing approaches.

Factors Affecting Exercise Performance

It may seem obvious that more severe cardiac defects are associated with greater compromise to exercise performance. As a guideline, the fewer the cardiac chambers and great arteries, the worse the exercise performance will be. Table 12.1 illustrates the ranges of aerobic capacity for some of the more common types of congenital cardiac defects.

Table 12.1   Exercise Capacity in Congenital Heart Disease

Lesion	Aerobic capacity
Simple shunt lesions	Normal
• Small ASD or VSD, unrepaired	Normal
• ASD or VSD, repaired	Normal or mildly reduced
• Large ASD or VSD, repaired	Normal or mildly reduced
• Large ASD, unrepaired	Mildly reduced
Simple ventricular outflow tract obstruction	Normal, except for severe cases
TOF, repaired	Mildly to moderately reduced
TGA, arterial switch	Normal or mildly reduced
TGA, atrial switch	Moderately reduced
Functional single ventricle, Fontan completion	Moderately to severely reduced

ASD = atrial septal defect. TOF = tetralogy of Fallot. TGA = transposition of the great arteries. VSD = ventricular septal defect.

The more complex lesions generally produce lower maximal aerobic capacity. While this decrease in aerobic performance is often a result of impaired cardiac output (as a consequence of decreased stroke volume, chronotropic response, or both), this is not always the case. Noncardiac abnormalities of the pulmonary or musculoskeletal system may limit exercise performance even in those patients with significant cardiac defects. Formal cardiopulmonary exercise testing may often be the only way to sort out those factors that limit exercise performance in any individual patient.

Table 12.1 illustrates that the presence of four normal cardiac chambers and normal great arteries often results in normal or near-normal exercise performance. Usually this is the case for isolated septal defects and mild-to-moderate ventricular outflow tract obstruction. However, the presence of severe residual lesions such as valvular stenosis or regurgitation may result in significantly abnormal exercise performance even with an otherwise morphologically normal heart and great arteries. This is due to the abnormal stress placed on the myocardium from long-standing abnormal hemodynamics caused by the residual valvular defects. In addition to the number of ventricles, the morphology of the systemic ventricle is important in determining the cardiovascular response to exercise. Cardiac defects that result in a morphological right ventricle serving as the systemic ventricle (the ventricle responsible for pumping blood to the body as opposed to the lungs) invariably have decreased aerobic capacity when compared to those defects with a systemic morphological left ventricle. In lesions that are so complex that a two-

ventricle repair is not feasible, the palliative surgical approach is to perform a Fontan-type surgery, essentially creating a functional single ventricle. The impact of residual abnormal hemodynamics as it pertains to simple and complex two-ventricle systems and single-ventricle physiology are discussed more extensively in subsequent sections.

Exercise Testing

The types of protocols commonly used in the exercise physiology laboratory are discussed elsewhere in this text. The choice of treadmill testing or cycle ergometry depends on the type of information desired. The cycle ergometer has practical advantages over treadmill testing: It is less expensive to purchase and maintain, it is safer, and it is less intimidating to exercising subjects. One of the most important advantages of cycle ergometry is decreased artifact when measuring ECG and systemic blood pressure, physiologic measures that are crucial in the congenital heart disease population. Physical working capacity can easily be quantified with modern electronically braked cycle ergometers, but it is very difficult to assess by treadmill because of the differences in walking or running economy in children of different sizes and gait patterns (87). If the desired information from the exercise test is peak oxygen consumption, however, treadmill testing is recommended because in running both upper and lower peripheral muscles consume oxygen; during cycling, it is primarily the lower limbs that are involved. Table 12.2 provides some examples of

Table 12.2.   Exercise Protocols to Match the Desired Information

Inquiry focus	Example	Recommended protocol	Reasoning
Aerobic fitness	Patients with established or suspected heart disease	Treadmill	Subjects will attain a higher $\dot{V}O_2$ value with treadmill as compared to the cycle protocol
Simple shunt lesions	Repaired ASD, VSD	Treadmill or cycle	Arrhythmia assessment
Obstructive lesions	Coarctation of the aorta	Cycle	BP assessment more accurate in the cycle protocol
Complex two-ventricle defects	Repaired TOF	Cycle	Arrhythmia assessment is critical measure, easier in cycle protocol
	Repaired TGA	Cycle	Ischemia and arrhythmia ECG assessment are easier in cycle protocol
Single-ventricle physiology	Fontan repair	Treadmill or cycle	Dependent upon query
Primary arrhythmias and channelopathies	Arrhythmia assessment or long QT	Cycle	There is less motion artifact on the cycle protocol ECG
Acquired heart disease and cardiomyopathies	Pre- or post-repair coronary arterial anomaly	Cycle	Ischemia ECG assessment is easier in cycle protocol

ASD = atrial septal defect. BP = blood pressure. ECG = electrocardiography. $\dot{V}O_2$ = oxygen consumption. TOF = tetralogy of Fallot. TGA = transposition of the great arteries. VSD = ventricular septal defect.

Adapted from Stephens, McBride, and Paridon 2010.

selected congenital heart defects and the assessment modality that would likely provide the most useful information to the referring physician.

Simple Two-Ventricle Defects

Isolated shunt lesions include such common congenital defects as atrial septal defects (ASD), ventricular septal defects (VSD), and patent ductus arteriosus (PDA). Although these defects can occur as part of complex defects, they also occur often as isolated anomalies. These defects share in common an increased pulmonary blood flow as a consequence of the shunting of blood from the high-resistance systemic circuit to the low-resistance pulmonary circuit across the communicating lesion. When exercise capacity is measured by formal exercise testing in patients with either small, hemodynamically insignificant, or repaired shunt lesions, many have normal exercise performance, but a considerable number have a low aerobic capacity (79, 80). While residual cardiac disease or, more rarely, a persistent degree of elevated pulmonary vascular resistance with exercise may be the cause of exercise limitation in a few cases (80), the cause in most patients with these shunt lesions is a sedentary lifestyle with physical inactivity. Regular exercise, exercise training, and in many cases competitive sport participation can be beneficial for most of these patients.

Atrial Septal Defect

In patients with ASD, blood flows left to right across the atrial defect during diastole as a consequence

Etiology of Impaired Chronotropy in Congenital Heart Disease

The chronotropic capacity of the heart (ability to augment heart rate) affects both cardiac and exercise performance. In the normal heart, stroke volume increases approximately 30% to 40% over the resting value during upright progressive exercise. Any further increase in cardiac output in excess of this amount is accomplished by augmenting the heart rate. Therefore, in order to achieve a normal maximal oxygen consumption, an adequate chronotropic response to exercise is necessary. In patients with congenital heart disease, the ability to increase stroke volume in a normal fashion is often limited or absent. Thus, any impairment in the chronotropic response even more significantly decreases cardiac output during exercise. The causes of chronotropic impairment vary, and more than one cause may be at work in any given patient.

Sinus node dysfunction is the most common explanation for chronotropic impairment with exercise. Abnormal function of the sinus node can have several causes, including surgical (extensive intracardiac repair or denervation) or pharmacological etiologies (adrenergic blockade treatment). During surgical procedures, the sinus node is vulnerable to direct damage. Surgical procedures that require extensive atrial manipulation or creation of extensive atrial suture lines result in the highest risk for postoperative sinus node dysfunction. Such defects include D-transposition of the great arteries requiring atrial baffling (Mustard or Senning) or Fontan operations using an atrial lateral tunnel approach. The cause of chronotropic impairment in the Fontan repair for single-ventricle physiology using an extracardiac conduit is unclear and subject to ongoing debate. In earlier surgical eras, tetralogy of Fallot was associated with significant incidence of chronotropic impairment. This was most likely a combination of direct damage to the sinus node from atrial cannulation during cardiopulmonary bypass and significant residual right-sided hemodynamic abnormalities. In the current era significant chronotropic impairment is much less common.

Chronotropy can also be impaired as a result of surgical disruption of the autonomic innervation of the heart during an operation. The most extreme example of this process would be the complete denervation that occurs with heart transplantation. Less severe, but still significant, disruption of autonomic innervation is seen in those defects that require transsection of the aortic root or significant manipulation of the great arteries, such as the arterial switch operation for transposition of the great arteries or the Ross procedure for aortic valve stenosis.

Chronotropy with exercise can also be impaired by both congenital and surgically acquired heart block. In the current era, surgically acquired heart block is, fortunately, uncommon. However, when it does occur, the atrioventricular node assumes the role of the ventricular pacemaker. This results in a loss of synchronous contraction of the atria and the ventricles, causing a decrease in stroke volume and significantly impaired chronotropy. Resynchrony through the use of pacemaker technology has been shown to improve cardiac output and exercise performance in adults; however, there are significant limitations when applied to small patients with complex cardiac lesions. Often the upper rate limit of these pacemakers is too low to allow normal chronotropic response, even if the sinus node is undamaged.

Exercise testing of patients with congenital heart disease provides the clinician with several potential end points by which clinical decisions can be made that are otherwise unrecognized in the resting state. An understanding of the potential mechanisms of impaired exercise performance, including mechanisms of chronotropic impairment, is but a small portion of the clinical acumen needed for those responsible for assessing exercise capacity in children, adolescents, and young adults with congenital heart disease. However, recognizing the mechanisms of potentially impaired chronotropy, including congenital, surgical, and pharmacological etiologies, provides a basis for understanding potential limitations to exercise performance and creating the rationale for planning further medical interventions.

of the greater right ventricular compliance. The total amount of blood across the shunt is negligible for small defects, and these patients generally have a normal exercise capacity. However, with larger defects the greater shunt size leads to right ventricular volume overload that could cause pulmonary hypertension during exercise (80). In addition, this shunt flow may limit preload to the left ventricle at higher heart rates. If this occurs, this may lead to a mildly reduced exercise capacity. With aging, left ventricular compliance may decrease. This may also help to limit left ventricular preload and thus cardiac output during exercise.

In young patients after ASD closure, nearly all have normal or near-normal exercise capacity. Nowadays repair is almost always done in early childhood. Residual pulmonary hypertension or atrial arrhythmias are very rare during childhood or at any age following early childhood repair. Exercise data for children with ASD prior to repair is scant, mainly because this lesion is often identified early in childhood. From the available information, aerobic capacity is mildly impaired prior to surgical repair (54). As stated previously, cardiac output with exercise may be limited due to a decrease in left ventricular preload. This is a consequence of left-to-right atrial shunting of blood flow preferentially to the more compliant right ventricle. Following surgical repair, exercise performance improves significantly even in older patients with long-standing right ventricular dysfunction (54, 78). However, exercise performance may not normalize if right ventricular dysfunction persists. In addition, long-standing atrial septal defects may result in atrial arrhythmias and chronotropic insufficiency, which may limit cardiac performance during exercise. In those patients with atrial septal defects repaired during infancy, exercise performance is usually excellent and indistinguishable from their healthy peers.

Routine exercise testing in this population repaired in early childhood is not generally needed. However, it can be useful in some circumstances. It may help identify the rare patient with significant chronotropic limitation or arrhythmias. It may also help in distinguishing noncardiac from cardiac-exercise-related symptoms. Exercise intolerance and persistent symptoms are much more likely to be found in patients whose defects were repaired in adolescence or adulthood. Exercise testing in this population is often useful to assess cardiopulmonary performance following repair.

Ventricular Septal Defect

A simple ventricular septal defect results in left-to-right shunting of blood flow. The amount of this shunting is usually measured with noninvasive procedures such as echocardiography or cardiac magnetic imaging using pulmonary-to-systemic flow ratio (Q_p/Q_s). In patients with restrictive VSDs and a Q_p/Q_s of <1.5/1, the shunt from the left ventricle to the right ventricle is relatively small and the impact on exercise performance is negligible. This is because the drop in systemic vascular resistance results in a majority of stroke volume being delivered to the aorta and the systemic vascular system rather than across the VSD. In patients with larger VSDs (Q_p/Q_s 1.5-2), there is usually low pulmonary artery pressure and resistance and only mild left ventricular volume overload. Under these circumstances, dynamic exercise is also usually well tolerated. However, isometric exercise, which increases systemic afterload much more than pulmonary afterload, results in an increase in both pulmonary flow and Q_p/Q_s, making isometric exercise somewhat less well tolerated. Those patients with large unrepaired VSDs (Q_p/Q_s >2), despite normal pulmonary pressures and resistances, often display exercise hemodynamics similar to the moderate-sized defects. However, when the VSD is large enough to allow significant transmission of the systemic pressure to the pulmonary vascular bed, pulmonary hypertension usually results, and neither dynamic nor isometric exercise is well tolerated. These patients are often symptomatic at rest or with minimal activity. This is due to pulmonary edema that results from the high pressures in the pulmonary arterial bed. However, current protocol dictates that these patients almost always undergo surgical repair in infancy. Therefore, the assessment of exercise performance is usually not contraindicated in this population.

Results evaluating maximal exercise performance in patients with a repaired or a residual VSD have been conflicting. Most, however, do show a mild decrease in aerobic capacity (78, 79, 105). Reasons for impaired aerobic capacity vary among patients, but all are related to impaired hemodynamics, including myocardial dysfunction, chronotropic impairment, and elevated pulmonary vascular resistance. However, patients in many of the studies evaluating exercise performance following VSD repair were repaired at older ages according to earlier practice, when the strategies of myocardial preservation were not as sophisticated as they are today.

Exercise testing in the current era for VSDs is usually performed for reasons similar to those described for ASDs and other simple shunt lesions. Performance during cardiopulmonary exercise testing should be essentially normal in a patient who underwent repair without complications in infancy or early childhood. Electrocardiographic conduction abnormalities are often reasons for exercise testing in this group. Bundle branch blocks (usually right) are common, especially in those patients repaired in earlier eras. Atrioventricular node conduction abnormalities are less common but can significantly impair chronotropic response when present. For this reason, exercise testing is often useful for routine screening in patients prior to their participation in competitive and recreational activities.

Obstructive Lesions

Lesions that result in either right or left ventricular outflow tract obstruction may occur as isolated defects or as part of a more complex cardiac abnormality. The most common site of outflow tract obstruction for either ventricle is at the semilunar valve, although obstruction both above and below the valves occurs on either the right or left ventricles.

As a consequence of outflow tract obstruction, resistance is increased, which increases the work required to maintain sufficient cardiac output. This results in myocardial hypertrophy of the ventricle to generate greater force during systole. In mild-to-moderate obstructive disease, cardiac output is generally well preserved both at rest and during exercise. In fact, congenital pressure overload appears to cause supranormal cardiac contractility. Not surprisingly, exercise performance under these circumstances is often normal. Therefore, aerobic capacity and physical working capacity are usually within the normal range for mild-to-moderate aortic and pulmonary valve stenosis. In the case of severe outflow tract obstruction, exercise capacity is often impaired due to the inability of the ventricle to maintain stroke volume despite compensatory ventricular hypertrophy. As a

result, cardiac output decreases during increased work. This implies that during exercise, cardiac output is below the expected value for any given level of oxygen consumption ($\dot{V}O_2$).

Aortic Stenosis

Aortic stenosis can be categorized under three major subtypes: subvalvar, valvar, and supravalvar. Subvalvar disease is a result of a subaortic muscular ridge, a fibromuscular ridge or tunnel, or a discrete subaortic membrane. Subaortic obstruction causes a high-velocity outflow jet and turbulence, resulting in chronic aortic valve leaflet damage and resultant aortic regurgitation. Congenitally stenotic aortic valves can be isolated (unicommissural or bicommissural lesions), or they can be found in association with other lesions, such as VSD, mitral valve abnormalities, hypoplasia of the aortic arch, or coarctation of the aorta. Supravalvar aortic stenosis at the sinotubular junction is typically seen in patients with Williams syndrome, in familial supraaortic stenosis (1), or as spontaneous mutations in otherwise normal individuals. Except in severe cases or in the presence of other significant defects, exercise performance is usually normal or near normal.

In severe left ventricular outflow obstruction, ventricular hypertrophy places the myocardium at risk for exercise-induced myocardial ischemia. This is due to insufficient capillary development to meet myocardial demand, as well as decreased cardiac output, which limits coronary flow reserve during exercise (68). Myocardial ischemia may be seen as electrocardiographic changes (ST segment depression) during exercise (46). Interestingly, in many cases these changes do not correspond to any exertion symptoms. Chest pain and syncope are relatively rare events in children, even with severe aortic stenosis. Therefore, limitations in exercise performance with severe obstruction are often due to the inability to deliver adequate blood flow and oxygen to the peripheral skeletal muscle rather than myocardial ischemia.

Guidelines that grade the degree of stenosis have been used to make recommendations about sport participation in competitive athletes. However, these guidelines are admittedly conservative, based upon scant literature, and do not account for the heterogeneity across the various subtypes of stenotic lesions (36). A recent long-term follow-up of sudden cardiac death (SCD) after aortic valvuloplasty calls these recommendations into question. In a large single-center report, there was only one case of SCD. This occurred during sleep. No deaths occurred with activity, and there was no difference in survival based on whether or not the patient had been activity restricted (9). While most of the previously reported patients with aortic stenosis who died suddenly had a high incidence of ECG abnormalities, and left ventricular hypertrophy or strain should be assessed when making recommendations for sport participation, the usefulness of these findings is limited. Resting ECGs are often abnormal in patients with aortic stenosis even in nonsevere cases. Varying degrees of left ventricular strain are often present and may persist or worsen with exercise. The ability to assess the presence of subendocardial ischemia in the presence of these preexisting ECG abnormalities is limited. While graded exercise testing may be helpful in unmasking important findings, such as blunted blood pressure response, ventricular ectopy, and in certain cases, myocardial ischemia in asymptomatic patients, care should be taken in interpreting the results. A positive finding on an exercise test is often a useful supportive test in patients who are close to meeting criteria for intervention by other testing modalities. A positive exercise test alone should be viewed with caution for patients who by all other criteria would not be candidates for intervention.

Bicuspid Aortic Valve

Bicuspid aortic valves are the most common type of congenital heart malformation, estimated to occur in 0.5% of the population, with males being affected three times more often than females (3, 63). When a female is found to have a bicuspid aortic valve in association with aortic coarctation, Turner syndrome or Turner mosaicism should be strongly suspected. The typical bicuspid aortic valve has two recognizable lines of cusp apposition. Fusion of the right and left, or right and non-coronary, cusps results in valves that are prone to either regurgitation or stenosis or both. Abnormalities of the aortic root, sinotubular junction, and ascending aorta occur as part of this lesion (87, 90, 97). Even in patients who do not have stenosis or regurgitation, dilation of the root and ascending aorta is common. Although extremely rare in the pediatric population, spontaneous rupture may occur in these patients but with much less frequency than in Marfan syndrome patients. There is a tendency for aortic root dilation to occur in patients with fusion of the right and left coronary cusps, while dilation of the ascending aorta appears to be more common in patients with fusion of the right and noncoronary cusps (25, 89).

Exercise performance in children and adolescents with bicuspid aortic valve mirrors that of their normal peers, and regular physical activity or competitive sports are not restricted in those patients without significant aortic root dilation. Although there is information to suggest that regular athletic training may increase aortic dimensions, the actual risk associated with this progression is unknown (25). There is evidence that endurance training may improve the elastic properties of the aorta (10, 47, 48, 95, 96). Close follow-up is warranted, and annual echocardiography may be helpful, but it should be individualized to the patient. Intense, repetitive isometric activities may enhance aortic stiffness and dilation; however, in the absence of aortic root dilation, isometric activities are currently acceptable (2, 5, 6). Controversy around this issue is likely to persist in the foreseeable future. Exercise testing is seldom needed for simple bicuspid aortic valve.

Coarctation of the Aorta

Coarctation of the aorta describes the narrowing of the aortic isthmus, the segment of the aorta between the origin of the left subclavian artery and the ductal ampulla or ligament in a left aortic arch (99). It is commonly associated with abnormalities of the aortic valve and mitral valves, the presence of a VSD, hypoplasia of the aortic arch, and even hypoplastic left heart syndrome. Histological abnormalities involving the elastic media at the site of the coarctation are integral to this lesion. Both unoperated and operated older patients are also at risk for the development of and rupture of intracranial aneurysms. Although rare after early successful palliation, increased risk for coronary artery disease, stroke, heart failure, aortic and cerebral aneurysmal rupture, and SCD persist (100). Exercise capacity is often reduced in these patients despite the adequacy of the repair (18, 37). Chronically elevated systolic blood pressure may play a role in cardiovascular morbidity and mortality. Endothelial dysfunction, reduced vessel elasticity, and enhanced baroreceptors may all play a role in the development of chronic systolic hypertension and the commonly found systolic hypertensive rise to graded dynamic or isometric exercise (4, 15, 38).

Medical and surgical history, including the presence or repair of associated lesions, four extremity blood pressures, and a resting ECG are essential in the initial evaluation of patients with coarctation of the aorta. Previous sites of cardiac catheterization are also important, particularly in patients who have had balloon dilation of native coarctation or dilation of recurrent or residual coarctation. The presence of an upper-to-lower-extremity blood pressure gradient should alert the physician to the presence of a possible residual coarctation.

Laboratory studies include the baseline resting ECG to assess for the presence of left ventricular hypertrophy. Echocardiography is very useful in the detection of residual gradients, associated lesions, and the site of the residual obstruction, if present, as well as left ventricular hypertrophy. Magnetic resonance imaging (MRI) with three-dimensional reconstruction offers exquisite detail of aortic anatomy and may be useful in determining the site of residual obstruction. Maximal exercise testing can be useful to assess the blood pressure response to exercise in these patients. This response may be abnormal even in the absence of a residual coarctation, and any abnormal findings may be related to residual abnormal vascular reactivity in these patients. Interestingly, even patients with excellent repairs and no residual gradient at rest may develop quite large upper-to-lower-extremity blood pressure gradients with exercise. These gradients may occur even in the presence of a normotensive upper extremity response to exercise. The etiology of this blood pressure response is uncertain but may well be related to the abnormal elasticity of the aorta at the site of the coarctation repair.

Pulmonary Stenosis

The most common type of right ventricular outflow obstruction is pulmonary stenosis (PS), which is caused by fused leaflets in most patients. The degree of obstruction is variable but is typically mild. More advanced obstruction results in right ventricular hypertrophy or strain. If left untreated, it can result in exercise intolerance (63) or atrial arrhythmias secondary to right atrial dilation. Most patients with advanced obstruction benefit from intervention, typically balloon valvuloplasty. Mild PS (peak gradient <30 mmHg) does not appear to significantly affect exercise performance. However, moderate (30-50 mmHg peak gradient) or severe (>50 mmHg) stenosis impairs performance, but this typically improves after intervention. Chances are good that reintervention will be unnecessary and exercise capacity will increase; however, the long-term impact of chronic pulmo-

nary regurgitation as a result of the intervention remains to be seen (11, 84). Long-term pulmonary insufficiency is increasingly being associated with impaired exercise performance. This has led to more efforts to quantify right ventricular volume, regurgitant fraction, and ejection fraction in hopes of better defining the optimal timing of pulmonary valve replacement. Limited data suggest that serial exercise testing may be useful as part of the evaluation of the timing for intervention. A decrease in either aerobic performance or pulmonary efficiency as measured by the $\dot{V}CO_2$ slope, for example, may be a useful indicator of increasing right ventricular compromise. More research in this area is needed, especially as the development of nonsurgical implantable valves makes this therapy a more attractive clinical option.

Complex Two-Ventricle Defects

The most common complex two-ventricle defects are tetralogy of Fallot and transposition of the great arteries (TGA). Other, less common, defects are Ebstein's anomaly and truncus arteriosus.

Tetralogy of Fallot

Tetralogy of Fallot (TOF) is a common, complex cardiac defect consisting of four major malformations in cardiac anatomy:

1. A large anterior malalignment ventricular septal defect
2. Overriding of the intraventricular septum by the aortic valve
3. Varying degrees of pulmonary valvular stenosis
4. Right ventricular hypertrophy, usually with significant subvalvular pulmonary obstruction

Repair of this defect is accomplished by surgical closure of the large VSD, simultaneously shifting the aorta into a left ventricular outflow position and relieving pulmonary outflow obstruction.

As a relatively common cardiac lesion with a long history of surgical repair, TOF has probably the most extensive number of studies of exercise performance of any complex two-ventricle defect. Prior to the last decade, most studies of exercise performance in TOF had found mild-to-moderate reduction of aerobic capacity compared to normal subjects (43, 50, 58, 59, 79, 86, 99, 102, 103). In the current era, however, one study found these patients to have nearly normal aerobic and physical working capacities (53).

Reasons for decreased indices of exercise performance in these patients are multifactorial. Residual right-sided lesions have been the focus of most attention. Free pulmonary insufficiency, residual pulmonary stenosis, and right ventricular dilation seem to be key factors in limiting aerobic capacity in this population. Residual stenosis, regurgitation, and branch pulmonary artery stenosis have all been independently associated with diminished exercise performance and inefficient ventilation during exercise. The latter is manifested as high ventilatory equivalents for carbon dioxide (minute ventilation is high when compared to carbon dioxide excretion) as well as a steep rise in the slope of minute ventilation relative to carbon dioxide production (24, 30, 31, 97).

Due to the residual cardiac lesions, patients surgically repaired for TOF have aerobic capacities that are generally around 80% to 85% of healthy matched peers (50, 78, 79, 86, 99, 103). However, exercise capacity varies widely in this population. It may range from severely depressed to, in some cases, well above normal. This heterogeneity in exercise performance reflects both the heterogeneity of the defect itself and the broad spectrum of residual disease seen following operative repair. Those patients with significant pulmonary regurgitation accompanied by biventricular systolic dysfunction appear to have the lowest exercise capacity (30). These are often young adults with long-standing residual right-sided abnormalities. Patients with restrictive right ventricular mechanics may not develop significant right ventricular dilation despite severe pulmonary regurgitation, and they often appear to have more preserved exercise capacity at long-term follow-up (26, 27). Prolongation of the electrocardiographic QRS is associated with dilated right ventricles, and the absolute measurement (>180 ms) as well as the rate of change of the QRS duration may be important predictors for the risk of sudden cardiac death during exercise. Other factors associated with ventricular arrhythmias and possibly with SCD include older age at repair, residual right ventricular outflow tract obstruction with increased right ventricular systolic pressure, RV dilation and dysfunction, and left ventricular systolic dysfunction (28, 29, 93). Premature atrial and ventricular ectopy are commonly observed on exercise testing and can be seen in as many as 50% of patients. However, fast atrial or ventricular couplets or runs of arrhythmias are not common during exercise testing and are likely a cause for concern.

Abnormalities of pulmonary function are also quite common following repair of TOF (70, 86, 102). Most commonly, patients present with restrictive airway patterns associated with single or multiple sternotomies. However, obstructive airway patterns have also been observed. Pulmonary mechanics of these patients tend to result in lower breathing reserves and higher ventilatory equivalents for oxygen and carbon dioxide. However, they are seldom enough to limit exercise performance (58, 86, 102). It is unclear whether the etiology of abnormal pulmonary function in these patients is an intrinsic pulmonary abnormality or secondary to abnormal pulmonary vasculature, especially pulmonary regurgitation.

It would appear that just as TOF is a heterogeneous cardiac defect, there are multiple causes of exercise impairment in this population. These patients often require more individual assessments than other cardiac patients, not only to determine their degree of exercise impairment but also the specific cause of their impairment. Regular exercise testing, including measurements of oxygen consumption and pulmonary function, in this population is necessary to assess causes of any exercise limitation. Exercise testing combined with imaging studies, such as an MRI, may be helpful in identifying patients who could benefit from pulmonary valve replacement; this may be especially useful in the late adolescent and young adult age group. This strategy has taken on more importance over the last decade because there has been an increased focus on earlier pulmonary valve replacement in the second and third decades of life in an attempt to preserve and rehabilitate right ventricular function. This trend is likely to accelerate in the near future as nonsurgically implanted pulmonary valves become routinely available. Changes in aerobic capacity and ventilatory efficiency may be useful markers of early right ventricular dysfunction, and they may be used to help determine the optimal timing for valve placement. The frequency of exercise testing in this population should be greater than in the pre-adolescent population, but the value for pediatric TOF patients remains a point of ongoing research.

D-Transposition of the Great Arteries

In dextro-transposition of the great arteries (D-TGA), the aorta arises from the right ventricle and the pulmonary artery from the left ventricle. This results in systemic venous return bypassing the lungs and being transported back to the systemic arterial circulation. Pulmonary venous return is pumped by the left ventricle back through the lungs. Without early surgical intervention, this type of circulation is usually not compatible with survival.

Until about 25 yr ago, D-TGA was repaired using the atrial switch operation, either the Mustard or Senning, by which baffles were constructed in the atria that channel systemic venous blood to the left ventricle and pulmonary venous blood to the right ventricle. These types of repairs result in extensive suture lines and a right ventricle that must function as a systemic pump.

Exercise performance in this population is usually moderately compromised, with maximal $\dot{V}O_2$ in the range of 60% to 70% of predicted. Many patients may be limited even in performing activities of daily living to varying degrees. The reasons for poor exercise performance are multiple and may include poor systemic right ventricular function, chronotropic impairment, tricuspid valve regurgitation, and ridged atrial baffles that limit augmentation of ventricular preload (17, 57, 65, 71, 85).

Today, patients with the atrial switch operation will be almost exclusively adults, and most will have some degree of significant cardiac dysfunction. A thorough evaluation is warranted, such as that previously described for TOF patients with significant residual defects. Routine exercise testing is used to assess for the presence of arrhythmias and chronotropic response. This testing is also useful for judging exercise capacity and potential limitations. The information gathered can be used in counseling patients about jobs, daily living activities, recreational sports, and the need for physical rehabilitation, when necessary.

Because of the long-term problems with the atrial switch operations, the current approach has been to perform an arterial switch operation whenever possible. In this procedure, the aorta and pulmonary arteries are transected and moved to their correct ventricles. This requires moving and reimplanting the coronary arteries. This has raised concern that torsion and kinking of the coronary arteries during surgery might limit coronary blood flow reserve and place children at risk for stress-induced left ventricular myocardial ischemia (69).

After the arterial switch operation, most patients, especially preadolescents, have normal or near-normal exercise performance (on average, 87% of predicted maximal $\dot{V}O_2$) (16, 23, 32, 44, 52, 74, 81). However, as this population ages there is evidence to suggest that their exercise performance declines somewhat (16, 23, 74). The reasons for this decline are unclear, but they may be related at least in part to a lack of physical activity and rising rates of obesity (75, 76). Complete or partial occlusion of coronary vessels, myocardial perfusion imaging defects, wall-motion abnormalities on stress echocardiography, and diminished coronary reserve have been noted in approximately 10% to 12% of patients following the arterial switch operation (32, 40, 44, 52, 67, 74). Although rarely causing symptoms, these conditions raise concerns about the risk of highly competitive sports in these patients as well as the potential risks of acquired atherosclerotic coronary disease and subsequent activity-related myocardial ischemia as this population ages. In addition, dilation of the aortic root is seen with increasing frequency as this population ages. Despite these concerns, sudden death after this operation is uncommon. To date, the need for additional intervention has been quite low, and indications for reintervention are not clear. The risk for aortic rupture or dissection during activity in this population is not known, but given the presence of scar tissue surrounding the aortic root and the presence of a suture line, it is probably much less than with aortic root dilation from connective tissue diseases. The effects of aortic root dilation on the coronary arteries are also unknown (7, 13, 60).

Exercise testing may help to unveil hidden residual hemodynamic abnormalities or arrhythmias and to evaluate for the presence of electrocardiographic changes suggestive of myocardial ischemia. Myocardial perfusion imaging during exercise stress testing, stress echocardiography, and MRI may also be considered if abnormalities are suspected, especially in cases where resting conduction abnormalities on the ECG limit the interpretation of ST segment changes. Some authorities believe that ECG, echocardiography, and stress testing should be repeated every 2 yr; however, there are no reliable data that this approach to screening frequency is superior to individual clinician judgment.

Ebstein's Anomaly

Ebstein's anomaly is characterized by apical displacement of the septal and posterior leaflets of the tricuspid valve, often accompanied by a secundum atrial septal defect, and it varies in its degree of severity. In patients with mild leaflet displacement and little or no insufficiency, symptoms may be absent and may not require any intervention. In more severe cases, cyanosis is often observed as elevated right atrial pressure results in right-to-left shunting of blood flow at the atrial level.

Few studies have been performed to examine exercise risk or performance in patients with Ebstein's anomaly. Heterogeneity in this patient population is great and will vary with the severity of the valvular dysfunction, in addition to the presence and degree of atrial shunting. Oxygen consumption is often reduced compared to normal individuals; however, patients have improved exercise performance after the tricuspid valve is palliated and the ASD is repaired (51). Patients repaired at a young age who have lower cardiothoracic ratios on chest X-ray at the time of intervention appear to have better outcomes than mid-term repairs (101). Preoperative patients are often cyanotic, which worsens with exercise. After surgical intervention most patients still have fairly limited exercise capacity, with an average maximal $\dot{V}O_2$ of 50% of predicted (101). Significant intracardiac repair often results in chronotropic insufficiency. Reduced right ventricular and left ventricular stroke volumes may limit cardiac output and therefore exercise performance, even in adult patients who are fully saturated (101). Similar to patients with other forms of complex two-ventricle physiology, such as tetralogy of Fallot, ventilatory inefficiency during exercise can be significant as a result of pulmonary abnormalities as well as right-to-left intraarterial shunting if an ASD is present.

Truncus Arteriosus

Truncus arteriosus is characterized by a single arterial vessel that arises from the base of the heart and gives origin to the coronary, pulmonary, and systemic arteries. Similar to other conotruncal defects, there is a great degree of variability in the severity of this disease. Early signs and symptoms may range from mild cyanosis to heart failure due to excessive blood flow through the pulmonary vascular bed. Therefore, early surgical intervention has been the treatment of choice.

Surgical repair involves separating the pulmonary arteries from the common trunk and connecting them to the right ventricle with a conduit. In patients who undergo surgical correction during the early years of life, eventually the smaller conduit will be outgrown. As with other rare two-ventricle defects, data are scant on exercise performance with isolated truncus arteriosus. Regular evaluation of exercise performance to assess the degree of exercise impairment may be beneficial to determine the need for reintervention, such as valve replacement during adolescence or adulthood similar to patients with tetralogy of Fallot.

Single-Ventricle Physiology

The Fontan procedure is the most common palliation for single-ventricle physiology (72). Here, the pulmonary and systemic circulation are separated, with venous return directed passively to the pulmonary vascular bed through a series of three surgical procedures, often before age 2. In this way, the functional single ventricle receives only pulmonary venous return, which is then pumped to the systemic circulation via the aorta.

Exercise performance is significantly impaired in patients with single-ventricle physiology after surgical palliation. Patients with unoperated single-ventricle physiology also have reduced exercise performance, primarily due to right-to-left intracardiac shunting that sends mixed arterial-venous blood flow to the common ventricle, resulting in systemic hypoxemia. During progressive exercise, hypoxemia occurs secondary to a fall in systemic vascular resistance, and resistance to blood flow through the pulmonary vascular bed remains relatively high in these patients (69). The decrease in systemic vascular resistance results in limited pulmonary blood flow compared to that in normal persons. As the system tries to meet the demands of exercising muscles and eliminate carbon dioxide, a larger portion of cardiac output is shunted away from the pulmonary vascular bed, limiting oxygen diffusion. To eliminate the greater fraction of carbon dioxide present in the blood and to maintain pH homeostasis, unoperated single-ventricle patients hyperventilate more than healthy peers and those with acyanotic congenital heart lesions. This pattern of breathing is associated with the marked limitations on exercise performance experienced by unoperated single-ventricle patients (20). Poor ventricular function and chronotropic incompetence have also been implicated as causes of impaired exercise response (21). Fortunately, in the current surgical era it is rare to find patients who have unrepaired single-ventricle physiology.

Exercise performance following Fontan palliation for single-ventricle physiology has been the focus of many research studies. Improvement following surgical repair is attributed to a significant reduction or complete elimination of intracardiac shunting. However, aerobic exercise performance in these patients is persistently impaired, ranging from 60% to 75% of normal values for age and gender (21, 22, 33, 39, 62, 72). Several factors have been implicated in the abnormal response during follow-up after the Fontan palliation. The primary determinant is the inability to maintain stroke volume during exercise in a normal fashion in these patients, which is multifactorial. In a few patients, impaired exercise performance has been attributed to poor systolic function. These patients are often very symptomatic. More commonly, systolic function is preserved, and exercise stroke volume appears to be restricted by an inability to maintain ventricular preload and adequate stroke volume. This is a result of the limited ability to maintain pulmonary blood flow in the absence of a subpulmonary ventricle and in some cases ventricular diastolic dysfunction.

Before Fontan surgery, patients hyperventilate more than their peers at rest and during exercise as a consequence of persistent systemic cyanosis. This results in elevated ventilatory equivalents for oxygen and carbon dioxide and overall poor ventilatory efficiency. Following surgical repair, the ventilatory equivalents for oxygen and carbon dioxide often improve, but they remain abnormal during exercise secondary to intracardiac or intrapulmonary shunting. The direct cause of abnormal ventilatory function during exercise after surgical repair is not clear, but it may be due to several factors, such as

- a small vital capacity following repeated palliative surgeries, resulting in uneven lung inflation secondary to adhesions after surgery,
- accelerated ventilation due to arterial oxygen desaturation and increased dead-space-to-tidal-volume ratio, and
- an inability to increase effective pulmonary blood flow in response to the increasing demands of the exercising muscle (21, 66, 69).

Exercise impairment in children with Fontan circulation is often related to both central (oxygen delivery) and peripheral (oxygen extraction) factors or both. While chronotropic impairment and intracardiac shunting affect Fontan exercise capacity as central cardiovascular factors, the importance of peripheral musculoskeletal factors should not be overlooked. Lean muscle mass and focused exercise training are important modulators of exercise capacity in both congenital and adult-onset heart disease, and they may be important to daily activities and overall exercise capacity in the Fontan population. In a recent multicenter study, central cardiovascular factors accounted for less than 40% of the variance in maximal $\dot{V}O_2$ in the Fontan population (72).

From a central cardiovascular standpoint, the absence of a subpulmonary pump (venous ventricle) results in passive pulmonary blood flow dependent on the pressure difference between the central venous pressure (CVP) and the systemic ventricular end-diastolic pressure (EDP), as well as relatively low pulmonary vascular resistance (PVR) (35). Given the limited ability to increase this CVP driving pressure due to the lack of a prepulmonary pump, any change in PVR or EDP can have a profound impact on limiting ventricular preload and overall stroke volume (35). Even with normal PVR and EDP, the inability to increase the driving force of blood flow through the pulmonary vascular bed results in decreased preload and, therefore, decreased stroke volume (35). These limitations, while often seen in the resting state, are more pronounced during exercise, limiting cardiac output to at most a two- to threefold increase rather than the three- to fivefold seen in normal two-ventricle physiology (35).

In a study by the Pediatric Heart Network (PHN), $\dot{V}O_2$ at the ventilatory anaerobic threshold was significantly better than at maximal exercise (78% vs. 65% of predicted, respectively) (72). Therefore, ability to maintain a higher submaximal $\dot{V}O_2$, as measured by the ventilatory anaerobic threshold, relative to the maximal $\dot{V}O_2$ with Fontan physiology does suggest some positive benefits in that sustained physical activities that are shorter in duration and usually performed below the anaerobic threshold will be better tolerated. These include almost all the activities of daily living, recreation, and short-burst competitive sports. Patients with Fontan physiology may in fact be more functional during submaximal levels of activity than would be expected on the basis of the maximal values obtained in the exercise physiology laboratory.

Pediatric Heart Network studies and others suggest that central cardiovascular factors are not the only limitation to exercise performance in the Fontan population. Peripheral factors can also play a critical role in the limitations on exercise performance in these patients. Factors such as lean muscle mass and exercise conditioning appear to be at least as important in determining exercise performance in this population as central factors. Children and adolescents with Fontan physiology have deficits in lean mass and skeletal muscle that may affect their ability to extract oxygen in a normal fashion. As a result, even when oxygen delivery is adequate from a central standpoint, the ability of the peripheral musculoskeletal system to meet metabolic demand may be compromised.

Primary Arrhythmias and Channelopathies

Approximately one-third of all sudden cardiac deaths that occur during athletic activities in the pediatric and young adult age range are not associated with any evidence of structural heart disease or cardiomyopathy (55). These deaths are most likely due to arrhythmias. Defects in the cardiac sodium and potassium channels are the types most often associated with SCD in athletics. These include prolonged QT syndrome and Brugada syndrome. In normal persons, the QT interval shortens with exercise, but the corrected QT interval (QTc) changes very little. In patients with prolonged QT syndrome the QTc may prolong during intense work, potentially leading to a life-threatening dysrhythmia. A formal exercise test may therefore be useful in helping to exclude prolonged QT syndrome in selected patients (see chapter 5). The authors have found that an exercise test, although not always diagnostic, can be useful in providing additional information when attempting to stratify risks for certain patients and assessing the adequacy of treatment strategies such as beta-blocker therapy.

Acquired Heart Disease and Cardiomyopathies

Diseases of the heart muscle can be acquired after birth or inherited. This section will discuss Kawasaki disease, anomalous coronary artery disease, and cardiomyopathies. Kawasaki disease affects the blood vessels and can lead to coronary artery disease after treatment of the immediate symptoms. Anomalies of the coronary arteries can be a primary disease or a structural form of congenital heart disease. Cardiomyopathies can also be acquired or inherited, the five types being dilated, hypertrophic, restrictive arrhythmogenic, right ventricular, and unclassified (mixed) physiology. The most common is hypertrophic cardiomyopathy, which is usually inherited and affects heart muscle proteins.

Kawasaki Disease

Kawasaki disease is the most common cause of acquired heart disease in children (61). Ischemic heart disease that can lead to myocardial infarction and SCD can occur due to coronary aneurysms, together with progressive coronary artery stenosis (49). The risk of SCD due to coronary artery disease appears to change over time. For the first 20 yr after the onset of Kawasaki disease, patients without evidence of coronary aneurysm or with initial transient dilation on echocardiography appear to have no greater risk for ventricular tachyarrhythmias and SCD than the general population (49). Those with aneurysms that regress to normal lumen diameter may have persisting structural and functional coronary abnormalities (49). Suda and colleagues recently reported the long-term prognosis (median 19 yr follow-up) of patients with giant coronary aneurysms that did not regress and, instead, remodeled over time, leading to intimal thickening and risk of ischemic heart disease. Of the 76 patients initially followed, 7 died and 1 underwent heart transplantation. In addition, there were many catheter and surgical coronary interventions with cumulative coronary intervention rates of 28%, 43%, and 59% at 5, 15, and 25 yr after disease onset, respectively (94). Certainly, in patients with Kawasaki disease, risk associated with physical activity and exercise depends on the degree of coronary involvement. Paridon et al. reported on 46 children and adolescents with a history of Kawasaki disease and showed that maximal oxygen consumption was within normal limits without a difference based on coronary artery status (i.e., none, regressed, or current aneurysms) (73). Another study focusing on children with persistent coronary aneurysms also showed normal peak oxygen consumption, workload, and anaerobic threshold when compared to control subjects (82).

Children with Kawasaki disease should have, in addition to a physical examination and ECG, a resting echocardiogram to evaluate heart function and determine the presence and size of coronary aneurysms. A maximal graded exercise test in conjunction with nuclear myocardial imaging or stress echocardiography can be helpful in assessing for evidence of ischemia, wall-motion abnormalities, and the presence of exercise-induced arrhythmias.

Anomalous Coronary Artery Disease

Coronary artery anomalies often result in an increased risk of myocardial ischemia and sudden cardiac death. Those in which the vessel begins from the inappropriate sinus of Valsalva and courses intraarterially and intramurally between the aorta and pulmonary artery, with the anomalous left coronary artery emerging from the right sinus, carry the greatest risk of a SCD. These anomalies are the second leading structural cause of SCD in young athletes in the United States (55). Discovery of coronary anomalies is difficult because many people may not experience exertional symptoms prior to a cardiac event. A resting screening ECG is almost always normal in these patients. Although chest pain both at rest and with exertion may often be associated with these coronary abnormalities, it does not confirm their presence. Chest pain, which is typical for true cardiogenic anomalies, is ubiquitous in the pediatric population and is very rarely cardiac in origin. Syncope during or immediately after strenuous athletic activity is a much more ominous sign that cardiac pathology may be involved. Transthoracic echocardiography with color Doppler should be performed to demonstrate coronary anatomy. However, when an anomalous coronary artery is suspected, advanced imaging such as cardiac MRI, ultrafast CT, and coronary angiography can be performed to provide confirmatory evidence. In patients where the left main coronary artery arises aberrantly, surgery is often recommended. There are no defined guidelines for the management of asymptomatic patients with anomalous right coronary artery, with some opting for surgery and others opting for conservative management, with or without exercise restriction. The relative risks and benefits of surgical versus nonsurgical management of these defects are far from clear and are evolving as more data about the risks of these anomalies and the short- and long-term risks of interventions become available.

Children suspected to have a congenital coronary anomaly should have a baseline physical examination and ECG, with a resting echocardiogram to delineate coronary anatomy, evaluate heart function, and assess for atrioventricular and aortic valve regurgitation as well as wall-motion abnormalities. A maximal graded exercise stress test, usually with nuclear myocardial perfusion or stress echocardiography, should be performed. Besides measuring aerobic and physical working capacities, the stress test will help assess for evidence of ischemia, exercise-induced symptoms, and exercise-induced arrhythmias. Routine follow-up with exercise testing should be performed at regular intervals both to assess operative results in patients in whom these procedures are performed and to follow nonsurgically managed patients for any changes that might indicate an increased risk for exercise-induced ischemia.

Hypertrophic Cardiomyopathy

Hypertrophic cardiomyopathy (HCM) is characterized by a thickened, nondilated left ventricle in the absence of other cardiac diseases such as aortic stenosis or systemic hypertension (56). Hypertrophy of the myocardium may be diffuse and may involve the left ventricle and ventricular septum. Obstruction of the left ventricular outflow tract by the thickened ventricular septum and the mitral valve causes increased impedance to left ventricular cardiac output, producing markedly increased ventricular pressures. This in turn may impair left ventricular function by increasing myocardial wall stress and oxygen demand. As a result, regional myocardial ischemia may occur during exercise in HCM. However, quantification of the extent of ischemia is often problematic. Exercise testing with the use of stress echocardiography can provide valuable insight into cardiac physiology during activity with the potential for detecting subclinical or occult pathology. The presence of left ventricular outflow tract obstruction at rest or with provocation can guide the choice of medication and can make a patient a potential candidate for septal reduction therapy if symptoms and gradient persist after maximal medical therapy. Some data suggest that a blunted blood pressure response to exercise in the pediatric population may indicate an increased risk for sudden death and that routine screening for this is therefore indicated. Some recent data call into question whether detecting dynamic outflow obstruction during exercise has value in distinguishing relatively mild HCM from the healthy population. More research is needed on this subject (104).

Other Cardiomyopathies

Less common cardiomyopathies include dilated, restrictive, or mixed physiology. These cardiomyopathies arise from a variety of etiologies such

Heart Transplantation

Exercise capacity, as measured by both maximal oxygen consumption and measures of musculoskeletal strength, such as manual muscle testing, decreases significantly in the pediatric population following heart transplantation. The causes of these limitations are multifactorial and include pretransplant deconditioning and malnutrition losses in muscle mass as well as post-transplant bed rest. The limited reports on exercise testing in pediatric heart transplant patients have reported aerobic capacities, as measured by maximal $\dot{V}O_2$, of 50% to 60% of healthy age- and sex-matched peers (14, 19, 77), similar to values reported in the adult heart transplant population. The causes appear to include both central and peripheral factors that combine to impair cardiovascular and musculoskeletal exercise capacity. Especially in early post-transplant periods, stroke volume is limited. This may be due to systolic impairment but more importantly to diastolic dysfunction with high cardiac filling pressures. Abnormalities of autonomic innervation and function also affect stroke volume, ultimately limiting cardiac output during exercise in the presence of impaired chronotropy. At the start there is a loss of autonomic innervation to the heart. This significantly decreases chronotropic reserve and blunts the time course of the chronotropic response. There is some evidence for reinnervation and improved chronotropy late after transplant in some patients or as a response to cardiac training. In addition to the cardiac effects, autonomic tone is abnormal in the peripheral vasculature. Brachial reactivity is impaired, and systemic vascular resistance is increased (8, 41, 83, 92).

Serial studies of exercise performance following pediatric heart transplantation are scant in their findings. The most robust studies from Davis et al. and Dipchand et al. are conflicting (14, 19). Both show early improvement in aerobic capacity and working capacity in the early recovery period. However, Davis saw a decline after about 3 yr of improvement, while Dipchand's population remained steady, with some patients showing a decrease associated with the onset of graft vasculopathy. The explanation for these results is unclear but is likely the combined improvement of systolic and especially diastolic function in the immediate post-transplant period as well as the longer-term improvement in musculoskeletal conditioning, even in the absence of formal rehabilitation. Improved chronotropy suggests that at least some patients benefit from autonomic reinnervation of the donor heart. The etiology of reinnervation is unclear.

Exercise testing in this population is useful for identifying causes of decreased performance as a way to target post-transplant rehabilitation. Long-term testing is often used to screen for the presence of ischemia due to progressive immune-mediated coronary vascular disease. Classical exertional angina may not occur in a denervated heart. Exercise testing that assesses electrocardiographic changes combined with nuclear or echocardiographic imaging may be useful. However, there are little data that objectively evaluate their utility.

as genetic, chemical or toxic, and postinfectious. Exercise performance in this population may vary from severely limited to normal, depending on the degree of ventricular dysfunction. Because of this wide variation, it is not possible to make general recommendations about physical activities and sport participation or the routine use of exercise testing.

Limited data on exercise performance are available in some of these groups. Patients who have received anthracycline therapy as part of a chemotherapeutic treatment may develop a slowly progressive, dose-dependent, dilated cardiomyopathy as children or young adults. Exercise performance may be normal in patients who are only mildly affected, and this may remain stable for many years (42, 90). However, symptoms will often occur in patients with significant, progressive decline in cardiac function. The presence of a restrictive physiological component, especially in the presence of pulmonary hypertension, may greatly increase the risk of both regular activities and exercise testing in the post-anthracycline population.

Left ventricular noncompaction cardiomyopathy is a rather recently recognized entity (12, 45). The incidence of this type of cardiomyopathy is unknown. Patients may present with severe heart failure but are often completely asymptomatic and are identified by an echocardiogram performed for an unrelated cause. For this reason, the risks of exercise and the utility of exercise testing in this population are largely unknown, but research is currently focused on algorithms for risk stratification.

Conclusion

The importance of exercise testing in the pediatric congenital heart disease population cannot be overemphasized. Given that most diagnostic tests of cardiac structure and function offered to the pediatric cardiologist require the patient to be in the resting state, exercise testing can provide greater insight into the physiological and hemodynamic changes that may help in clinical decision making, and it more closely reflects habitual physical activity and recreational and competitive sport.

Early diagnosis and early repair, combined with improved surgical techniques, have resulted in normal or near-normal exercise performance in most patients with simple cardiac defects. The challenge for the future is to determine the causes of impaired performance in those patients with complex defects and hemodynamics. This understanding will hopefully lead to modification of our current practices, resulting in improved exercise performance and better guidance for safe participation in physical activity and sport.

Exercise-Induced Dyspnea

Steven R. Boas, MD

Case Presentation

Quinta is a 14-year-old female competitive cross country runner who presents with a 3 mo history of dyspnea with exertion. She states that prior to the onset of symptoms, she had never had any difficulty with running. The first episode occurred during a varsity high school cross country meet, and she has since had symptoms with each meet. Quinta notes an abrupt onset of biphasic dyspnea, although the inspiratory phase appears more problematic. No associated wheeze or cough is appreciated, but a harsh, high-pitched inspiratory noise is heard on an inconsistent basis. She experiences upper chest and throat tightness. She usually tries to "run through" her symptoms without success. Recently, she started to note paresthesia in both arms with some numbness present. She also describes a presyncopal sensation during these episodes but has never lost consciousness. Her past medical history is otherwise unremarkable. Quinta is self-described as competitive and gets anxious around the time of her athletic performance. She is a straight A student in the ninth grade, taking all honor classes.

Dyspnea is defined as the subjective feeling of breathlessness. It is a term used to describe the perception of uncomfortable respiratory sensations (27). Dyspnea with exertion may signify the early stages of pulmonary or cardiac disease that is not yet apparent at rest, or it may represent a mild condition that only presents with the increased work of breathing associated with exercise.

Differential Diagnosis

The following list outlines the principal pulmonary conditions associated with exercise-induced dyspnea in the pediatric population. The reader is referred to other chapters in this book for discussion on dyspnea and abnormalities associated with nonpulmonary conditions.

Respiratory Causes of Exercise-Induced Dyspnea

- Airways
  - Exercise-induced bronchospasm
  - Exercise-induced laryngeal obstruction
  - Exercise-induced tracheobronchomalacia, excessive dynamic airway collapse

- Breathing mechanics and control of breathing
  - Exercise-induced hyperventilation
  - Anxiety
  - Physical limitation
  - Deconditioning
- Restrictive limitations
  - Obesity, chest wall abnormalities, muscular dystrophy, myopathies
  - Interstitial lung disease

Exercise-Induced Bronchospasm

Exercise-induced bronchospasm (EIB) results from a narrowing of the airways during or after exercise. The National Heart, Lung, and Blood Institute (NHLBI) expert panel notes that "exercise may be the only trigger of asthma for some individuals, and its presence is considered a marker of inadequate asthma control" (31). EIB occurs in up to 90% of children with asthma (9). EIB involves both bronchospasm and the release of inflammatory mediators. Although many mechanisms have been proposed for EIB, cooling and drying of the exercising airway resulting from increased ventilation has been the most widely accepted theory. EIB presents with bronchoconstriction, most notably 5 to 10 min following peak exercise, although some people experience symptoms during exercise. Symptoms typically resolve 15 to 30 min later. A refractory period lasting between 45 min and 3 h has been noted in 50% of children with EIB (29). The NHLBI expert panel states that a history of cough, dyspnea, chest pain or tightness, wheezing, or endurance problems during exercise suggests EIB (31). Children are often poor perceivers of EIB, with about 50% reporting no symptoms despite positive challenge tests (29).

Physical examination is often normal unless EIB is part of the more typical presentation of asthma. Signs and symptoms of allergies, hyperinflation on lung examination, and wheezing upon auscultation are some of the more common physical exam findings of asthma.

If EIB represents one aspect of more persistent asthma, then resting pulmonary function testing may indicate airway obstruction on spirometry, increased residual volumes on body plethysmography, and increased resistance, reactance, and impedance measurements on impulse oscillometry assessment. The exercise provocation is considered the hallmark for diagnosing EIB. For children and adolescents, the treadmill is commonly used. Running is performed at about 80% of maximal capacity with the subject exercising for 5 to 8 min. Exercising in a cool, dry environment will typically elicit the best response. Spirometry is performed before exercise as well as immediately after exercise and again at 5, 10, and 15 min postexercise. What constitutes a significant decline in the forced expiratory volume in 1 s (FEV_1) following exercise has varied among different authors. However, declines of 12% to 15% are typically used for the EIB diagnosis (34). Other challenge protocols have been used to diagnose EIB, including eucapnic voluntary hyperpnea, mannitol, methacholine challenge, and cold air challenges.

Exercise-Induced Laryngeal Obstruction

Since the sentinel work by Christopher and colleagues in 1983 describing vocal cord dysfunction (VCD), many published studies have followed using over 40 different names for this condition (8, 30). A recent international consensus conference has proposed that the condition be called "inducible laryngeal obstruction (ILO) causing breathing problems" (7). Multiple nonexercise causes of ILO exist; see the excellent review articles for further information (30). Exercise as

a cause of laryngeal obstruction can be divided into two types. There is a primary type where the act of increased ventilation during exercise induces laryngeal obstruction at the supraglottic level or at the glottic level with partial or complete adduction of the vocal folds. A secondary type of exertional ILO occurs when specific factors place the glottis and supraglottic structures at higher risk of exertional closure. These associated factors include gastroesophageal reflux and chronic aspiration, laryngeal irritants such as smoke, chemicals, dust, and chlorine, anxiety, sinusitis, rhinitis, vocal cord paralysis or paresis, and anatomical obstruction (e.g., large polyps, nodules).

The history obtained in children with exercise-induced laryngeal obstruction typically includes an early adolescent onset with symptoms that are intensity related. Personality features may include a high academically performing person who demonstrates symptoms predominantly with exercise (25). While a female preponderance has been cited in the literature, the condition often exists in adolescent males as well. Dyspnea is the most common presenting symptom (73%-95%) followed by a wheeze, stridor, or cough (30). Physical examination is often normal at rest, with signs only presenting during highly intense activity. The use of accessory breathing muscles with mild, prompted deep inhalations is commonly seen. Stridor or wheeze does not usually occur at rest but may present during exercise. The sounds may be audible or only appreciated by auscultation. The quality of the stridor is usually high-pitched and best heard during inhalation over the trachea. Wheezing may also be heard as a primary sign of ILO or as a manifestation of concomitant exertional asthma. The presence of wheezing and stridor are consistent with ILO but are not required for the diagnosis.

Pulmonary function testing may reveal a truncated or flattened inspiratory limb of the maximum flow volume curve. This finding is often intermittent and is only suggestive of, but not diagnostic of, symptomatic ILO (40). Although ILO is thought of as predominantly an inspiratory disorder, expiratory involvement may occur as well (10). A change in the expiratory portion of the flow-volume curve is not inconsistent with ILO and may erroneously lead to a false diagnosis of asthma. Lack of reversibility in the abnormal expiratory flows upon administration of a bronchodilator should be noted. The use of impulse oscillometry can assist in the diagnosis of ILO by demonstrating higher inspiratory amplitude (19).

Exercise testing using a sufficiently high intensity is an essential part of the evaluation. A treadmill test is used to reproduce the patient's symptoms either at a preset intensity level or in an individualized symptom-limited fashion. When the symptoms of dyspnea and wheeze appear, inspiratory limitation of the maximal flow-volume curves can be seen, with some patients showing expiratory flow limitation as well. Observing the pattern of breathing during symptoms with excessive accessory muscle use, listening for audible inspiratory stridor or wheeze, and auscultating the trachea and upper chest can aid in the diagnosis of laryngeal obstruction. Breath-by-breath gas analysis can be performed during exercise testing to demonstrate changes in tidal volume during laryngeal obstruction and to identify other conditions with associated ventilatory abnormalities. Tidal loop abnormalities obtained during and immediately after exercise may also support the diagnosis of ILO (5). Postexercise spirometry can evaluate for exercise-induced asthma as a separate or concomitant condition.

Visualization of the laryngeal structures will help determine which type of ILO is present. Inspection of the anatomy of the glottic and supraglottic structures, evaluation of the laryngeal mucosa, and assessment of laryngeal movements during various breathing maneuvers (i.e., panting, phonation, and deep breathing) are recommended. The classic finding seen for ILO at the glottis may include adduction of the anterior two-thirds with an associated posterior diamond-shaped chink (8). Supraglottic closure on a partial or complete basis can also be observed with these breathing maneuvers (21). Laryngoscopy has been employed pre- and postexercise, during exercise, and following a positive exercise provocation (30).

Treatment for ILO varies based on the type of inducer identified. Medical management for reflux, sinusitis, or allergies, if present, should be initiated. Self-hypnosis, relaxed throat breathing techniques, biofeedback, and psychological counseling have all been effectively used and should be chosen on an individual basis (4, 41, 44, 48).

Exercise-Induced Tracheobronchomalacia or Excessive Dynamic Airway Collapse

Airway limitation during exercise may result from dynamic collapse of central airways. Data for exercise-induced tracheobronchomalacia in children are limited to anecdotal pediatric reports. Recently, exercise-associated excessive dynamic airway collapse (EDAC) of the trachea has been described in adults with associated symptoms of tachypnea, respiratory distress, and abnormal respiratory sounds (3, 46). Dyspnea has been reported in 59% of children with primary airway tracheobronchomalacia and reduced exercise tolerance in 35% (6), with more symptoms likely to occur as the severity of airway collapse increases (11). During resting breathing, the tracheobronchomalacia segment is quiet and produces minimal or no symptoms. With increased respiratory effort or exertion, the pressure surrounding the tracheobronchial segment becomes positive. In combination with the Bernoulli effect, these forces cause the membranous portion of the tracheobronchial tree to collapse, obscuring the lumen of the airway. On physical examination during exertion, a harsh, monophonic expiratory wheeze or stridor may be heard. The sound is usually prominent in the central airways. Some patients may also experience cough as the airway collapses and prevents the clearance of distal secretions. As airway collapse progresses, the individual may feel presyncopal. Bronchoscopic evaluation with high-intensity exercise (cycle ergometry or treadmill) remains the definitive diagnostic procedure. Assessment of lung function during exercise may demonstrate a flattening of the flow-volume loops at high lung volumes with preservation of lung function at low lung volumes (11). Careful consideration of whether a bronchodilator should be used prior to exercise testing as the use of a beta agonist may cause further floppiness of the airways. Whether the mechanisms explaining exertional tracheobronchomalacia in an otherwise healthy athlete are similar to that seen in the infant and childhood form of malacia is unknown.

Exercise-Induced Hyperventilation and Anxiety

The normal ventilatory response to the demands of exercise includes increasing both breathing frequency and tidal volume to varying degrees, resulting in increasing minute ventilation. This increased ventilation matches the physiological needs of the exercising muscle, ensuring adequate oxygen uptake and carbon dioxide elimination (28). This "heavy" or "rapid" breathing occurring during exercise is a normal response.

When ventilation exceeds physiologic demand during exercise, then exertional hyperventilation (EIH) occurs with an associated decline in P_aCO_2. The true incidence of this entity in children is unknown, although it was reported as 34% in one clinical series (2, 16). Exercise-induced hyperventilation may occur in isolation (primary) or in response to a trigger such as laryngospasm, laryngeal obstruction, or an asthma attack (secondary). A person with EIH experiences mainly rapid breathing with intermittent deep breathing and an associated sensation of shortness of breath. Musculoskeletal tension from the exercising muscle may be perceived as chest tightness. As arterial carbon dioxide becomes further reduced, dizziness or syncope may ensue. The exact P_aCO_2 that results in paresthesia and dizziness is variable, but it has been reported in healthy subjects to occur near 20 mmHg (12, 36). Numbness or tingling of the extremities, sweating of the hands, and spasm of the muscles in the hands and feet (tetany) resulting from alkalosis are also common complaints (15, 43). Physical examination at rest is usually normal. The presence of deep inhalation and exhalation with resultant associated dizziness during auscultation of the lungs may occur in healthy people and does not necessarily suggest a predisposition to exertional hyperventilation. During an acute episode, the increased exertional ventilation is usually readily observed, although it is not diagnostic for EIH. The use of accessory respiratory muscles, pallor, diaphoresis, and cold skin may be observed as the process progresses. Tetany of the extremities may be experienced. While demonstrating a decline in P_aCO_2 by direct measurement of arterial blood is the gold standard for diagnosing hyperventilation, noninvasive measurements of CO_2 production such as end-tidal CO_2 collection during exercise testing with breath-by-breath gas analysis is widely used and provides a good estimation of arterial CO_2 in the absence of underlying lung disease (12). The exercise protocol can be individualized based on the circumstances around which the hyperventilation is suspected. Ventilatory equivalents, $\dot{V}_E/\dot{V}CO_2$ and $\dot{V}_E/\dot{V}O_2$, are elevated during hyperventilation and significantly correlate with the decline in P_aCO_2 ($r = -.71$) (18). Oxygen saturations by pulse oximetry are unchanged. If arterial or arterialized blood is obtained during EIH, an elevated pH and low P_aCO_2 consistent with an acute uncompensated respiratory alkalosis will be measured. While the cause of EIH is poorly understood, it has been speculated that neurologic dysfunction of central or peripheral oxygen receptors during exercise may occur with associated abnormal ventilatory homeostasis (16).

A panic reaction or anxiety reaction may cause exercise-induced dyspnea associated with hyperventilation (45). The onset of symptoms is abrupt. Increased breathing frequency and associated decline in end-tidal PCO_2 can ensue with associated clinical manifestations of hyperventilation. For some individuals, anxiety or panic may occur during rest in anticipation of exercise. Low end-tidal CO_2, high breathing frequencies, high respiratory quotients, and high oxygen levels are associated with anxiety at rest or at low exercise levels. Once the demands of exercise increase, these ventilatory and physiological values return to near baseline values and undergo the normal physiological response with exercise. Anxiety can also present during exercise as evidenced by an irregular breathing pattern or breath holding. These breathing patterns can be observed during exercise testing (treadmill or cycle ergometry of varying intensities) as well as during breath-by-breath gas analysis with blood gas measurement. With breath holding, a respiratory acidosis can rapidly occur in addition to the metabolic acidosis seen at intense exercise. Hypoxemia can also be seen during exertional breath holding (26).

Physical Limitation

It is a normal physiological response to increase minute ventilation as exercise demands increase. For athletes who may push themselves in order to reach or surpass their physical limits, the perceived dyspnea may in fact represent the normal respiratory discomfort in breathing fast or deep. This physiological limit reflects the high levels of ventilation needed to meet the self-imposed metabolic demands. This ventilation is exaggerated as one passes the ventilatory threshold with lactate accumulation accompanying increased ventilation and hypocapnia (low P_aCO_2).

Defining this physiological limit is difficult and somewhat a diagnosis of exclusion. In one study, this physical limitation was determined by the level of baseline conditioning associated with parameters obtained on a breath-by-breath gas analysis during exhaustive exercise (1). For the low-conditioned person as defined by a $\dot{V}O_{2max}$ <80% of predicted, symptoms during testing would need to occur in association with one other indicator of poor condition such as low anaerobic threshold, decreased oxygen pulse, increased ventilator equivalents for oxygen, or greater-than-expected heart rate versus work rate slope. A normally conditioned person was defined as a $\dot{V}O_{2max}$ between 80% and 120% of predicted without any of the preceding indicators. Above-normal condition was defined as a $\dot{V}O_{2max}$ >120% of predicted without any abnormal indicators. Using this model, the authors suggest that 52% of children referred for evaluation of EIB had actually reached their normal physiological limit, with two-thirds having above-normal or normal conditioning (1).

In contrast, Mahut and colleagues defined a normal physiological response with breathlessness related to intense exertion as a normal performance with normal ventilatory response and demand or a low-performance individual (<83% $\dot{V}O_2$ predicted) with a ventilatory response related to poor effort without evidence of poor conditioning (22). In their study, 67% of patients with exertional dyspnea demonstrated a normal physiological response to exercise testing. The tested subjects were children and adolescents with exertional dyspnea of at least 4 wk duration, without a response to asthma therapy, and were considered nonathletes. In the study by Abu-Hasan and colleagues, only a third of those with physiological limitations were considered highly conditioned (1).

Because the mechanism of excessive dyspnea as one tries to surpass physiological limits is not precisely defined, the true prevalence of exertional dyspnea due to this "supraphysiologic" state remains unknown. Clinically, these individuals feel short of breath with more intense exertion. In contrast to laryngeal obstruction, symptoms will occur when one has reached ventilatory threshold, while in ILO, symptoms often start abruptly long before ventilatory threshold has occurred. No symptoms of hyperventilation, restriction, or obstruction should be present. Physical examination findings are limited to signs of increased work of breathing (i.e., accessory muscle use) in order to maintain ventilation at this high level. No wheeze, stridor, or adventitial sounds are present. On breath-by-breath gas analysis during high-intensity exercise, normal ventilatory parameters should be present without signs of airway obstruction by either tidal loop analysis or flow volume loops. Treatment includes aerobic conditioning and training to reduce the ventilatory demands imposed by high-intensity exercise.

Deconditioning

Poor conditioning or deconditioning can be defined as reduced exercise capacity due to disuse in the absence of any underlying primary disease (42). The lowered exercise capacity is associated with decreased peripheral oxygen extraction.

Criteria for this diagnosis include a decreased peak oxygen pulse <80% of predicted, a decreased ventilatory threshold <44% of peak $\dot{V}O_2$ predicted, and a decreased peak $\dot{V}O_2$ predicted <83%. In these individuals, lactate accumulation and increased $\dot{V}_E$ occur with lower levels of exercise (22). In Abu-Hasan's study, 17% of subjects were considered to have decreased conditioning (1), and in the report by Seear et al., 23% of those previously diagnosed with EIB were considered "unfit" (38).

Individuals who are deconditioned or recently have become deconditioned may perceive the increased ventilation at lower exercise intensity as abnormal or as dyspnea. Physical exam findings are essentially normal, with signs of increased work of breathing without hyperventilation, restriction, or obstruction. A maximal graded exercise tolerance test on a cycle ergometer with breath-by-breath gas analysis and heart rate monitoring can help delineate this diagnosis. Physiological data were described previously with an increase in heart rate in response to exercise workload. Recovery data may also be beneficial in establishing a diagnosis of deconditioning, with a slow return to baseline heart rate noted. Treatment includes aerobic exercise training.

Obesity

Exercise-induced dyspnea commonly occurs in obese persons; it has been reported to affect as much as 80% of this population in association with various physical activities (13, 23). Respiratory mechanics are altered in the obese person (33). Chest wall and respiratory system compliance are lower, and airway resistance is higher. Total lung capacity, functional residual capacity, expiratory reserve volume, and maximal voluntary ventilation are reduced, due in part to mechanical forces displacing the diaphragm higher into the chest cavity. An exercise tolerance test with a maximal graded protocol with breath-by-breath gas analysis will assist in this diagnosis. Obese persons commonly exercise at lower lung volumes than those seen in nonobese persons. Measures of mechanical efficiency such as $\dot{V}_E/\dot{V}CO_2$ are decreased in obese children. Total oxygen consumption at peak or anaerobic threshold in obese children has been reported as similar in some studies and decreased in others when compared to nonobese children (32, 37, 49). It appears that absolute $\dot{V}O_2$ in L/min is similar between obese and nonobese individuals but when normalized by weight, the values in obese children are reduced. Corrections of $\dot{V}O_2$

by a fat-free mass factor may be more beneficial in this population, although normative values are not readily available. Following exercise training, exertional dyspnea decreases; fitness (peak $\dot{V}O_2$ and at anaerobic threshold) improves; breathing frequency at peak exercise decreases and efficiency improves (17, 35).

Chest Wall Abnormalities

Abnormalities in the chest wall configuration can result in exercise-induced dyspnea. The deformities include scoliosis, pectus excavatum, pectus carinatum, and other rib anomalies. Depending on the severity of the deformity, a restrictive process can ensue. A comprehensive evaluation of the impact of pectus deformities on exercise is reviewed in chapter 17.

Scoliosis may reduce total lung capacity and vital capacity. The angle of scoliosis deformity (the Cobb angle) has been correlated with the degree of lung impairment for individuals with a thoracic curve. Exercise-induced dyspnea and reduced exercise tolerance are common. With thoracic involvement, nonhomogeneous ventilation may occur with associated $\dot{V}_A/Q'$ mismatch (20). Due to the limited vital capacity, the ability to increase tidal volume with exercise is compromised, with tidal volumes approaching the vital capacity. $\dot{V}O_2$ at peak is lower due to these ventilatory limitations. A high heart rate reserve and low breathing reserve are observed during exercise (24, 39, 45).

Interstitial Lung Disease

While chest wall deformities and obesity are relatively common, other pediatric restrictive lung disease is relatively rare; this includes pediatric interstitial lung disease. While exercise-induced dyspnea is widely recognized for adult restrictive conditions such as idiopathic pulmonary fibrosis, limited data exist in the pediatric population. In adult interstitial lung disease, there is a combination of ventilatory impairment from loss of functioning alveoli and a functionally reduced pulmonary capillary bed, leading to a low $\dot{V}O_2$ peak. The ventilatory abnormalities lead to a high tidal volume-to-inspiratory capacity ratio with associated increased breathing frequencies. Breathing reserves are usually decreased as well. The ensuing ventilation perfusion mismatch leads to increased dead space ventilation. As a result of the reduced capillary bed and the failure to increase pulmonary circulation with exercise, hypoxemia develops with exercise. P_aO_2 decreases

with increasing exercise workload. Arterial and end-tidal CO_2 are no longer similar, with higher arterial levels of CO_2 accumulating, and a gradient develops. Higher ventilatory equivalents are seen at ventilatory threshold. Exercise tolerance testing serves a useful role in both diagnosing interstitial lung disease and monitoring the extent of progression (45). Muscular dystrophy and other myopathies can also be associated with restrictive lung disease due to respiratory muscle weakness.

Evaluation

While there is a wide spectrum of conditions that can result in exercise-induced dyspnea, an organized diagnostic approach can help determine the cause for a given person. Weiss and Rundell have proposed a flow diagram algorithm to address the evaluation process and differential diagnosis for exercise-induced dyspnea (47).

The initial part of any evaluation for exercise-induced dyspnea begins with obtaining a detailed exercise history (see *History Taking for Exercise-Induced Dyspnea*). The elements of an exercise history include duration of symptoms, context of when the symptoms occur, specific sports or activities that are associated with the dyspnea, the intensity required to elicit the symptoms, timing of symptoms, and modifying factors. Combining these elements with other standard aspects of history taking such as a past medical history, current medication usage and allergies, family history, and psychosocial history can assist in focusing on the possible etiology of the dyspnea.

The physical examination in a person with exercise-induced dyspnea is often normal (see *Physical Examination for Exercise-Induced Dyspnea*). However, certain abnormal findings may suggest an underlying chronic condition such as allergies, chronic lung disease, chest wall disorders, or cardiac disease. The relevance of any examination abnormalities in the context of the exercise-induced dyspnea should be considered and appropriate evaluation undertaken.

Performing a resting pulmonary function test (PFT) in all pediatric patients who present with exercise-induced dyspnea is an important part of the evaluation. PFTs include the assessment of lung volumes to determine whether any lung restriction is present, diffusion capacity and oxyhemoglobin saturation by pulse oximetry to assess oxygen delivery, and spirometry and impulse oscillometry to assess for airway obstruction.

History Taking for Exercise-Induced Dyspnea

Duration

How long have the symptoms been present? _____

Context

What were the circumstances around the time when symptoms first occurred? _____
_____

What has happened over time? Better? Worse? _____

Sport or Activities

What sports or activities cause the symptoms to occur? Competitive or recreational? _____
_____

Practices versus competition _____

Early in the activity or later on _____

How often do the symptoms occur with the activity? _____

Intensity

What intensity level is required to elicit the symptoms? _____

Symptoms

What symptoms are present? What is the most common one? _____

Location and description of symptoms. _____

Short of breath, pain, tightness _____

Noises – stridor, wheeze, cough _____

Phase of breathing – inspiratory, expiratory, biphasic _____

Alteration in voice _____

Dizziness, syncope _____

Heartburn, dysgeusia _____

Headaches, muscle cramps _____

Tingling in arms, hands, feet, or legs _____

Tetany _____

Timing

Gradual or abrupt onset _____

During or after exercise_____

At peak, submaximal, warm-up, recovery _____

Duration of symptoms _____

Morning or night _____

(continued)

History Taking for Exercise-Induced Dyspnea (continued)

Timing (continued)

Relationship to eating or drinking _____

Presence or absence of a refractory period_____

Modifying Factors

Impact of temperature (cold vs. hot) _____

_____

Impact of humidity (low vs. high) _____

_____

Impact of rest, fluids, medications, altered breathing_____

_____

History Reports Conducted

☐ Prior evaluation

☐ Medical history

☐ Current medications and allergies

☐ Family history

☐ Psychosocial history

Environment exposures _____

_____

Presence of anxiety, life stressors _____

_____

Academic performance _____

_____

From T.W. Rowland, American College of Sports Medicine, and North American Society for Pediatric Exercise Medicine, 2018, *Cardiopulmonary exercise testing in children and adolescents* (Champaign, IL: Human Kinetics).

Both inspiratory and expiratory portions of the spirometric flow-volume loops should be obtained with the tracings inspected for signs of blunting, truncation, and concavity. Use of a bronchodilator post-testing may help to determine the presence of any reversibility of airway obstruction. Nonexercise diagnostic studies may then be selected based on the initial resting PFT data. Eucapnic hyperventilation, cold air challenges, and methacholine provocation are some of the common techniques used to determine the presence or absence of airway hyperreactivity.

Exercise Testing

The choice of the appropriate exercise test to perform to determine the cause of exercise-induced dyspnea is not well standardized. Different pulmonary physiology laboratories use different testing protocols. Despite this lack of standardization, three basic protocols with varying components emerge as the most common ones used in the pediatric population.

An EIB challenge is typically performed on a treadmill with speed and grade adjusted to

Physical Examination for Exercise-Induced Dyspnea

General

Mood and affect: ☐ anxious ☐ flat ☐ depressed

Body weight: ☐ thin ☐ athletic ☐ overweight ☐ obese

Body type: ☐ Marfanoid

Head, Eyes, Ears, Nose, and Throat

General: ☐ allergic shiners ☐ allergic creases

Nasal mucosa: ☐ pale ☐ inflamed

Orophranyx: ☐ cobblestoning ☐ tonsillar enlargement

Respiratory

Inspection: ☐ chest wall anomalies ☐ accessory muscle use
☐ abdominal breathing ☐ rib expansion ☐ rate of breathing

Palpation: ☐ rib pain ☐ costochondritis

Percussion: ☐ hyperinflation ☐ diaphragm movement

Auscultation: ☐ wheezes ☐ stridor ☐ cough
☐ crackles ☐ inspiratory ☐ expiratory

Cardiac

Heart rate: ☐ fast ☐ slow

☐ Presence of murmurs

Abdomen

☐ Organomegaly ☐ Obesity ☐ Epigastric tenderness

Extremities

☐ Digital clubbing ☐ Perfusion

Musculoskeletal

☐ Scoliosis ☐ Pectus excavatum ☐ Pectus carinatum

☐ Flexibility

From T.W. Rowland, American College of Sports Medicine, and North American Society for Pediatric Exercise Medicine, 2018, *Cardiopulmonary exercise testing in children and adolescents* (Champaign, IL: Human Kinetics).

achieve an asthmagenic state as previously described (14). Spirometry is performed before and after exercise at varying time points. While considered less asthmagenic, a cycle ergometry protocol can be used as an alternative if the treadmill protocol is not practical. Given the degree of hyperpnea generated by this protocol, ATS standards for exhalation may not be reached. It is important to view the tracings of the tidal loops and flow-volume loops from the spirometry, especially during and after exercise, in conjunction with the numeric results.

In contrast to the EIB challenge, where the intensity of the provocation reaches about 70% to 80% of maximal capability, many of the conditions discussed require a higher-intensity provocation to elicit symptoms. The intensity of these protocols can be predetermined (e.g., maximal predicted heart rate) or symptom-limited with the subject having input into the appropriate intensity. The use of a treadmill is more provocative in eliciting symptoms than cycle ergometry, although the choice of modality is often determined based on the clinical situation. Breath-by-breath gas analysis is often used as an adjunct during this high-intensity protocol so that ventilatory, cardiac, and gas exchange parameters can be assessed. These additional parameters are often needed to differentiate the various conditions associated with exercise-induced dyspnea.

A third protocol employs the use of a cycle ergometer with breath-by-breath gas analysis and is performed as a maximal incremental protocol to exhaustion. This is often called an exercise tolerance study. The strength of this exercise protocol compared to the others described is in its ability to directly determine work capacity and analyze the progression of physiological parameters with data generated in a smooth, continuous manner. This type of protocol is generally not symptom limited.

While not comprehensive, normative values exist for most of the parameters generated during an exercise tolerance test.

The use of a flexible laryngoscope can provide information in addition to the data obtained in the preceding protocols. Direct views of the laryngeal structures at rest, during provocative maneuvers, and immediately after exercise can indicate whether obstruction is present or absent. The use of a flexible bronchoscope during an appropriate exercise provocation may be necessary to help diagnose tracheobronchial abnormalities.

Conclusion

Quinta's history supported an upper airway source of her dyspnea. Physical examination was essentially normal at rest, with increased accessory muscle use upon prompted deep breathing. Subsequent PFTs showed a prominent blunting over the inspiratory portion of the flow-volume curve. Dynamic treadmill exercise testing with breath-by-breath gas analysis successfully provoked her symptoms and demonstrated extrathoracic airflow limitation. A flexible laryngoscopy was performed and confirmed inducible laryngeal obstruction at the supraglottic location induced by high-intensity exercise. Multiple therapeutic modalities, including biofeedback, relaxed throat breathing, and counseling were initiated.

In summary, multiple causes of exercise-induced dyspnea exist in the pediatric population. While exercise-induced bronchospasm has traditionally been considered the most common cause of dyspnea with exercise, many other conditions exist that mimic EIB. A careful exercise history and physical examination along with appropriate diagnostic studies can help differentiate these other conditions.

Chest Pain With Exercise

Julie Brothers, MD

Case Presentation

Nathan is a 10-year-old male who plays on the traveling soccer team. He presents with complaints of several episodes of chest discomfort during both practice and games over the past 2 mo. The pain is described as sharp, is located over his sternum, and increases with inspiration. There have been no nausea, dizziness, wheezing, or palpitations. When he has the pain he stops his exercise, the discomfort lasts approximately 30 s, and then he resumes play. He has been seen in the emergency department of the local hospital twice for these episodes, where electrocardiogram, chest X-ray, and blood studies were all normal. His school is requiring that he be seen by a physician for clearance for gym class or sports.

Chest pain at rest or with exertion in children and adolescents is a common presenting complaint to the pediatrician and emergency department and often results in referral to a pediatric cardiologist (6, 11, 35, 37). In fact, there are nearly three-quarters of a million such visits per year for the evaluation of pediatric chest pain (4, 35). Chest discomfort in the young is nearly always noncardiac in origin (33), in contrast to adults, where chest pain is commonly a manifestation of heart disease. Because of this association with cardiac disease in adults, chest pain in a child or adolescent, especially when it occurs during exercise, often causes anxiety and distress in the patient and the parents, and this can lead to unnecessary restriction of physical activities (24). The complaint should therefore be taken seriously, especially when chest pain is associated solely with exertion, in which case it is likely to reflect pulmonary, gastrointestinal, or, rarely, cardiac disease. Exercise testing can be a valuable part of the evaluation of these patients.

A complaint of recurrent episodes of brief chest discomfort occurring at rest is common in children and adolescents and rarely reflects underlying disease. Typically, this idiopathic pain is pleuritic

and is not accompanied by any associated symptoms such as dizziness, palpitations, dyspnea, or nausea. While the origin of this type of pain, often called a precordial catch, is obscure, some have suggested that transient nerve entrapment in the intercostal region could be responsible (30). In some cases, recurrent pain of this nature is believed to have a psychogenic basis (28). In the otherwise healthy patient without associated symptoms, further evaluation for this type of transient chest pain limited to rest is often not necessary. Exercise testing is not commonly indicated for brief chest discomfort that occurs at rest.

Differential Diagnosis

It is more common that an underlying cause can be discovered for chest pain that is triggered by exertion. In the pediatric age group, such episodes can be due to a variety of conditions, which are listed next (31, 34). Findings on exercise testing can be valuable in sorting out this differential diagnosis, particularly with the goal of ruling out the rare but serious cardiac abnormalities that can present with this symptom.

Causes of Chest Pain With Exercise in Children and Adolescents

- Musculoskeletal Abnormalities
  - Costochondritis
  - Idiopathic ("stitch")
  - Chest wall strain or overuse injury
  - Direct trauma or contusion/rib fracture
- Pulmonary Diseases
  - Asthma
  - Vocal cord dysfunction
- Gastrointestinal Abnormalities
  - Gastroesophageal reflux
  - Esophagitis or gastritis
  - Esophageal spasm
- Cardiac Anomalies
  - Coronary artery anomalies: congenital or acquired
  - Hypertrophic cardiomyopathy
  - Pulmonary hypertension
  - Dilated cardiomyopathy
  - Mitral valve prolapse
  - Severe pulmonary stenosis
  - Left ventricular outflow obstruction
  - Tachyarrhythmias
  - Pericarditis
  - Myocarditis
- Other
  - Certain drugs
  - Anxiety

Musculoskeletal Abnormalities

Musculoskeletal disease is considered the most common cause of exertional chest pain in children and adolescents (28, 35). Costochondritis is characterized by pain in the upper two or more costochondral joints; the discomfort is exacerbated by the deep breathing that occurs with exercise. The pain is short-lived. Chest wall tenderness can often be reproduced by palpating the painful area, but inflammation is absent. This pain resolves spontaneously but may recur intermittently throughout adolescence. A stitch is a cramping pain that occurs with exercise, typically located laterally and inferiorly beneath the left rib cage. The discomfort increases with deep breathing and often becomes severe enough to make

exercise impossible. Manual palpation increases pain, but there are no signs of inflammation. Its cause is unknown. Originally considered to be a manifestation of trapped gastrointestinal gas or splenic engorgement, most now believe this pain is related to spasm or strain of musculoskeletal structures surrounding the diaphragm. Chest wall strain or overuse injury mainly occurs in children who participate in gymnastics and weightlifting. This may limit them during exercise and cause localized pain and swelling or erythema where the injury occurred. Xiphodynia is an uncommon cause of musculoskeletal chest pain. The diagnosis is suggested by reproducible pain to light pressure over the xiphoid process (16).

Pulmonary Diseases

Airway issues are a common explanation for chest pain with exercise in youth (6, 26). Exercise-induced asthma is the most frequent among these, even in those without evidence of wheezing or rales. Indeed, in a study by Wiens and colleagues (39), the researchers found that 73% of children who had chest pain with exertion had evidence of exercise-induced asthma, which is a much greater percentage than previously believed. In most cases the chest pain is described as a "tightness" and is associated with cough, wheezing, and excessive breathlessness with exercise. Vocal cord dysfunction arises from abnormal adduction of the vocal cords during exercise and is often misdiagnosed as exercise-induced asthma because children complain of exertional chest pain and shortness of breath. This diagnosis should be suspected if the patient has trouble breathing in (asthma creates difficulty with expiration), exhibits stridor with exercise, or gets no symptom relief from asthma medication. Airway problems with exercise are described in detail in chapters 11 and 13.

Gastrointestinal Abnormalities

Gastrointestinal causes of chest pain occur in about 8% of patients (9). Most commonly, gastroesophageal reflux may induce chest pain with vigorous physical activity, particularly running. The diagnosis should be considered in a patient complaining of midsternal pain with exercise, sometimes accompanied by nausea, cough, vomiting, or an acid taste in the back of the throat. While reflux discomfort has often been described as "burning," this pain can also be sharp. Some patients with this problem will describe a history of "heartburn" following heavy meals.

Cardiac Anomalies

Cardiac causes of exertional chest pain in children are rare (13, 35). Still, for several reasons, ruling out the existence of a cardiovascular abnormality is a central focus in the evaluation of an ostensibly healthy young person with a complaint of chest pain associated with exercise:

1. Whether or not expressed, patients and their families normally have concerns about a cardiac cause because this symptom is an expression of serious heart disease in adults.
2. Cardiac abnormalities that cause chest pain with exercise are important and often life-threatening.
3. Angina symptoms may be the only expression of these rare cardiac abnormalities.

Exertional chest pain in patients with known cardiac disease should be taken seriously, and further evaluation is warranted to investigate the cause.

Coronary Artery Anomalies

Children with congenital coronary artery abnormalities or acquired disease are at risk for myocardial ischemia that can lead to infarction, notably during or just after exercise. The physical examination in these patients is almost always deceptively normal. In those with anomalous left coronary artery from the pulmonary artery (ALCAPA), children will usually have had surgical repair during infancy. However, there are cases where older children or adolescents are not diagnosed during infancy and may present with ischemic chest pain with exertion. Postoperatively, too, these patients remain at risk for exertional chest pain from ischemia due to impaired coronary flow reserve, even in patent grafts, resulting in myocardial ischemia during times of increased oxygen demand (36). Anomalous origin of the left coronary artery (AAOLCA) or the right coronary artery (AAORCA) from the wrong sinus of Valsalva can lead to true anginal chest pain, generally with exertion. These congenital coronary anomalies are the second leading structural cardiac cause of sudden death behind hypertrophic cardiomyopathy. Sudden death with exercise may be the first symptom (10). Such tragedies are hypothesized to occur because of decreased coronary blood flow through the anomalous vessel, resulting in myocardial ischemia or ventricular tachyarrhythmias (21). This diminished blood flow likely results from an anatomical malformation of the anomalous vessel, which may include one or more of the following: acute angle takeoff from the aorta creating a slit-like orifice that easily collapses, the presence of an ostial ridge, or a proximal intramural course that gets compressed within the aortic wall between the great arteries. Kawasaki disease can be complicated by coronary artery aneurysms, which are associated with rupture, thrombosis, or stenosis, all of which can cause myocardial ischemia and chest pain with exercise (32).

Children with genetic dyslipidemias, primarily familial hypercholesterolemia, may experience anginal chest pain with exertion and ultimate myocardial infarction; however, it is extremely rare for true ischemia to occur during childhood and adolescence with the common heterozygote form (approximately one in 250-300 people in the United States). The rare patient (one in 1,000,000) who has homozygous familial hypercholesterolemia is at risk for coronary artery disease in the first two decades of life (20) and could present with exercise-induced chest pain. Coronary vasculopathy can occur any time after heart transplantation, making these children at risk for developing myocardial ischemia. Indeed, chest pain, especially during or just after exercise, can be the first sign of rejection and the start of progressive coronary vasculopathy (15). Children and adolescents who have undergone the arterial switch operation for transposition of the great arteries as neonates are at risk for coronary ostial stenosis as older children and adolescents. This is a rare but extremely serious complication in the years following the procedure (2). Chest pain with exertion may be the first symptom in a patient with a coronary artery occlusion but some may not experience any pain despite coronary artery stenosis or occlusion.

Cardiomyopathies

In children with hypertrophic obstructive cardiomyopathy, characterized by a dramatic thickening of the ventricular walls, chest pain from myocardial ischemia during exercise can be the presenting complaint. This disorder is inherited in an autosomal dominant fashion, and there may be a family history of the cardiomyopathy or of sudden cardiac death. Still, the diagnosis is often missed on routine physical examination. These patients may have a heart murmur that is louder in the standing position or after Valsalva. Similarly, dilated cardiomyopathy may present with exertional chest pain and fatigue; there may also be a family history of this disorder.

Other Abnormalities

Children with mitral valve prolapse may present with chest pain that, while sometimes occurring with exercise, is more commonly nonexertional (1). Patients with pulmonary hypertension can experience chest pain with exercise, and this may be an initial symptom in those with the idiopathic form. Severe pulmonic stenosis may produce squeezing or pressure-type chest pain with exertion that may be caused by true ischemia; these children have almost always been diagnosed before presenting with pain. Left-sided obstructive lesions, including aortic, supraaortic, and subaortic valvular stenosis, can present with exertional chest pain, dizziness, and fatigue. These children typically have a harsh ejection murmur, sometimes accompanied by an ejection click from a bicuspid aortic valve. Children with pericarditis usually experience a sudden onset of sharp chest pain, often during or just after a viral illness. The pain is usually better in the upright position and worse in the supine position and with deep inspiration. Patients may also present with fever, palpitations, and shortness of breath. Children with myocarditis often present with stabbing chest pain that often accompanies pericarditis; this is termed myopericarditis.

Supraventricular and ventricular tachycardia usually present with palpitations that may be exacerbated by exercise, but these arrhythmias can also cause chest pain, often brief and sharp.

Cocaine use or the use or overdose on other sympathomimetic drugs can lead to exertional chest pain, either from arrhythmias or from true myocardial ischemia.

Evaluation

A great deal of insightful information can be gained by taking a carefully obtained, thorough history from a patient who experiences chest pain with exertion (see *Chest Pain History in the Pediatric Patient*). One needs to understand the setting or settings in which the pain occurs (during competitive sports, with certain body position, etc.), its characteristics (location, duration, radiation, severity), and any associated symptoms (dizziness, palpitations, dyspnea, cough, wheezing, vomiting). A complete past medical, surgical, and family history can help with the diagnosis. Genetic disorders, such as Marfan or Turner syndrome, should be considered. Asking about a history of trauma, recent psychological stress, and a history of drug use (inquire without parents in room if possible) may also aid in the diagnosis (12).

Once a thorough history has been performed, a physical examination should be completed. The examination should always begin with a review of vital signs and anthropometric measurements. Extreme height may be indicative of Marfan syndrome. Dysmorphic features, such as those that might cause one to suspect Marfan syndrome, should be noted. Chest wall inspection should include evaluation of past surgical scars, evidence of pectus excavatum or carinatum, or other bony abnormalities. A breast inspection should look for evidence of puberty and any warmth or erythema around the nipple. Palpation of the chest wall should be performed to elicit reproducible pain. Lung examination should focus on crackles, wheezes, or decreased aeration. Cardiovascular examination should focus on rhythm, heart sounds (distant or not), and any clicks, murmurs, gallops, or rubs to suggest a cardiac etiology of the chest pain. Femoral pulses should be palpated to assess for weak pulses or radiofemoral delay (31).

Based on the history and physical examination, a decision can be made about the need for further investigations and tests (25). The extent of such studies will depend on the situation. A patient with known asthma who is experiencing chest tightness with exercise associated with cough and wheezing needs medical treatment but probably no further diagnostic testing. On the other hand, an adolescent distance runner who complains of chest pain associated with palpitations and near-syncope during a competitive race deserves a complete cardiac assessment, which might include echocardiography, electrocardiogram, and treadmill exercise testing. That is, in such a setting, these symptoms could reflect a cardiac disorder, the nature of which would only be revealed through a comprehensive laboratory evaluation.

In a study by Kane et al. (19), the authors concluded that in over 4,400 outpatient visits for chest pain, the 32 patients who had true cardiac pathology would have been diagnosed using a good history, physical examination, and ECG. However, if there is concern about a possible cardiac etiology based on history or abnormal findings on physical examination, an echocardiogram is usually performed to evaluate for structural cardiac disease as the cause of exertional chest pain. This is important because there are some cardiac causes of exertional chest pain, such as coronary anomalies, in which patients have a normal ECG and physical examination, and to

Chest Pain History in the Pediatric Patient

Description

Duration (circle):   acute versus chronic

Frequency of occurrence _____

Length of time pain occurs _____

Location on chest: point with 1 finger

Severity: rank on a scale of 1 to 5 _____

Radiation to other areas (circle):   neck   arm   back   upper abdomen

Occurs with or just after exercise _____

Precipitating factors (circle):   deep inspiration   certain positions   eating

Relieving factors_____

Associated factors (circle):   recent infection   fevers   syncope   nausea

palpitations   headaches

Medical History

Cardiac disease or surgeries_____

Chest or abdominal surgery_____

☐ Asthma

☐ Sickle cell disease

☐ Autoimmune disease

☐ Marfan syndrome

☐ Turner syndrome

☐ Trauma

☐ Drug use and abuse

Recent psychological stressors _____

Psychiatric diagnoses _____

Family History

☐ Sudden cardiac death, especially with exertion

☐ Early death from unknown cause

☐ Cardiac disease

☐ Cardiomyopathy

☐ Arrhythmia

☐ Familial hypercholesterolemia or other genetic dyslipidemia

☐ Marfan syndrome

From T.W. Rowland, American College of Sports Medicine, and North American Society for Pediatric Exercise Medicine, 2018, *Cardiopulmonary exercise testing in children and adolescents* (Champaign, IL: Human Kinetics). Adapted from Reddy 2010.

diagnose the abnormality would require further imaging studies.

Exercise Testing

If the patient's pain is considered to be musculoskeletal in origin, an exercise stress test is unlikely to contribute any further information to the diagnosis. In a study by Kyle et al. of children with structurally normal hearts (22), the authors found that nearly half of the 433 patients referred to the outpatient pediatric cardiology clinic had an exercise stress test performed, and 79% of these patients had chest pain with exercise during the test. There were only four abnormal exercise studies, and none were felt to be related to the chest pain. Of 12 patients who had structural abnormalities found by echocardiogram, 11 had a normal exercise test. Still, demonstrating that measures of blood pressure, heart rate, and electrocardiogram are normal at a time when a patient is experiencing chest discomfort during an exercise test can be helpful, reassuring information.

The exercise stress test is valuable in examining possible cardiac and pulmonary etiologies for chest pain with exercise. Specifically, the assessment of responses of heart rate and blood pressure, electrocardiographic evidence of ischemia, arrhythmias, and exercise-induced asthma can be key elements in the differential diagnosis process (27).

Children and adolescents with certain congenital coronary artery abnormalities, Kawasaki disease, and genetic dyslipidemias are at risk for limitations of coronary artery blood flow and myocardial ischemia during exercise. A maximum exercise stress test in these patients is useful for examining ST segment changes with exercise as evidence of ischemia. Criteria for positive ST segment changes include ST segment depression with a horizontal, upward, or downward sloping ST segment occurring at least 80 ms after the J point, that is, at least 1 mm or more below the baseline, and this has to occur for at least three to five QRS complexes in a row (29). Both false negative and false positive results can occur (7). Certainly, though, any evidence of ST segment depression in children with exertional chest pain warrants further investigation with nuclear imaging, stress echocardiography, or cardiac catheterization.

Hypertrophic cardiomyopathy patients with exertional chest pain should have an exercise stress test to help in the prediction of disease severity, prognosis, and future clinical management. This disease predisposes the patient to myocardial ischemia, arrhythmias, and abnormal blood pressure responses to exercise. Exercise-induced hypotension in patients with hypertrophic cardiomyopathy has been correlated with a high risk for sudden death (14). Thus, close attention should be paid to electrocardiographic ST segments, heart rhythm, and blood pressure during exercise testing.

Significant pulmonary stenosis, in the presence of an intact ventricular septum and normal-functioning tricuspid valve, can lead to abnormalities on the exercise test. In those with exertional chest pain, close attention should be paid to the blood pressure and ST segments because many patients can have elevated systolic pressure with exercise. As the pulmonary valve annulus area decreases, this can lead to elevated right ventricular end-diastolic pressure and diminished stroke volume index; this can lead to myocardial ischemia in the right ventricle. The ECG may show ST segment depression in the mid-precordial and inferior leads (18).

An exercise stress test should be performed if a patient with aortic stenosis (including valvar, subvalvar, and supravalvar obstructions) complains of exertional chest pain, since this can be a harbinger of progression of obstruction. During a maximum exercise test, there should be particular attention on blood pressure, aerobic and physical working capacities, and ST segment changes. Studies have shown that the severity of left ventricular outflow tract obstruction has a negative correlation with exercise tolerance. Systolic blood pressure response to exercise in patients with left ventricular outflow obstruction is also important. In those with mild to moderate stenosis, blood pressure responses are usually normal, but in patients with severe obstruction, the exercise systolic blood pressure may be blunted or even lower than resting blood pressure, a sign of impaired left ventricular performance (17). ST segment depression can also occur due to subendocardial left ventricular ischemia that is caused by a mismatch between myocardial oxygen supply and demand (38).

Children and adolescents with known or suspected supraventricular or ventricular tachycardia who have exertional chest pain, if their hearts are otherwise normal, can be evaluated with a maximum exercise stress test (see chapter 5). Most children with rare or occasional arrhythmias will have a normal heart rate and blood pressure response and a normal working capacity in a maximum exercise stress test. The exercise test may trigger supraventricular or ventricular tachycardia, and this would help in the diagnosis and in evaluating any abnormal physiological responses to the arrhythmia (8).

As described in chapter 13, exercise testing is useful in the evaluation of exertional chest pain in a patient with a structurally normal heart. One should consider the possible diagnosis of exercise-induced asthma or vocal cord dysfunction. An exercise test may be best used to evaluate for these diagnoses by incorporating a specific "exercise-induced asthma" protocol. This protocol starts the exercise abruptly with minimal warm-up, targeting the heart rate at 180 bpm at least and exercising for 6 to 8 min. Pulmonary function tests should be obtained pre-exercise and at 2, 5, 10, 15, 20, and 25 min afterwards. A decrease in the forced expiratory volume in 1 s (FEV_1) or peak expiratory flow (PEF) of at least 15% when compared to pre-exercise value is considered positive for exercise-induced asthma (39). With this protocol, vocal cord dysfunction can be diagnosed in two ways: truncation of the inspiratory flow loops and documentation of abnormal adduction of vocal cords using flexible fiber-optic rhinolaryngoscopy while symptomatic (23).

Exercise-induced gastroesophageal reflux is generally a diagnosis by exclusion in the exercise testing laboratory, since all physiological and electrocardiographic findings will be normal. However, this diagnosis should be entertained in any patient who exhibits sharp retrosternal discomfort during exercise testing with such normal findings. Further assessment by a gastroenterologist may be considered, which could include esophageal pH probe recordings during exercise.

Conclusion

Complaints of chest pain are common in children but only very rarely reflect underlying cardiac disease. Most otherwise healthy young patients whose intermittent episodes of pain occur only at rest do not need exercise testing. Exercise testing should be considered for those who experience chest discomfort with exercise, however, as a means of discriminating between musculoskeletal, pulmonary, gastroesophageal, and cardiac causes.

Nathan was referred to a pediatric cardiologist. The physical examination was normal, without any murmurs heard. There was no chest wall pain to palpation or inspiration. Further questioning revealed he had used an albuterol nebulizer as a toddler when he had upper respiratory tract infections, but his mom felt he had outgrown this, and he has not used albuterol for the past several years.

Although Nathan was referred to a pediatric cardiologist, the chest pain was noncardiac and occurred solely with exercise, so he was referred to a pediatric pulmonologist for further evaluation. Given the past history of possible asthma, an exercise stress test was ordered using an exercise-induced asthma protocol. This showed normal baseline pulmonary function tests. During exercise, he experienced chest pain and shortness of breath, which improved after 2 to 3 min in recovery. His pulmonary function tests showed a 22% decrease at 10 min after exercise. He was given albuterol, and the pulmonary function tests then improved back to baseline and his symptoms improved. He was diagnosed with exercise-induced asthma and was given albuterol to be taken prior to exercise. His chest pain did not recur.

Nathan's case study demonstrates the importance of taking a good history and tailoring the evaluation to each patient.

Presyncope and Syncope With Exercise

Julie Brothers, MD

Case Presentation

Rebecca is a 17-year-old varsity cross country runner whom you have followed in your pediatric practice since she was a young child. She presents to you today with complaints of multiple episodes of dizziness with exertion, the first event having started approximately 1 yr ago. She describes the dizziness as "spinning," and it occurs not during running but just after she has finished. It usually occurs after running at a high-tempo pace for several miles and usually in the summer months. She denies frank syncope, chest pain, shortness of breath, or palpitations. There was a documented blood pressure of 62/40 at one time when paramedics were called while the event was occurring. She has been seen in the emergency room twice for the complaint; the most recent was 2 d ago. In the emergency room, she had a normal heart rate and blood pressure, normal physical examination, and normal laboratory values for anemia and thyroid. Her chest radiograph and electrocardiogram (ECG) were unremarkable. She has been restricted from exercise until she is seen by you. She is requesting exercise clearance.

Syncope is a pathophysiologic process whereby a person experiences loss of consciousness and postural tone as a result of cerebral hypoperfusion. The events are short-lived, and the person awakens spontaneously when blood pressure increases. There are no neurologic sequelae (16, 17). Presyncope is defined as a sensation of light-headedness that precedes or nearly results in collapse but does not result in actual loss of consciousness (46). If the person assumes the supine position when this sensation occurs, true syncope often is avoided.

Syncope with or without exertion is quite common in the general population, occurring in up to 40% of people over a lifetime (33, 36). In fact, syncope accounts for anywhere between 1% and 3% of all emergency room visits (31, 33). Most cases of syncope and presyncope are benign (8), including those that occur at rest and after the cessation of exercise (2, 24). These episodes are generally known as reflex or vasovagal syncope (17). However, exertional syncope or presyncope may be caused by an underlying cardiac disorder and may be related to future sudden cardiac death (32, 39, 47). In a study by Colvicchi and colleagues in Italy, of 7,568 athletes undergoing preparticipa-

tion screening, 6.2% (11) had experienced syncope within the past 5 yr. Of those, most were not related to exercise, 12% experienced postexercise syncope, and 1.3% had syncope with exercise. The latter group was ultimately diagnosed with right ventricular outflow tract tachycardia, hypertrophic cardiomyopathy, and exercise-induced neutrally mediated syncope (10).

Differential Diagnosis

The following list presents the main causes of syncope and presyncope. This chapter will focus on exertional causes of syncope and presyncope, which can be separated into noncardiac and cardiac causes. In determining the cause of exertional syncope or presyncope, a thorough history is extremely important, and it should focus on symptoms before, during, and after the event (7, 36, 44). In general, syncope that occurs during exercise is a greater cause for concern than syncope that occurs after exercise, and it has been linked to several cardiac causes (5, 24, 25, 43). Therefore, anyone presenting with presyncope or syncope with exercise should have a complete work-up and should be restricted from athletics until a complete evaluation has been performed (14).

Noncardiac Causes of Exertional Syncope and Presyncope

- Vasovagal syncope
- Postexertion collapse
- Hyperthermia or dehydration
- Hyponatremia
- Hyperventilation
- Seizure disorder

Cardiac Causes of Exertional Syncope and Presyncope

- Structural heart defects
  - Left ventricular outflow tract obstruction
    - o Hypertrophic cardiomyopathy
    - o Severe aortic stenosis
  - Right ventricular outflow tract obstruction
    - o Severe pulmonic stenosis
    - o Primary pulmonary hypertension
  - Coronary artery anomalies
  - Arrhythmogenic right ventricular cardiomyopathy
  - Dilated cardiomyopathy
- Commotio cordis
- Arrhythmias
  - Bradyarrhythmias: Mobitz type II and complete heart block
  - Tachyarrhythmias: supraventricular tachycardia
  - Long QT syndrome
  - Brugada syndrome
  - Ventricular tachycardia

Noncardiac Causes

Most exercise-induced syncope is benign, although any young person with syncope during or after exertion, as opposed to those who experience syncope at rest, should be evaluated. A thorough history and physical examination are extremely important in helping to differentiate between noncardiac and potential cardiac causes. This will also influence what additional testing is recommended for further evaluation.

Vasovagal Syncope

Vasovagal syncope is sometimes called vasodepressor or neurally mediated syncope. This form of syncope usually occurs after exercise, not during it, and it often follows an abrupt cessation of activity, such as during a huddle in a football game or at the end of a running race (2, 21, 22). Vasovagal syncope occurs when the person is upright. Symptoms include palpitations, lightheadedness, warmth, and nausea. The person falls to the ground slowly in contrast to a quick drop. If patients can assume a supine position, they may often avoid complete syncope (23). Most people awaken rapidly (5-30 s) but may continue to feel fatigued, sometimes for hours after the event (2, 23). The pathophysiology of postexercise vasovagal syncope is not completely understood but is likely a combination of mechanisms, including metabolic, neural, and flow mediated (9, 49).

Postexertion Collapse

Postexertion collapse is a syndrome that generally occurs after significant exertion, commonly after a competitive race (26). Like vasovagal syncope, patients slump to the ground and protect themselves; however, the distinction between the two is that in postexertion collapse, patients do not lose consciousness (26). In one study, 59% of marathon runners who sought medical attention had experienced postexertion collapse over a 12 yr study period (48). It was previously thought that postexercise collapse was due to biochemical derangements, such as dehydration, hyperthermia, and hypernatremia. However, in a study by Holtzhausen and colleagues, the researchers found no differences in body temperature, plasma volume, or sodium levels between those who collapsed and control runners who did not collapse during that race (26). This entity is thought to be related, in part, to the increase in heart rate, contractility, and stroke volume as well as skeletal muscle vasodilation that occurs with exercise. During exercise, cardiac output increases and is dependent on preload, and the peripheral muscle activity must help deliver venous blood back to the heart. When vigorous exercise stops suddenly, the peripheral resistance is still low, causing pooling of blood in the lower extremities and decreased venous return to the heart; this leads to reduced stroke volume and cardiac output (2, 24, 26). Postexertion collapse is benign and should be differentiated from other significant causes of syncope during or after exercise.

Hyperthermia or Dehydration

Heat-related syncope is considered part of a spectrum called heat illness and is diagnosed by a core temperature of at least 40 °C in conjunction with central nervous system dysfunction (2, 45). This form of syncope and presyncope most likely occurs during prolonged exercise in the heat. These individuals can experience syncope either during or after exercise. Similar to postexertion collapse, they may not lose consciousness. They often have symptoms prior to syncope of confusion and delirium, and this may continue until the hyperthermia is reversed. Immediate medical attention is necessary (10).

Dehydration is usually found with heat-related syncope, but it can occur without hot climate conditions and without coexistent central nervous dysfunction. It usually occurs during warm temperatures (e.g., playing football twice daily at the end of the summer in full uniform), but it may also be due to limited water intake, either with warm climates or with inadequate fluid intake during exercising. For children, limited fluid intake may be by their own choice (e.g., the patient may not like to urinate during school hours so he or she does not drink all day) or by athletic coaches who may limit fluid breaks during practice. Also, some people sweat heavily and are not adequately replacing their insensible losses. The diagnosis of dehydration is suggested when the history reveals a limited amount of fluid intake or few trips to the restroom during the day, hot weather conditions with a significant amount of exercise performed, or documenting orthostatic vital signs on examination.

Hyponatremia

Similar to heat-related syncope, hyponatremia-related presyncope or syncope is likely to occur during prolonged exercise in the heat. This form of exertional syncope occurs due to excessive sweating with sodium loss through the sweat while drinking water, which may exacerbate the hyponatremia (50). The hyponatremia can potentially lead to central nervous system derangements and occasionally presyncope or syncope (2). Immediate medical attention is needed to prevent serious adverse events (10).

Hyperventilation

Hyperventilation during exercise, especially during competition, can lead to dizziness and, occasionally, overt syncope. This is generally psy-chogenic in nature. Clues to this diagnosis include overbreathing or air hunger, light-headedness, or mouth dryness in association with paresthesias of the face, hands (fingers), and feet (toes).

Seizure Disorder

Seizures, or tonic-clonic activity, may occur whenever there is decreased blood pressure and the cerebral perfusion drops. If a "seizure" occurs during exertional syncope, it does not necessarily mean that epilepsy is the true cause of syncope. Other clues, such as urinary incontinence and a prolonged ictal state, should clue the practitioner in to the possibility of a seizure disorder as the cause. If this is suspected based on the history, a referral to a neurologist and a possible electroencephalogram, or brain imaging, is warranted.

Cardiac Causes

Cardiac causes of exertional presyncope and syncope account for approximately 1% of syncope in athletes and may be due to structural or arrhythmogenic causes (11). With syncope from a cardiac cause, there are generally minimal or no prodromal symptoms; if there is a prodrome, it may be light-headedness or vision changes. Because unconsciousness often happens before collapse, the patient is at risk of injury. Generally, episodes are brief, and the patient usually feels well after the event and may not seek medical attention (2).

Left Ventricular Outflow Tract Obstruction

Left ventricular outflow tract obstruction is associated with exercise-induced syncope. Hypertrophic cardiomyopathy (HCM), also known as idiopathic hypertrophic subaortic stenosis (IHSS), is the leading cause of sudden cardiac death in athletes (43), and this diagnosis should always be considered in any exercise-associated syncope. HCM is characterized by a thickened left ventricle, often with a disproportionately thick septum. With HCM, exercise-induced syncope occurs through two potential mechanisms: structural and arrhythmic. During exercise, the dynamic outflow gradient (increases with increased contractility or decreased preload or afterload) can become great enough to actually obstruct forward flow, leading to syncope (2). HCM also places the patient at risk for ventricular arrhythmias, especially ventricular tachycardia, which can lead to exercise-induced syncope (2). In those with severe aortic valve stenosis, syncope with exertion is associated with a more serious prognosis (13). Exercise-induced syncope in severe aortic stenosis is thought to be caused by at least one of the following: ventricular arrhythmias, inability to generate sufficient cardiac output due to the fixed obstruction, or an imbalance of myocardial oxygen supply and demand during exercise (23).

Right Ventricular Outflow Tract Obstruction

Severe pulmonic stenosis is a cause of exercise-induced syncope; however, because most children and adolescents will have had surgery or balloon catheter dilation of the stenosis, this is an extremely rare cause of syncope. Syncope with exercise can be an indicator of previously undiagnosed primary pulmonary hypertension. Initial symptoms may be shortness of breath on exertion; however, because the pulmonary vascular bed is constricted, these patients cannot adequately increase their cardiac output during exercise and may present with exertional presyncope or syncope (23).

Coronary Artery Anomalies

Coronary artery anomalies, specifically those in which one or both coronary arteries arise anomalously from the wrong aortic sinus, are the second leading structural cause of sudden cardiac death in young persons in the United States, comprising up to 17% of cases (5, 39, 40, 55). The highest-risk lesion is an anomalous left coronary artery from the right sinus of Valsalva with an interarterial, intramural course (3, 5). Sudden cardiac death is hypothesized to occur from decreased blood flow through the anomalous coronary, resulting in myocardial ischemia or ventricular tachyarrhythmias. These probably result from an anatomical malformation, such as a slit-like orifice, an initial course in the wall of the aorta that becomes compressed, or an ostial ridge (3, 51, 54). Patients are often asymptomatic, but anyone with presyncope or syncope in association with exercise should be evaluated for an anomalous coronary artery because these symptoms could be predictive of a future risk of sudden death (3).

Arrhythmogenic Right Ventricular Cardiomyopathy

Arrhythmogenic right ventricular cardiomyopathy (ARVC) or arrhythmogenic right ventricular dysplasia (ARVD) is a rare, genetically inherited myocardial disease that can be associated with fatal arrhythmias. It is the cause of 4% of all cases of sudden cardiac death in athletes in the United States (12) and 22% of sudden cardiac death in athletes worldwide, with an increased prevalence in northeast Italy (4). A family history of syncope, ventricular tachycardia, or sudden death, notably in young males during exercise, should raise the question of ARVC (15). In ARVC, the normal heart tissue is progressively replaced with fat and fibrous tissue, almost always in the right ventricle. While this tends to occur predominantly in the right ventricle, there may be left ventricular involvement as well. The abnormal myocardium predisposes these patients to ventricular tachycardia that originates from the right ventricular outflow tract. Presyncope and syncope with exertion can occur due to the ventricular tachycardia. Diagnosis is suggested by the electrocardiogram, which often shows T wave inversion in the right precordial leads, V_1 to V_3, in adolescents and adults (38).

Dilated Cardiomyopathy

Dilated cardiomyopathy usually has been diagnosed prior to exercise-induced symptoms because patients generally have signs and symptoms of congestive heart failure. However, syncope associated with dilated cardiomyopathy is associated with increased mortality (52). Presyncope and syncope are likely caused by runs of ventricular tachycardia; sudden death can occur if these do not terminate (54).

Commotio Cordis

Commotio cordis is one of the leading causes of sudden cardiac death in young athletes and is most common in males involved in high-impact sports, such as baseball, boxing, and hockey (2, 42). Following a blunt, nonpenetrating trauma to the chest, a nonsustained ventricular arrhythmia can occur, causing syncope. If the impact leads to ventricular fibrillation, then sudden cardiac arrest is often the result. Thankfully this is a rare event; multiple variables need to be present in order for a potentially lethal arrhythmia to ensue. At time of impact, these factors include location and rate of speed, object shape and firmness, and timing in the cardiac cycle (36, 37).

Mobitz Type II and Complete Heart Block

When bradyarrhythmias cause exertional symptoms, it is usually from an acute episode and not a chronic condition. First-degree atrioventricular (AV) block and Mobitz type I (or Wenckebach) are considered benign conditions that may be found in athletic children and adolescents and do not cause exertional presyncope or syncope (57). However, higher-grade block, including Mobitz type II and complete heart block, are not normal findings and need further evaluation. Presyncope and syncope during exercise may occur because the heart rate cannot increase appropriately, leading to decreased cardiac output. Some people with congenital complete heart block may have an adequate junctional escape rhythm; however, if they become symptomatic with syncope or ventricular ectopy, then pacing is required (57).

Supraventricular Tachycardia

Supraventricular tachycardia rarely causes exertional syncope but may cause presyncope, especially if there is a drop in blood pressure with the tachycardia (36). Much more often, the patient experiences palpitations. The two most common supraventricular tachycardias are AV nodal reentrant tachycardia and Wolff-Parkinson-White (WPW) syndrome, which is an atrioventricular reciprocating tachycardia. WPW can be diagnosed based on a resting ECG that shows a short PR interval and a QRS that has a slurred initial portion (delta wave). Both types of supraventricular tachycardia are amenable to oral beta-blockade as well as radiofrequency catheter ablation (36).

Long QT Syndrome

Long QT syndrome (LQTS) is a group of genetically inherited disorders that can be responsible for sudden death in young athletes (41). It is usually diagnosed by finding a significantly prolonged QT interval on a baseline ECG, often >500 ms. LQTS affects the cardiac ion channels, resulting in prolonged ventricular repolarization. This can lead to rapid ventricular tachycardia, often in a polymorphic pattern known as torsades de pointes (19). Patients may have presyncope or syncope, often during exertion, which is generally caused by the development of torsades de pointes (19, 29). If the ECG does not show QT interval prolongation or only mild prolongation, then a high index of suspicion is necessary for LQTS, and further questioning should focus on family history, symptoms, and the ECG. All patients who present with exercise-induced presyncope or syncope should have an ECG with the QT interval, or corrected QT interval, noted.

Brugada Syndrome

Brugada syndrome is almost always a genetically inherited condition that can cause syncope, either with exertion or at rest, and sudden cardiac death.

It can often be diagnosed by ECG abnormalities, including ST elevation in leads V_1 to V_3 with a right bundle branch block appearance; a prolonged PR interval may also be seen (56). Unlike ventricular arrhythmias that have other causes, those caused by Brugada syndrome are increased by vagal tone, so events may happen during sleep; hyperthermia also induces arrhythmias (1). Athletic young people may be more prone to exercise-induced syncope caused by ventricular arrhythmias because of their higher vagal tone and because of body temperature increases during exercise (1).

Right Ventricular Outflow Tract Tachycardia

Right ventricular outflow tract tachycardia is the most common type of ventricular tachycardia in athletic patients, and it may present as presyncope and syncope during exercise. Sudden death is rare, however. The mechanism is considered to be intracellular calcium overload; because exercise increases levels of cyclic adenosine monophosphate, this leads to an increase in intracellular calcium, which can lead to ventricular tachycardia that may be sustained (35).

Catecholaminergic Polymorphic Ventricular Tachycardia

Catecholaminergic polymorphic ventricular tachycardia (CPVT) is a genetically inherited disorder that can cause recurrent exertional presyncope and syncope with a high rate of sudden cardiac death (34). It is caused by cardiac electrical instability through activation of the sympathetic nervous system. The average age of diagnosis is between 7 and 9. Because the ECG is usually normal, a high index of suspicion for CPVT is warranted for anyone with recurrent exertional presyncope or syncope at a young age.

Evaluation

Any young person with presyncope or syncope during or just after exercise should be seen by a physician, and a detailed history and physical examination should be performed to help elucidate the cause (table 15.1) (24). One of the main elements in the history to clarify is whether the event occurred during or immediately after exercise. Questions should focus on the presence of a prodrome and whether the loss of consciousness was abrupt or gradual. Generally, a prodrome of presyncope, warmth, and nausea may be neurally mediated, whereas a lack of prodrome or an abrupt

Table 15.1   Causes of Exertional Presyncope or Syncope in the Young

	Vasovagal	Postexertion collapse	Hyperthermia or dehydration	Arrhythmic
Association with exercise	After activity while standing	After extreme exertion	After extreme exertion, hot weather	With exertion
Recurrence	Frequent	Rare	Rare	Rare
Prodrome	Warmth, dizziness, nausea	Dizziness, tiredness	Confusion	Usually none
Injury	Rare	Rare	Rare	Common
Postevent symptoms	Fatigue	Fatigue	Confusion	None
Cardiac disease	Rare	Rare	Rare	Common

Adapted from Bader and Link 2013.

onset of palpitations may be arrhythmic. If the patient was injured while fainting with limited prodrome, it is more likely cardiac and less likely benign. If there is a post-ictal state with bowel or bladder incontinence, then a seizure should be high in the differential diagnosis. Fatigue after the event can be caused by vasovagal syncope or postexertion collapse. Confusion or delirium as a prodrome that persists may be due to hyperthermia or hyponatremia. A history of chest wall trauma may suggest commotio cordis. Long QT syndrome is suggested if the event happened with loud noises or upon entering cold water.

Further insight into the event can be determined by other aspects of the patient's history. A story of recurrent exertional presyncope and light-headedness is less likely to be cardiac in origin than occasional true syncope with exercise. A medical history should be obtained, with any chronic diseases and any medications noted, including prescription and over-the-counter. The patient should be asked about use of illicit drugs and performance-enhancing substances; this should be done without the parents in the room, if possible. A careful family history should include questions about sudden cardiac death or any unexplained deaths in the young as well as death from a "heart attack" or "seizure" at a young age, which may have been an arrhythmia that was misdiagnosed (24).

A physical examination should be performed, including vital signs with orthostatic blood pressure and heart rates. Close attention should be paid to the cardiac examination, listening for murmurs, especially those that increase with standing or during Valsalva (hypertrophic cardiomyopathy). Carotid and radial pulses should be felt for a bifid pulse seen with hypertrophic car-

diomyopathy or a slow rising pulse of aortic stenosis. Delay between carotid and femoral or weak femoral pulses is suggestive of aortic coarctation.

After a thorough history and physical examination are complete, decisions should be made about further diagnostic evaluation. In patients with classic vasovagal syncope, postexertion collapse, or hyperthermia or hyponatremia after exercise, no further testing is necessary. However, for those whose symptoms may have a cardiac cause, further diagnostic testing should be considered, including a 12-lead electrocardiogram, echocardiogram, and exercise stress test (EST). An ECG can be useful for diagnosing many arrhythmic causes, including long QT syndrome, Brugada syndrome, WPW, heart block, and ARVC. ECG abnormalities are noted in the majority of patients with hypertrophic cardiomyopathy, including ST segment depression, left ventricular hypertrophy, left axis deviation, and abnormal Q waves (18). While an ECG may be helpful in noting some major causes of exertional presyncope and syncope, the ECG is generally normal in patients with coronary artery anomalies and CPVT. Because the ECG is only a screening tool, for anyone whose condition is suspected of having a cardiac cause, an echocardiogram should be obtained to look for structural and functional contributions to the patient's symptoms. An exercise stress test may also be performed as part of the patient's work-up, especially if the symptoms are occurring in association with exercise.

Exercise Testing

In those patients with exertional presyncope or syncope, the exercise test is useful for trying to reproduce the patient's symptoms and evaluating whether the symptoms correlate with any blood pressure or electrocardiographic abnormalities. It can also be used to demonstrate arrhythmias with exercise and evidence of ischemia in those patients where this is a concern. However, the exercise test is short and probably at a different level of intensity than the activity in which the patient experienced symptoms. The test can be adapted, when possible, to simulate the patient's usual exercise to help elicit the symptoms. For instance, if the syncope only occurs after a long, competitive race, then the patient could run on the treadmill for a distance and speed similar to the race pace; the person who experiences symptoms while sprinting might undergo repetitive sprints during testing (57). Because the exercise test has a low sensitivity for arrhythmic causes of syncope, a normal test may not preclude the need for other investigations (30). If the symptoms cannot be elicited during the exercise test, it may be necessary to monitor the patient in other ways, such as with a looping event recorder (53).

For patients with known or suspected hypertrophic cardiomyopathy who report exertional presyncope or syncope, the exercise stress test can provide valuable information. Because hypertrophic cardiomyopathy can cause myocardial ischemia, arrhythmias, and abnormal blood pressure responses to exercise (20), any one of these can be the cause of the patient's symptoms. Exercise-induced hypotension, which can lead to exertional presyncope or syncope, correlates with a risk of sudden death in patients with this diagnosis (20).

Severe left ventricular outflow obstruction (aortic stenosis) can lead to exertional symptoms. During the exercise test in these patients, attention should be paid to blood pressure response, physical working and aerobic capacities, and ECG abnormalities. With severe aortic stenosis, physical working capacity may be markedly diminished (27). Systolic blood pressure also may fall, leading to dizziness or frank syncope. Ischemia leading to ST segment depression greater than 2 mm during exercise in those with severe stenosis may also be the cause of the patient's symptoms.

Patients with severe pulmonary stenosis may be noted to have suprasystemic right ventricular pressures and elevated systolic blood pressure during exercise as well as decreased stroke volume index, which can lead to potential ischemic changes in the right ventricle. This can cause exercise-induced symptoms. Pay particular attention to the blood pressure and ECG, notably looking for ST segment depression in the mid-precordial and inferior ECG leads (28). Exercise testing is important in children with pulmonary hypertension, especially if they are experiencing exercise-induced presyncope or syncope. The test should focus on oxygen saturation at rest and during exercise, blood pressure response to exercise, and aerobic performance. Some children with moderate to severe pulmonary hypertension may benefit from a six-minute walk test instead of a traditional cycle or treadmill test.

If a patient with complete heart block is experiencing exertional symptoms of presyncope or syncope, an EST is warranted to evaluate the patient's exercise performance, heart rate response, focusing on the atrial rate, and ventricular ectopy. However, a patient with exertional presyncope or syncope with congenital or acquired complete heart block also must be considered for a pacemaker placement before further exercise can be allowed (57). Exercise testing in children and adolescents with coronary artery anomalies is important because they are at risk for myocardial ischemia during exercise, which may show itself as presyncope or syncope. Close attention should be paid to ST segment changes with exercise as evidence of ischemia; criteria for electrocardiographic evidence of ischemia were discussed in chapter 5.

Ventricular tachyarrhythmias may be noted during the exercise stress test. The cause of the patient's exertional symptoms may be CPVT, which can be induced during exercise.

An exercise test is important for patients with suspected or known LQTS, especially if they have had exertional symptoms. During exercise, changes in the QTc interval should be noted as well as the appearance of any dysrhythmias that might cause the patient's symptoms. The corrected QT interval as defined by Bazett's formula (6) is

QTc = QT interval divided by the square root of the preceding RR interval in seconds

For example, if the QT interval is 0.400 s and the preceding RR interval is 0.880 s, then

$$QTc = \frac{0.400}{\sqrt{0.880}}$$
$$QTc = 0.426 \text{ s}$$

The exercise test is useful in measuring the QTc response to exercise, evaluating the chronotropic response both before and after starting oral beta-blockade, and identifying any ventricular arrhythmias. In people without LQTS, the QT interval should shorten with exercise. In patients with LQT1, the QT interval often does not demonstrate such shortening, and there is pronounced lengthening in recovery when the heart rate decreases. They also often demonstrate chronotropic impairment. Those with LQT2 tend to have a decrease in the QT interval with exercise but a significant lengthening during late recovery; they have a normal chronotropic response to exercise. Patients who have LQT3 have a significant decrease in their QT interval with exercise when compared both to other patients with LQTS and to controls (2) (see chapter 5).

Conclusion

A young patient who experiences exertional presyncope or syncope deserves a careful medical evaluation. A thorough history and physical examination should help to distinguish those with benign vasovagal syncope, postexertion collapse, or exercise-induced hyperthermia or hyponatremia from those in whom a cardiac cause is suspected and who need further diagnostic testing. An EST should be part of that evaluation because it can provide valuable information.

Turning back to our case study, on further questioning it is revealed that Rebecca drinks 20 oz of water daily and one cup of coffee in the morning. She often skips breakfast as well because she does not have time to eat. She sometimes experiences dizziness in the morning when she stands up while getting out of bed. Rebecca's physical examination included orthostatic vital signs. Her heart rate increased from 52 bpm supine to 90 bpm in the upright position. Her blood pressure decreased from 116/68 to 108/64 from supine to upright position, respectively. Her physical examination was otherwise unremarkable.

Rebecca had an ECG performed, which was normal. Although her symptoms were classic for dehydration as well as a vasovagal component, because she was a competitive athlete, the cardiologist opted for an echocardiogram, which was normal, and an exercise stress test to try to induce her symptoms.

Rebecca performed a maximal treadmill exercise stress test. She had a high-normal aerobic capacity. There were no arrhythmias or ischemic changes. There was a normal heart rate and blood pressure response to exercise. She developed dizziness in early recovery with a 30 mmHg drop in her blood pressure. After she drank a cup of water and was placed in the supine position, her blood pressure improved within 2 min.

For Rebecca, her history, physical examination, and ECG were reassuring that this was unlikely to be cardiac-related exertional syncope. Her exercise stress test was able to document dizziness associated with a drop in blood pressure that responded to fluids and supine positioning. She was counseled on improving her hydration and salt intake and not to skip meals. She returned for a follow-up visit 3 mo later and had not experienced any further dizziness or syncope.

Exercise Fatigue

Thomas W. Rowland, MD

Case Presentation

Sixteen-year-old Joseph has been swimming competitively since age 7. His performance improvement has been steady in the past but times in competitive meets this year have been deteriorating. Compared to the previous winter, his time has increased from 2:19.6 to 2:35.3 min in the 200 yd individual medley and from 24.3 to 28.8 s in the 50 yd freestyle. Joseph complains of marked muscle fatigue but denies shortness of breath or other symptoms. During the summer he maintained his training by bicycling 50 to 60 mi each week and exercising on a stationary cross-country skiing machine for an equivalent of 14 mi.

There is hardly a body system that does not contribute in some manner to exercise performance: The synchrony of muscle fiber contraction. The chain of elements required for oxygen delivery. The efficiency of the muscle cell's metabolic machinery. Provision for energy substrate. Neural electrical transmission by virtually instantaneous ion transfer across cell membranes. All of this regulated by extremely accurate "body clocks." Within the brain, complex neural connections permit visual-motor tracing, strategizing, and motivation. During all this, homeostasis must be maintained. So we have mechanisms for thermoregulation, the preservation of fluid–electrolyte balance, and the maintenance of blood glucose.

It is not hard to see that a breakdown in the effectiveness of any of the myriad components of this exercise "machine" will be expressed as a limitation of physical capacity. The proverbial chain is only as strong as its weakest link. Thus, when trying to identify any malfunctioning part, one must consider a range of diagnostic possibilities. Still, given an appropriate evaluation of patients with exercise intolerance, the clinician can often identify its underlying cause. Often, therapeutic interventions can successfully restore normal exercise capacity.

This chapter will examine the differential diagnosis and appropriate evaluation of the young patient who describes unexpected ease of fatigue with exercise. The discussion will then focus on the role of exercise testing in identifying the severity and etiology of exercise intolerance.

Differential Diagnosis

Fatigue with exercise that is considered severe enough to seek medical attention deserves a careful and thorough assessment. From the listing of diagnostic possibilities that follows, we can see that a wide range of illnesses can manifest as exercise intolerance. Thus the importance of this evaluation lies not only in restoring exercise capacity but also in ruling out significant underlying medical disorders.

Causes of Exercise Fatigue

- Cardiac abnormalities
- Pulmonary disease
- Anemia and iron deficiency
- Changes in body weight
- Post-viral infections
- Sedentary lifestyle or genetic limitations
- Emotional difficulties
- Medications
- Chronic disease
- Overtraining (athlete burnout)

Cardiac Abnormalities

Ease of fatigue with exercise is a common final symptomatic pathway for all forms of heart disease. Patients with dilated cardiomyopathies are limited by low cardiac output and pulmonary venous congestion, while those with hypertrophic cardiomyopathy often are limited with exercise by the diminished stroke volume related to a small left ventricular cavity as well as diastolic dysfunction. Among the forms of congenital heart disease, those creating large left-to-right shunts (such as ventricular septal defect or patent ductus arteriosus) decrease lung compliance via increased pulmonary blood flow as well as pulmonary hypertension. These are anomalies usually detected early in life by other findings, but certain anomalies, such as a large atrial septal defect or partial anomalous pulmonary venous return, may remain occult during childhood.

Significant obstructive lesions (such as aortic or pulmonary valve stenosis) limit stroke volume during physical activities. Right-to-left shunts are revealed by arterial desaturation that typically worsens with exercise, limiting performance by restricting oxygen delivery. Patients with third-degree (complete) heart block or sinus node dysfunction may not be capable of fully compensating for bradycardic responses to exercise by augmenting stroke volume.

Exercise-induced arrhythmias are unusual in young subjects, especially in the absence of underlying structural cardiac abnormalities. Still, easy fatigability and shortness of breath with physical activities can be the expression of episodes of atrial fibrillation, atrial flutter, supraventricular tachycardia, or ventricular tachyarryhthmias.

Pulmonary Disease

Young patients often cannot distinguish "shortness of breath" from "abnormal ease of fatigue" with exercise. The clinical approach to the patient whose complaint is specifically about respiratory limitations with exercise is outlined in chapter 13. Often, however, the clinician must approach the patient from a more global perspective of the differential diagnosis of exercise fatigue. Patients with concerns about ease of fatigue with exercise may be experiencing their limitation on a pulmonary basis. To review the discussion in chapter 13, then, respiratory limitations from problems such as exercise-induced asthma, vocal cord dysfunction, anatomic airway obstruction, and hyperventilation often need to be considered in the diagnostic evaluation, particularly if shortness of breath is a component of the patient's report about ease of fatigue with exercise.

Anemia and Iron Deficiency

Anemia, a depressed concentration of blood hemoglobin, has long been recognized to impair exercise performance. Oxygen delivery to exercising muscles is progressively impaired as hemoglobin level falls; even decrements in hemoglobin concentration as small as 1 g/L have been recognized to cause fatigue with exercise. Most cases of anemia in young persons relate to iron deficiency. Identification and treatment of anemia as the cause of exercise intolerance is important because such symptoms will normally resolve with the restoration of normal hemoglobin levels.

Overt anemia is a late sign of the depletion of body's iron stores, and it is possible to have significant iron deficiency (as indicated by a depressed level of serum ferritin, the storage form of iron) with normal hemoglobin levels. In animals, this condition of non-anemic iron deficiency has been clearly demonstrated to impair endurance exercise capacity because of the lack of iron needed for the function of the metabolic pathways (Krebs cycle, cytochrome chain) within the skeletal muscle cell.

Non-anemic iron deficiency (low serum ferritin level with normal hemoglobin concentration) is common in female adolescent endurance athletes, with frequencies reported as high as 25% to 45% (4). Menstrual losses, dietary iron insufficiency, and gastrointestinal losses are the most likely causes. But whether this condition impairs performance in humans is currently controversial, and there is little convincing experimental proof on either side of the argument. Still, most experts would agree that a low ferritin level documented in an athlete who exhibits unexplained decrement in sport performance should be treated with iron supplementation.

Changes in Body Weight

Most forms of physical exercise are weight bearing. It follows that a major determinant of performance on such activities will be dictated not only by the muscular capacities of the child but also by the load that must be transported. Changes in the "baggage" created by excess body fat, then, can explain worsening exercise intolerance in youth with increasing levels of obesity.

Equally important, excessive weight loss can result in decreased exercise capacity. This is seen in wrestlers who have manipulated food intake or body fluid losses to achieve weight classification goals or in patients with eating disorders such as anorexia nervosa. An optimal weight and body composition—difficult to define—appear to be important for maximizing sport performance in a way that varies between individuals and the type of sport involved (6).

The training regimens of highly competitive child athletes typically involve significant increases in energy demands. These must be met by increased caloric consumption. If this is not done, the young athlete is effectively undernourished, which can be expected to limit exercise performance. It is essential that this be avoided in growing children, given the caloric requirements for normal growth and development. A mismatch in energy expenditure with training and caloric intake is revealed by failure of the athlete to achieve normal weight gain, or in extreme cases, even weight loss.

Post-Viral Infections

Some patients can experience marked fatigue with exercise after having suffered a significant viral infection, a disability that persists well beyond the normal recuperation period (2). Termed post-viral infection chronic fatigue syndrome, this condition is devastating to athletes, who find it impossible to train or compete in their sport even many months after their illness. Infectious mononucleosis is most commonly implicated in such cases, but other infectious agents have also been described, including influenza, viral hepatitis, and enteroviral meningitis.

The cause of chronic exercise fatigue after a viral infection has not been determined. An abundance of hypotheses have been put forward, each with limited experimental or observational evidence, including persistent viral infection, psychological depression, altered muscle energy metabolism, and reduced threshold for sensory fatigue in the brain. For the most part, no insights have been forthcoming from exercise testing of adolescents with chronic fatigue syndrome, which generally reveals either no abnormalities or a depressed $\dot{V}O_{2max}$ or endurance capacity consistent with a physical deconditioning or a sedentary lifestyle (1, 5). Whether such findings in the latter case are the cause or the effect of the patient's complaints of excessive exercise fatigue is not clear.

Sedentary Lifestyle or Genetic Limitations

Young people with sedentary lifestyles, including low fitness or insufficient training, or inherent constraints on exercise capacity, will perform poorly on physical tasks that exceed their physiologic limitations. Not uncommon, too, is the scenario in which the child who was performing well on the soccer field when younger is now, having reached the early teen years, is finding it hard to keep up with his or her peers on the pitch. Interpreted by the child's coach and parents as "increasing exercise fatigue," this may be explained by the fact that competition normally sharpens when one reaches adolescence, and effective sport participation now requires training. Alternatively, he or she may have been an early-maturing child who reached the limits of his genetic potential for a certain activity at an age before that of his or her teammates.

Emotional Difficulties

It is unlikely that ease of fatigue with exercise would be an isolated expression of emotional disorders or mental stress. Such problems are more likely manifest as a global lassitude, which might include reports of general tiredness, behavior change, poor sleep habits (too much or too little), and a deterioration in school grades. Patients with emotional difficulties often experience other associated psychogenic symptoms, such as headaches, abdominal pain, and dizzy spells. Stressors such as serious family illness or death, parents' divorce, or breaking up with a girl- or boyfriend could explain limitations in ability to exercise, but this can usually be expected to occur within the context of more general fatigue in daily life.

Medications

Certain medications may depress exercise performance or at least create symptoms that are interpreted as limitations of interest or capacity in physical activities. These include beta-blockers, antihistamines, diuretics, anticonvulsants, sedatives, and muscle relaxants.

Chronic Disease

The onset of any of a number of chronic diseases involving noncardiopulmonary systems might be expected to present initially with fatigue triggered by exercise. These include a range of endocrine, autoimmune, neoplastic, and hematologic illnesses that should be considered in the child with unexplained onset of symptoms of easy fatigability. Some skeletal anomalies, such as pectus excavatum deformity of the chest (see chapter 17), may limit endurance fitness. Many neuromuscular diseases (muscular dystrophy, mitochondrial disorders) can initially manifest as weaknesses and inability to sustain normal levels of exercise.

Overtraining

Overtraining, or athlete burnout, describes a decrement of competitive performance as a result of excessive levels of training. Typically, athletes respond to such a situation by training even more heavily, resulting in a more precipitous drop in performance (7). Besides failing performance, athletes who are overtraining often describe other complaints, including frequent respiratory infections, sleep disorders, muscle soreness, loss of appetite, and mood disturbances. Whether this phenomenon reflects a physiological or psychological response to excessive training load is not clear. Despite considerable research, no specific biochemical or physiological diagnostic marker for this overtraining effect has been discovered. The diagnosis, then, is considered after exclusion of other causes of exercise fatigue, but burnout is to be suspected in the athlete with an overzealous training regimen that lacks periods of rest for recuperation.

Evaluation

Assessment of the child or adolescent who reports abnormal ease of fatigue with exercise normally requires a careful, multifaceted approach (see *Fatigue Assessment in the Pediatric Patient*). This includes a thorough history and examination coupled with a cardiopulmonary exercise test and certain pertinent laboratory tests.

History

A complete, well-thought-out history typically provides the sharpest diagnostic tool for discerning

---

Fatigue Assessment in the Pediatric Patient

History

Narrative _____

Documentation _____

Associated symptoms _____

Athletic history _____

Habitual physical activity _____

Nutrition_____

Weight changes _____

Psychosocial issues _____

Medical history _____

Family history _____

Physical examination

Laboratory tests

From T.W. Rowland, American College of Sports Medicine, and North American Society for Pediatric Exercise Medicine, 2018, *Cardiopulmonary exercise testing in children and adolescents* (Champaign, IL: Human Kinetics).

the etiology of exercise fatigue. Questions need to be directed to the patient and parents in light of the differential diagnosis outlined previously.

- *Narrative.* It is often useful to begin the history-taking by having the patient relate his or her complaint (without interruption) in his or her own words. This provides the examiner with not only the nature and scope of the problem but also, often importantly, the way the patient (or parent) perceives his or her symptoms and exercise limitation.

- *Documentation.* The details of the complaint of exercise fatigue need to be outlined: When the patient says "getting tired easily," what does he or she mean (shortness of breath, muscle weakness)? Who is concerned about it (child, parent, coach, gym teacher)? What is its duration (sudden, "for years," etc.)? What types of activity are involved (climbing stairs, one-mile run in gym class, recreational sports)? If the patient is an athlete, what is the documentation of performance decrement (finish times)? What types of sports are involved?

- *Associated symptoms.* Clues particularly to cardiac abnormalities (palpitations, chest pain, tachycardia, dizziness, syncope) or pulmonary disease (wheeze, cough, stridor, tightness in the throat) can be provided.

- *Athletic history.* If the patient is a competitive athlete, details of the training history (frequency, duration) are important. Is training year round? Is there any cross training with other sports? Is rest built into the training regimen? How competitive is the athlete—defending state cross country champion or back-of-the-pack runner enjoying the social benefits of team participation?

- *Habitual physical activity.* For the nonathlete, how much regular physical activity is he or she engaged in? Does the patient have interests that involve considerable sedentary time? Does he or she keep up with his or her peers in physical education class?

- *Nutrition.* A short dietary history should be obtained. What does the patient normally eat for breakfast? Lunch? Dinner? Snacks? If he or she is an athlete, what fluids are consumed with training and competition? Does the patient eat red meat? How often?

- *Weight changes.* Is there any pattern of excessive weight gain or of weight loss?

- *Psychosocial issues.* Is there a history of signs of global lassitude: personality change, sleep disturbance, decline in school grades, being "tired all the time"? Have there been any recent major emotionally upsetting events in the patient's life?

- *Medical history.* Has the patient experienced any medical illnesses, particularly significant viral infections, in the past 6 mo? Is there a past history of anemia, asthma, or heart murmur? If a female, is the volume of menstrual flow excessive? Does he or she take any medications or nutritional supplements?

- *Family history.* Identifying members of the family with serious cardiac or pulmonary disease may have a bearing on the patient's diagnosis. It may also be important to recognize family expectations—does the patient have parents or siblings who have been successful in sport competition?

Physical Examination

Given the broad differential diagnosis possibilities in the assessment of exercise fatigue, a thorough physical examination is clearly pertinent. The examination might be focused on cardiac and pulmonary findings, but attention to features as discrepant as skin color (pallor, cyanosis) and subjective assessment of muscle strength are important as well. Visual assessment of body fat content (does the patient appear obese or undernourished?) and muscularity (does he *look* like a healthy athlete?) is useful. In the encounter with the patient, does he or she appear depressed or withdrawn?

Laboratory Tests

A complete blood count, serum ferritin level, and sedimentation rate or C-reactive protein level are generally indicated for young patients complaining of ease of fatigue with exercise. Other tests can be considered in light of findings on the history and physical examination, such as electrocardiogram, echocardiogram, pulmonary function tests, chest X-ray, and thyroid function tests.

Exercise Testing

The principal role of cardiopulmonary exercise testing in assessing the patient who complains of exercise intolerance is to document the reality and severity of the exercise limitation. This is addressed by measures of treadmill endurance time on a standardized protocol or a size-related peak work capacity with cycle testing, but the most accurate means of assessing endurance capacity is by determination of $\dot{V}O_{2max}$. The results are best compared to norms established within the same testing laboratory. Differences in patient populations, laboratory personnel, and equipment make the use of published normal values questionable.

The central question here is whether the $\dot{V}O_{2max}$ value obtained (in consideration of variables such as body fat content, sex, and age) is lower than one would expect given the level of athleticism and regular physical activity of the patient. If the $\dot{V}O_{2max}$ of a healthy, active but nonathletic child is equal to the mean normal value, it is unlikely he or she is truly experiencing any abnormal exercise fatigue. Indeed, it is not unusual to find that a child whose parents are convinced he or she is tiring easily on the playing field exhibits normal endurance fitness on laboratory exercise testing.

Sometimes, however, the interpretation of a particular $\dot{V}O_{2max}$ value can be difficult. If a competitive soccer player complaining of deteriorating exercise performance is found to have a $\dot{V}O_{2max}$ just above the normal range, is this reassuring that there is no significant problem? Or does this value, although high, represent a major decline from his or her previous level of aerobic fitness? A one-time assessment cannot answer this question, but findings on serial testing might.

Experienced exercise testing staffs recognize that the patient's behavior in approaching and undergoing the exercise test can provide insights into his or her physical capacities, even if a true maximal effort is not provided on the test. Fit people typically enjoy the challenge of an exercise test and perform with motivation. Those who find physical activity abhorrent are likely to ask to stop exercising when it starts to become physically unpleasant, and while a true physiological assessment cannot be determined by such a submaximal test, the diagnosis of limited exercise capacity is not difficult.

Besides providing an objective measure of exercise capacity, treadmill or cycle testing can provide specific information about the differential diagnosis of exercise fatigue. The appearance of arrhythmias or ischemic ST changes on the electrocardiogram points to a possible cardiac etiology. A low ventilatory reserve (maximal voluntary ventilation at rest minus maximal minute ventilation) suggests a pulmonary etiology for exercise fatigue. Assessment of airway flow rates before and after exercise can provide insight into exercise-induced bronchospasm or vocal cord dysfunction. Dampened heart rate response to exercise could reflect a drug effect. Exercise limitation from a pectus excavatum deformity is expected to limit tidal volume and stroke volume during exercise. Low levels of oxygen saturation by pulse oximetry might reflect an occult right-to-left shunt.

Conclusion

The list of possible causes of abnormal exercise fatigue in young athletes and nonathletes is long. However, with a careful assessment, including exercise testing, the diagnosis can often be established. This comprehensive approach is important because (a) exercise fatigue can be the initial presenting sign of significant health issues, and (b) many of its underlying causes can be readily reversed with appropriate management.

The multitude of possible etiologies outlined in this chapter would need to be considered in the clinical approach to Joseph in the case report, who comes with a complaint of deteriorating swim performance despite a very high volume of training. It might be reasonable to expect that the answer might lie with overtraining. Joseph has missed the message that periods of rest are important in sustaining a positive sport training effect.

IV

TESTING SPECIAL POPULATIONS

For some members of the pediatric population, special care is needed when designing the exercise test. Part IV describes considerations that are important in the exercise evaluation of those with physical and intellectual limitations.

Pectus Excavatum

Thomas W. Rowland, MD

Pectus excavatum is a congenital deformity of the chest of unknown etiology in which abnormal formation of the osseus–cartilaginous junctions of the ribs and sternum create an anterior concave depression. Its incidence is approximately 1 in every 300 to 400 live births. Occurring predominantly in males, pectus excavatum is familial in approximately one-third of cases. It is not an infrequent finding in patients with Marfan syndrome, and some people demonstrate an associated scoliosis of the spine. However, pectus excavatum is usually an isolated malformation (9).

The severity of the chest wall depression can be assessed objectively by measuring an index obtained by dividing the inner width of the chest at the level of the pectus deformity by the anteroposterior distance between the posterior aspect of the sternum and the anterior surface of the spine by CT scan or chest X-ray. Normal people have a ratio of about 2.5; an index exceeding 3.1 is commonly considered severe and warrants surgery (9).

The original surgery for pectus excavatum deformity was the Ravitch procedure, which continues to be performed with certain modifications. This operation involves an anterior thoracotomy, with resection of abnormal cartilage and fixation of the sternum with a metal bar, which is removed after 1 yr. More recently in a less invasive approach, the Nuss procedure has been also used, in which a concave bar is passed behind the sternum and then flipped to project the sternum anteriorly. The bar is removed after 2 to 3 yr.

Physiological Implications

The pathological implications of a foreshortened anteroposterior diameter of the chest created by a pectus excavatum defect, as well as the response of any such abnormalities to surgical intervention, have long been debated. Does a pectus deformity impair cardiac and pulmonary function? Is physical performance compromised? Are psychological implications important? This uncertainty has challenged clinicians' ability to make the best decisions about surgical intervention. Cardiopulmonary exercise testing may help with this process.

Clinicians often hear complaints from patients with this deformity about shortness of breath and exercise limitations, and these symptoms are presumed to be related to cardiopulmonary dysfunction, particularly as a result of compression and lateral displacement of the lungs and the anterior, thin-walled right ventricle. The degree of exercise symptoms has not always matched clinical cardiac and pulmonary findings, however, and most patients with pectus excavatum deformity have demonstrated normal intracardiac pressures and cardiac output during cardiac catheterization (6). Some have suggested, in fact, that the limitations of exercise capacity in these patients might have an extra-cardiac basis (such as poor exercise habits), similar to that observed in adult patients with congestive heart failure, whose physical limitations cannot always be attributed to a decrease in myocardial function.

In 1972, a report by Beiser and colleagues in the *New England Journal of Medicine* described six adult patients with mild to moderate pectus deformity who performed submaximal supine and maximal upright exercise during cardiac catheterization (2). At equivalent low work intensities (defined by percent pulmonary artery saturation), cardiac output and stroke volume were significantly lower in the upright position than in the supine. The authors concluded that pectus excavatum deformity can reduce cardiac output by compression of the heart that occurs in the upright position: "[S]ince the sternal depression of patients with pectus excavatum is usually deepest just above the xiphosternal junction, cardiac compression or restriction might occur only with the patient in the upright position when the heart descends into the space most severely compromised by the pectus deformity."

That stroke volume and cardiac output during low-intensity exercise are expected to be lower in the upright than in the supine position in normal subjects (because of mobilization of blood from the lower extremities) was acknowledged by the authors. However, they described a 14% lower stroke volume with upright versus supine exercise in eight normal subjects but a 31% reduction in their patients with pectus excavatum.

Krueger et al. later would support the cardiac compression concept in these patients in their examination of ventricular chamber dimensions by transesophageal echocardiography before and after pectus surgical repair (8). Significantly greater postoperative values compared to those before surgery were found for right ventricular diameter (3.0 ± 0.9 vs. 2.4 ± 0.8 cm), area (18.4 ± 7.5 vs. 12.5 ± 5.2 cm^2), and volume (40 ± 23 vs. 22 ± 12 ml), respectively.

These studies notwithstanding, subsequent clinical studies of heart and lung function in patients with pectus excavatum have provided inconsistent findings. Some demonstrated limitations of cardiac output and stroke volume as well as pulmonary volumes (forced vital capacity, tidal volume) at rest (7, 15) while others did not (3, 10, 16). However, exercise studies more commonly have indicated diminished cardiopulmonary function in patients with a pectus deformity when compared to normal subjects (17, 21).

Characteristically, even those studies that demonstrate an average difference in physiological measures between patients with pectus deformity and healthy youths show high interindividual differences in the pectus subjects. Such differences have not consistently been related to the severity of the defect (19). The study of Rowland et al. comparing maximal cycle exercise findings in 12 boys with moderate-to-severe pectus deformity and 20 control subjects illustrates this issue (17). The patients with pectus deformity had significantly lower mean values for endurance fitness: peak work capacity (2.60 ± 0.28 vs. 3.11 ± 0.45 W/kg), maximal tidal volume (3.02 ± 0.44 vs. 3.46 ± 0.43 ml/kg), maximal stroke index (56 ± 9 vs. 62 ± 11 ml/m^2), and maximal cardiac index (10.61 ± 1.62 vs. 12.00 ± 2.20 L · min^{-1} · m^{-2}). As illustrated in figures 17.1 and 17.2, however, despite statistically significant mean differences, a broad overlap was observed in these

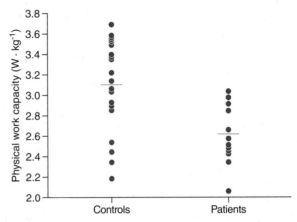

Figure 17.1 Physical work capacity expressed relative to body mass in patients with pectus excavatum deformity and healthy control subjects. Horizontal bars are mean values.

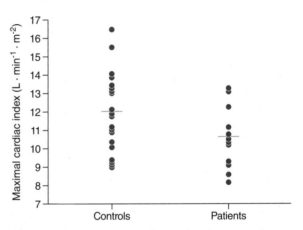

Figure 17.2 Maximal cardiac index during cycle testing of 12 patients with pectus excavatum deformity and 20 control subjects.

values between the two groups. In fact, the great majority of the patients with pectus excavatum demonstrated values within the "normal range." From a clinical decision-making perspective, then, to conclude that cardiopulmonary findings with exercise were an expression of limitations from the pectus deformity in any given child was problematic. That is, a physiological variable, such as maximal cardiac output, in a patient with a pectus deformity that is in the low range of normal for the healthy pediatric population could be normal or, instead, could represent a reduction caused by the chest wall deformity.

Surgical Results

Documenting improvements in cardiac or pulmonary function following surgery is not only important in

- confirming the value of surgery but also
- providing evidence of a true physiologic embarrassment caused by the pectus defect.

Subjective improvement in exercise tolerance has often been reported by patients following pectus repair, but, again, reports of the degree of physiological change have been variable.

In a meta-analysis of eight studies, Malek et al. found evidence of a moderate degree of improvement (6%-31%) in cardiovascular function at rest (cardiac dimensions, stroke volume) and with exercise testing (endurance time, $\dot{V}O_{2max}$, maximal stroke volume) following surgical correction of pectus excavatum [11]. Other studies, however, have failed to reveal any change in $\dot{V}O_{2max}$ after surgical intervention [1, 3, 4, 16, 20].

In 2006, Malek et al. performed a meta-analysis of 12 studies examining the effects of surgical intervention on 313 patients and concluded that surgical repair of pectus excavatum does not significantly improve pulmonary function (forced vital capacity, total lung capacity, maximal voluntary ventilation) [12]. Six years later, Chen et al. described a similar meta-analysis of 15 studies in the literature and found that resting forced vital capacity *decreased* within 1 yr after surgical correction using either the Nuss or Ravitch techniques, but improvement over preoperative values was noted by 3 yr after surgery [5].

It is important to recognize that

- a considerable heterogeneity of individual responses was observed among the investigations in these pulmonary meta-analyses, and
- only studies that examined resting ventilatory variables were included.

With exercise testing, some investigators have reported significant increases in maximal tidal volume and minute ventilation and a decreased infringement on respiratory reserve (percent maximum voluntary ventilation [MVV] of maximal minute ventilation) following pectus repair [3, 14, 18]. Others, however, have shown no changes in ventilation at maximal exercise after surgical intervention [1].

Pre- to postsurgical comparisons of markers of cardiac and pulmonary capacity in patients with pectus excavatum are limited by a major challenge—the timing of the postoperative assessment. Consider a patient with a pectus deformity who prior to surgical repair has a $\dot{V}O_{2max}$ during treadmill testing of $40 \text{ ml} \cdot \text{kg}^{-1} \cdot \text{min}^{-1}$. If one remeasures his aerobic fitness 2 to 3 wk postoperatively, it will be expected to decline, simply from inactivity following the surgery. This presumably explains the observation of Sigalet et al. that $\dot{V}O_{2max}$ had decreased by over 15% from preoperative values when 11 patients with pectus excavatum deformity were evaluated 3 mo after surgical repair [18].

As a postoperative patient gradually resumes normal activities, $\dot{V}O_{2max}$ will rise. But if one waits too long to perform postoperative testing, any number of possible extrinsic variables may influence the patient's peak aerobic power. He may begin to participate in athletic activities or choose to become more sedentary. It would not be unreasonable to expect that his body composition would change over time, particularly with a gain in body fat in the sedentary period immediately after surgery, which would artifactually deflate $\dot{V}O_{2max}$ values. The "correct" time to perform postoperative testing in such comparisons is therefore ill-defined. One could draw variable conclusions about the success or failure of the surgery in changing aerobic fitness simply based on the timing of the postoperative exercise testing.

Cardiopulmonary Testing

Exercise testing that records gas exchange variables may shed some light on adverse cardiac and pulmonary effects in pectus excavatum patients, thus adding physiological data for decisions about surgery [13]. However, interpreting these findings requires considerable caution.

Cardiac Effects

Any impact of a pectus deformity on cardiac filling should appear as a limitation of exercise stroke volume. In this scenario, low maximal stroke volume serves as the basis for diminished maximal cardiac output, and, according to the Fick equation, diminished $\dot{V}O_{2max}$. Assessment of these three variables—stroke volume, cardiac output, and $\dot{V}O_{2max}$—may provide evidence for a negative impact of the pectus deformity on cardiac functional capacity.

In a patient whose pectus deformity is depressing heart function, oxygen pulse, as a surrogate marker of stroke volume (see chapter 10), will be expected to be low at maximal exercise. The curve of heart rate plotted against oxygen uptake during the course of exercise may be displaced superiorly. Since myocardial inotropic and lusitropic function are presumably not disturbed in patients with a pectus defect, the pattern of stroke volume response (early rise, followed by relative stability to peak exercise) should be normal.

While low $\dot{V}O_{2max}$ and peak stroke volume during a standard exercise test should be characteristic of cardiac limitations imposed by a pectus deformity, a major diagnostic problem arises: These features are also typical of normal, healthy people who have low aerobic fitness because of genetic endowment or lack of habitual physical activity. Such findings must be considered in light of the patient's activity habits. That is, depression of stroke volume and cardiac output during an exercise test may reveal either the compressive effects of a pectus deformity or poor aerobic fitness in an inactive child.

Compounding this interpretive challenge is the possibility that the presence of a pectus excavatum deformity may, by itself, cause a person, because of embarrassment or fear, to choose a sedentary lifestyle. Given a finding of a low $\dot{V}O_{2max}$ and diminished maximal stroke volume with exercise, we cannot assume the physiological impact of a pectus deformity in a physically inactive child. We must ask about the reasons for a sedentary lifestyle in such a child.

It is also important to consider whether physiological findings on exercise testing are truly consistent with the patient's description of the degree of exercise limitation. It is unlikely, for example, that a $\dot{V}O_{2max}$ in the low normal range could be considered causal to a patient's complaints of being unable to climb a flight of stairs.

Pulmonary Effects

The chest wall disfiguration in patients with a pectus excavatum deformity may create a restrictive respiratory pattern on resting assessment that produces low maximal voluntary ventilation, low measures of vital capacity, and low tidal volume. Because of the tidal volume limitations, minute ventilation relies more heavily on breathing frequency. The findings of low tidal volume (per kg) at maximal exercise are consistent with the negative effect of a pectus deformity (3, 20, 21). At maximal exercise, too, a pectus deformity may cause an increase in the maximal minute ventilation as it relates to MVV measured at rest. In normal individuals the %MVV at $\dot{V}O_{2max}$ is 50% to 70%. Greater values indicate an elevated recruitment of breathing reserve typically seen in patients with pulmonary limitations to exercise, which might occur with a significant pectus deformity.

The same caveat holds for tidal volume at rest and during exercise in patients with pectus deformity as for cardiac findings: These results must be considered with respect to the typical activity level of the child. A sedentary lifestyle can create a data profile similar to that expected in patients with cardiopulmonary embarrassment from their pectus deformity. We must collect information about a child's athleticism and normal activity pattern before attributing test results to the effects of a pectus excavatum deformity.

Conclusion

Current research indicates that, when taken as a whole, patients with pectus excavatum deformity have lower endurance fitness than healthy youths, and this is associated with limitations of cardiac stroke volume and respiratory tidal volume during exercise. However, the research may provide little help for the practitioner in making individual recommendations about surgical intervention because there is broad overlap in the results for patients versus healthy control subjects. Among patients with pectus deformity there are those with little or no exercise impairment and there are others with significant cardiopulmonary limitations.

Cardiopulmonary findings on exercise testing may provide some guidance in making this distinction. During cardiopulmonary testing, findings of a low $\dot{V}O_{2max}$ with a depressed maximal stroke volume (or oxygen pulse) or limitations of tidal volume are consistent with the negative influences of a pectus deformity on cardiac and pulmonary function. It is important, however, to consider such findings in the context of the patient's athleticism and normal levels of physical activity because low fitness levels may produce these results in otherwise normal children.

Obesity

Laura Banks, PhD, and Brian W. McCrindle, MD, MPH

The high and increasing prevalence of childhood obesity has been well reported in the literature (13, 14, 32, 42, 43). Childhood obesity has pathophysiological consequences for cardiopulmonary function, both at rest and during exercise stress. Cardiopulmonary exercise testing of the obese child may require special considerations. This chapter will discuss physiological differences for obese children and adaptations and modifications for performing and interpreting the results of cardiopulmonary exercise testing.

Quantifying Childhood Obesity

Obesity is defined as abnormal or excessive adipose tissue accumulation resulting from an imbalance between energy intake and expenditure. Body composition can be determined noninvasively using hydrostatic weighing, dual X-ray absorptiometry, or air-displacement plethysmography. Air-displacement plethysmography provides a noninvasive, accurate assessment of fat and fat-free mass (i.e., protein, water, mineral, glycogen) based on measures of body volume and density (18). This method may be advantageous when compared to both hydrostatic weighing (where water displacement is required) and dual-energy X-ray absorptiometry (where radiation exposure is required). Furthermore, air-displacement plethysmography does not underestimate body fat percentage in children and adolescents (21). To date, air-displacement plethysmography has been validated against hydrostatic weighing in healthy young adults (8, 33). Nonetheless, additional studies are warranted to determine the validity of air-displacement plethysmography in younger children and adolescents.

The quantification of childhood obesity can be challenging in clinical practice. More rigorous assessments of adiposity, including dual-energy X-ray absorptiometry (DEXA) scanning, magnetic resonance imaging (MRI), and whole body plethysmography, are impractical for use in routine clinical practice. Simple anthropometric measures suffice provided that they are normalized according to the child's age and sex and such charts are readily available (10, 15, 27). Height and weight can be plotted on standardized sex-specific growth charts, and discrepancies between height and weight percentiles may indicate increased adiposity. Body mass index (BMI) is the most common measure used to indicate the level of adiposity; it is calculated from weight and height (weight in kilograms divided by height in meters squared, or kg/m^2). Since BMI varies by sex and age, it must be plotted and expressed as a percentile based on distributions in normal populations (figure 18.1 and table 18.1).

One of the limitations of using BMI to assess adiposity is that body composition is not evaluated (i.e., fat vs. fat-free mass components) (22). For example, BMI percentiles are inappropriate for use in children and adolescents with a high percentage of skeletal muscle mass because this fat-free mass component weighs more than fat mass; therefore, a young athlete may be incorrectly labeled as being obese based on having a relatively high BMI.

Measurement of waist circumference is a useful adjunct to BMI percentile, and it may be more reflective of cardiometabolically relevant abdominal, central, or visceral adiposity in children and adolescents (11, 26, 36). Waist circumference can be plotted on standardized sex-specific charts for age and expressed as a percentile. It can be normalized to body size more simply by dividing by height. A waist-to-height ratio <0.5 is considered normal, 0.5 to 0.59 indicates overweight, and ≥0.6 indicates obesity and significantly increased cardiometabolic risk. BMI and waist measures can be used together, so an athlete with increased muscle mass may have an increased BMI but a normal waist measure. Also, obese children who participate in an exercise intervention may at first

2 to 20 years: Girls
Body mass index-for-age percentiles

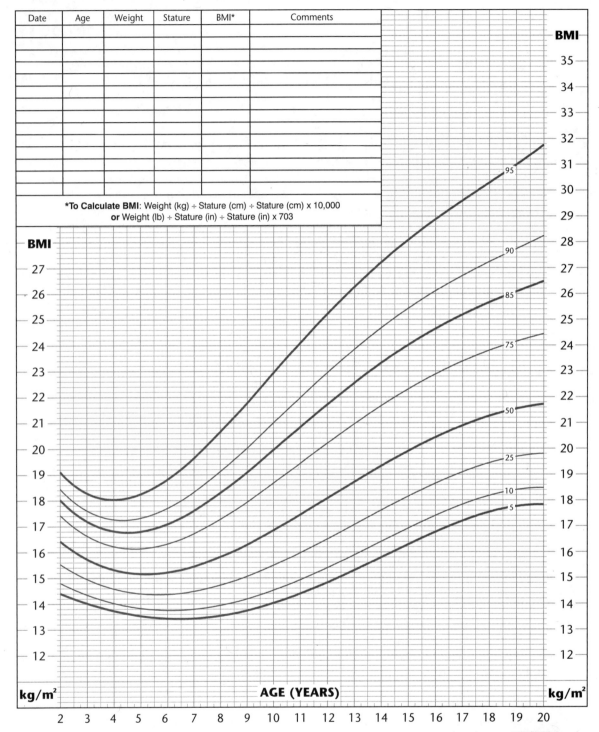

*To Calculate BMI: Weight (kg) ÷ Stature (cm) ÷ Stature (cm) x 10,000
or Weight (lb) ÷ Stature (in) ÷ Stature (in) x 703

SOURCE: Developed by the National Center for Health Statistics in collaboration with the National Center for Chronic Disease Prevention and Health Promotion (2000). http://www.cdc.gov/growthcharts

Figure 18.1 Body mass index growth charts (BMI for age) for girls and boys.

2 to 20 years: Boys
Body mass index-for-age percentiles

NAME _____

RECORD # _____

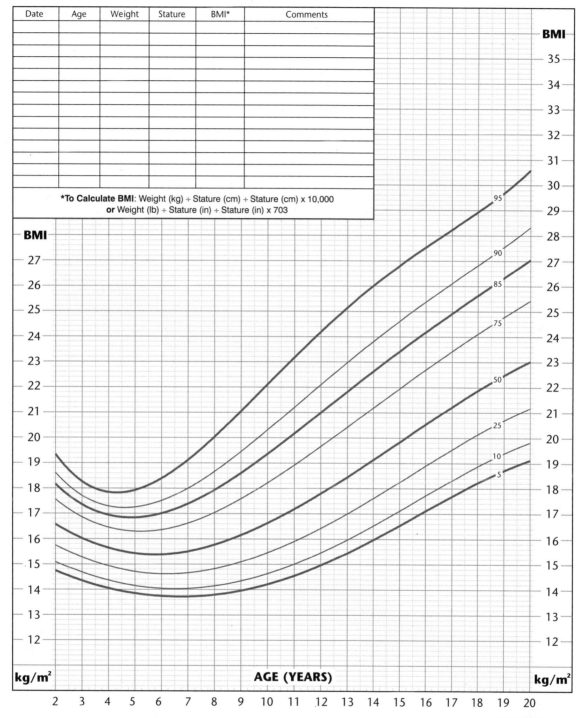

Date	Age	Weight	Stature	BMI*	Comments

*To Calculate BMI: Weight (kg) ÷ Stature (cm) ÷ Stature (cm) x 10,000
or Weight (lb) ÷ Stature (in) ÷ Stature (in) x 703

AGE (YEARS)

SOURCE: Developed by the National Center for Health Statistics in collaboration with the National Center for
Chronic Disease Prevention and Health Promotion (2000). **http://www.cdc.gov/growthcharts**

Figure 18.1 *(continued)*

Table 18.1 Classification of Body Mass Index Percentiles

Body mass index (BMI) percentiles (adjusted for age and sex)	Weight category
0 to <5th	Underweight
≥5th to <85th	Normal weight
≥85th to <95th	Overweight
≥95th to <99th	Obese
≥99th	Severely obese

Adjusted for age and sex in children and adolescents.

Data from Barlow 2007.

not change their BMI but may reduce their waist measure as they lose fat and increase muscle.

Skinfold thickness measurements have been used to determine body fatness in children because subcutaneous fat is highly correlated with total body fat. Measurement of skinfold thickness may be a useful adjunct to BMI percentiles in estimating body fatness among normal-weight children; however, measurements fail to improve the classification of obesity status beyond BMI percentiles in obese children (>95th percentile) (23, 34).

Physiological Adaptations

Obese children may exhibit cardiopulmonary adaptations and pathology that differ from their normal-weight peers. While cardiopulmonary function may be within the normative range for most overweight and obese children, several pathophysiological adaptations may be evident (figure 18.2).

Cardiac pathophysiological adaptations may include increases in blood volume, cardiac output, left ventricular end-diastolic pressure, blood pressure, and left ventricular mass. An increased rate of ventilation, greater metabolic demand associated with breathing, and reductions in respiratory muscle efficiency and compliance reflect pathophysiological pulmonary changes. Musculoskeletal pathophysiological adaptations associated with obesity may include a reduction in skeletal muscle mass, impairment in energy metabolism, accumulation of intramyocellular fat, and increased stress and pain in weight-bearing joints. Musculoskeletal conditions are common in the weight-bearing lower-body joints of obese children (45). Structural malalignments in the hip (e.g., reduced femoral anteversion during maturation and slipped capital femoral epiphysis) and knee joints (e.g., idiopathic tibial vara or Blount's disease) have been associated with obesity (24, 25, 38).

Ultimately, these pathophysiological changes and conditions may result in the clinical presentation of early cardiac dysfunction, type 2 diabetes, asthma, and orthopedic syndromes. The obese child's overall physical health should be evaluated before exercise testing because these changes may all contribute to performance impairments. Clinicians may need to adapt the cardiopulmonary exercise testing protocols for particular children or adolescents. An individualized approach to testing may improve compliance to test protocols and enable a child to more readily achieve a maximal effort.

Effects of Obesity on Physiologic Measures

Obesity may have significant implications for quantifying the physiological response to exercise. The scaling of peak oxygen consumption must be considered to account for potential differences in body composition. The following section will address the reporting of peak oxygen consumption data, including absolute peak $\dot{V}O_2$, ratio scaling $\dot{V}O_2$ to body mass, ratio scaling $\dot{V}O_2$ to fat-free body mass, and allometric scaling. The reporting of submaximal exercise responses will also be addressed, including ventilatory anaerobic threshold, blood pressure, heart rate, and economy.

Peak Versus Maximal Exercise Capacity

Maximal cardiopulmonary exercise performance in children is commonly defined by the highest rate of oxygen consumption obtained during the test. While most obese and nonobese children can

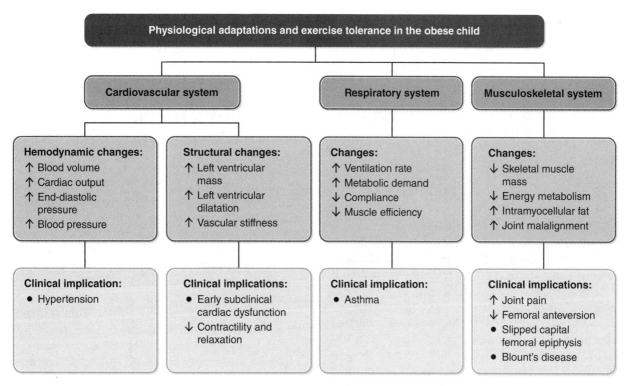

Figure 18.2 Physiological adaptations and clinical implications in the obese child and adolescent.

perform to maximal effort, they rarely demonstrate a plateau in oxygen consumption despite increasing workload (2, 4, 5). As described in chapter 7, children may more commonly achieve a peak $\dot{V}O_2$ ($\dot{V}O_{2peak}$) during cardiopulmonary exercise testing. A $\dot{V}O_{2peak}$ may still reflect a maximal effort on the part of the child, and it can be classified based on predefined secondary criteria for maximal heart rate and respiratory quotient. Armstrong et al. have demonstrated that the presence of a $\dot{V}O_2$ plateau for children performing exercise testing does not indicate superior maximal cardiopulmonary exercise results (including in $\dot{V}O_2$, heart rate, and blood lactate) when compared to children who did not achieve a $\dot{V}O_2$ plateau (2, 4). The absence of a $\dot{V}O_2$ plateau is particularly common in exercise testing results for the obese child. Breithaupt et al. evaluated the cardiopulmonary exercise response in 62 obese children and reported that fewer than 30% of them obtained a true $\dot{V}O_{2max}$ as defined by a plateau in values at high exercise intensities (12).

Absolute $\dot{V}O_{2peak}$

Notably, intra- and inter-individual variations in $\dot{V}O_{2peak}$ may be the result of several physiological factors, including

1. cardiopulmonary fitness,
2. oxygen extraction and utilization within muscle, and
3. changes in body mass (including fat and fat-free mass components).

Higher reported values of absolute $\dot{V}O_{2peak}$ in obese children and adolescents suggest that cardiopulmonary function is preserved and perhaps not influenced by increasing body fat mass (17, 28, 31, 46). Exercise intolerance in the obese child may be a result of the metabolic demand of moving a greater body fat mass rather than an intrinsic impairment in cardiopulmonary function (20, 29).

Ratio $\dot{V}O_2$ Scaling to Whole Body Mass

The interpretation of cardiopulmonary exercise performance requires special consideration in obese children to account for potential differences in whole body mass. Normalization of cardiopulmonary exercise data should account for differences in body size in order to take into account the effects of growth on physiological function in the context of obesity. This is particularly important when comparing obese and nonobese children because they often present with differences in both body fat mass and maturation status. In this case, the issue of scaling $\dot{V}O_{2peak}$ findings becomes increasingly important. In comparison to absolute $\dot{V}O_{2peak}$, relative $\dot{V}O_{2peak}$ has been most commonly reported in the literature. Relative measures of $\dot{V}O_{2peak}$ aim to normalize values and facilitate between-individual comparisons. A comparison of $\dot{V}O_{2peak}$ between obese and nonobese children has been reported in several studies (9, 17, 28, 31, 39, 46). A strong relationship has been noted between higher $\dot{V}O_{2peak}$ and increased whole body mass; physical activity levels and maturation may also be important (3, 9).

The interpretation of cardiopulmonary function during submaximal and maximal exercise has been skewed because most studies use a traditional ratio scaling approach. The traditional approach reports $\dot{V}O_2$ in a given unit of time as a function of whole body mass. Studies have reported a significant reduction in $\dot{V}O_{2peak}$ among obese children compared to nonobese children when this traditional approach was used (milliliters of oxygen per kilogram of body mass per minute, or $ml \cdot kg^{-1} \cdot min^{-1}$) (35). Ratio scaling based on whole body mass is subject to error because it assumes that the relationship between whole body mass and $\dot{V}O_{2peak}$ is linear. This approach tends to overestimate cardiopulmonary function in lean children and underestimate it in obese children. Thus, it is difficult to determine whether obese children have an actual reduction in cardiopulmonary function or whether differences in $\dot{V}O_{2peak}$ may reflect differences in body fat mass.

Ratio $\dot{V}O_2$ Scaling to the Fat-Free Component of Body Mass

Alternative methods have been suggested to provide a more accurate representation of $\dot{V}O_{2peak}$. Normalization of $\dot{V}O_2$ to fat-free body mass has been suggested as a simple alternative ratio scaling approach for use in cohorts where body fat mass may differ between groups (e.g., obese vs. nonobese) or where it may change considerably over time (e.g., weight loss) (16, 19, 29). Studies have reported no differences in $\dot{V}O_{2peak}$ between obese and nonobese children when measured as a function of fat-free body mass. Dencker et al. studied cardiopulmonary exercise testing in a larger population-based sample of 225 children (ages 8-11) (16). They observed that apparent differences in $\dot{V}O_{2peak}$ indexed to whole body mass disappeared when $\dot{V}O_{2peak}$ was scaled to fat-free

body mass (16). These findings further indicate that a ratio approach involving whole body mass may not be appropriate and, when more appropriate scaling approaches are used, obese children have normal cardiopulmonary function.

Allometric $\dot{V}O_2$ Scaling

The assumption of a nonlinear relationship between $\dot{V}O_{2peak}$ and whole body mass may ultimately provide a more conservative approach to reporting differences in $\dot{V}O_{2peak}$. Allometric scaling has been proposed as a promising method to accurately determine and compare $\dot{V}O_{2peak}$ values between cohorts where a large difference exists in whole body mass or maturation. An allometric scaling approach for the reporting of $\dot{V}O_{2peak}$ data uses an exponential function to normalize differences in whole body mass, resulting in a true interpretation of $\dot{V}O_2$. When this alternative scaling approach has been used, no differences in $\dot{V}O_{2peak}$ between obese and nonobese children are noted, although this approach has only been used in a small number of studies to date (35). The use of allometric scaling is warranted in future studies involving obese and nonobese children.

$\dot{V}O_2$ Scaling Limitations

A primary limitation of these alternative approaches, including indexing $\dot{V}O_{2peak}$ to either fat-free body mass or an exponential function of whole body mass, is that normative data have not been established. Future work in the interpretation of $\dot{V}O_{2peak}$ values is required for obese children to fully evaluate whether intrinsic impairments in cardiopulmonary function exist. Normalization of exercise testing data may help to determine whether cardiopulmonary function is modulated by the fat or fat-free body mass components.

The interpretation of $\dot{V}O_{2peak}$ data is particularly important when comparing obese and normal-weight children; it is less important when monitoring changes in $\dot{V}O_{2peak}$ within individuals. Nonetheless, $\dot{V}O_2$ scaling has direct implications for understanding the cardiopulmonary health of obese children beyond the laboratory environment. While it has been proposed that a population-based decline in high-intensity physical activity may be associated with a decline in $\dot{V}O_{2peak}$, this may be due to increases in whole body mass and not to an intrinsic decline in cardiopulmonary function (1).

Ventilatory Anaerobic Threshold

Ventilatory anaerobic threshold (VAT) is defined as the exercise intensity above which oxygen supply to and utilization within the working muscles is insufficient to meet energy demands (see chapter 8). VAT can be determined by

1. collecting ventilatory data using a metabolic cart during cardiopulmonary exercise testing and determining minute ventilation,

2. plotting a graph of minute ventilation versus $\dot{V}O_2$, and

3. identifying the $\dot{V}O_2$ value at the nonlinear inflection point where minute ventilation increases disproportionately to $\dot{V}O_2$ (7).

VAT provides an accurate, less effort-dependent measure of cardiopulmonary function. The submaximal exercise intensity from which VAT is derived may be advantageous when testing the untrained obese child. Data collected at the time of VAT, including heart rate and $\dot{V}O_2$, can be useful for determining exercise performance and prescription. Because VAT is commonly expressed as a function of whole body mass, it is susceptible to the same interpretation limitations as $\dot{V}O_{2peak}$ (as described previously).

The importance of scaling cardiopulmonary exercise data, including $\dot{V}O_{2peak}$ and VAT, has been previously reported. Zanconato et al. performed a matched case-control study to demonstrate how the expression of cardiopulmonary variables based on body size may lead to interpretations in cardiopulmonary exercise testing results (46). They observed that $\dot{V}O_{2peak}$ expressed relative to whole body mass was approximately 30% lower in young obese adolescents when compared to their normal-weight peers (males: 37 ± 7 vs. 54 ± 9 ml · kg^{-1} · min^{-1}; females: 38 ± 8 vs. 51 ± 9 ml · kg^{-1} · min^{-1}, $p < 0.001$ for both). Similarly, it was observed that VAT expressed relative to whole body mass was approximately 25% lower in young, obese adolescents when compared to their normal-weight peers (males: 32 ± 6 vs. 42 ± 7 ml · kg^{-1} · min^{-1}, $p < 0.01$; females: 29 ± 6 vs. 38 ± 5 ml · kg^{-1} · min^{-1}, $p < 0.001$). However, VAT expressed as a percentage of $\dot{V}O_{2peak}$ did not differ between obese and normal-weight subjects (males: 84 ± 10% vs. 78 ± 9%; females: 79 ± 10% vs. 78 ± 10%). This study demonstrates that cardiopulmonary function may be preserved in the obese adolescent. Alternative scaling approaches may also be suitable for determining intrinsic differences between obese and normal-weight children. Nonetheless, normative VAT data using alternative scaling methods to control for large variations in whole body mass have not been reported in the literature. VAT should be reported independent of whole body mass when evaluating cardiopulmonary function in both obese and normal-weight children.

Blood Pressure and Heart Rate Response

A linear, positive relationship between $\dot{V}O_2$ and heart rate is present during incremental exercise in the obese child. Maximal heart rate and systolic blood pressure responses are largely similar in obese and normal-weight children (19, 29, 46). Therefore, the same maximal heart rate criteria (outlined in chapter 5) for defining an exhaustive exercise effort apply when assessing whether a maximal or peak exercise effort was obtained (41). Reported differences in submaximal heart response between obese and normal-weight children are inconsistent, with some studies reporting a higher heart rate response and others reporting no significant difference in heart rate at a given exercise workload (30, 46).

Exercise Economy

A higher absolute $\dot{V}O_2$ and ventilation during submaximal exercise stages in the obese child may provide evidence of higher metabolic and ventilatory cost. While absolute $\dot{V}O_{2peak}$ (reported in ml/min) may in fact be similar in obese and normal-weight children, obese children may perform the cardiopulmonary exercise test with a shorter duration or a lower workload. This may indicate a reduction in mechanical efficiency in obese children (46). Norman et al. studied cardiopulmonary exercise test results from a large sample of 129 obese and normal-weight adolescents (37). They noted a greater metabolic cost during unloaded cycle ergometry exercise in obese adolescents than in those of normal weight, independent of both exercise workload and test duration. Further analysis showed that a greater metabolic cost was associated with reduced cardiopulmonary exercise performance. Thus, they hypothesized that a portion of the exercise-induced metabolic cost may be related to decreased mechanical efficiency during exercise. Furthermore, an increased whole body mass, but not relative body fat distribution (44), has been shown to be a significant predictor of greater energy expenditure during exercise (17). Poor exercise performance in the obese child may, therefore, be secondary to obesity-related changes in mechanical efficiency or energy expenditure rather than cardiopulmonary function.

Cardiopulmonary Exercise Testing Modifications

The physiological response to exercise in most obese children is not different enough from their normal-weight peers to warrant significant protocol modifications. Nonetheless, special consideration may be given to pretest participation screening, type of cardiopulmonary exercise testing protocol, exercise modality, and familiarization with the exercise test protocol. These modifications should be considered on an individual basis depending on the severity of obesity and the pathophysiological adaptations that may impair cardiopulmonary function during exhaustive exercise.

Pretest Participation Screening

Screening for contraindications to cardiopulmonary exercise testing should be performed to document any preexisting conditions or exercise symptoms. This may include chest pain, dizziness, prescription drug use, loss of balance, loss of consciousness, and joint conditions that may be worsened with physical exertion. These conditions are relative contraindications that should be assessed by a physician before cardiopulmonary exercise testing of the obese child. The Physical Activity Readiness Questionnaire (PAR-Q) provides a simple yet effective screening for physical activity participation in older adolescents and adults (aged 15-64). A similar questionnaire for younger and obese children has not been well validated, perhaps due to the low prevalence of adverse events in healthy children who perform cardiopulmonary exercise tests. Nonetheless, pretest screening of obese children is warranted given the physiological adaptations that may contraindicate maximal exercise.

Protocol

Cardiopulmonary exercise testing protocols may be chosen based on the physiological adaptations of the patient. A continuous ramp protocol may be preferable as this smaller, yet continuous, incremental approach to increasing exercise intensity may promote more steady-state increases in the cardiopulmonary response. A prolonged warm-up is recommended to enable the participant to reach a comfortable steady state and become familiar with the equipment and the testing environment. A prolonged cool-down should also be included to ensure an adequate recovery of cardiopulmonary variables, including heart rate and blood pressure, to near resting levels.

Exercise Mode

Cardiopulmonary exercise testing is most commonly performed with treadmill or cycle ergometry. A treadmill-based protocol may pose challenges for obese children because they have a larger weight-bearing load. Significant differences in body fat mass may place the obese child at a disadvantage compared to normal-weight peers. However, a weight-supported treadmill harness may reduce the weight-bearing load while promoting improved balance and safety during exercise testing. Furthermore, treadmill exercise protocols engage larger muscle mass than cycle ergometry-based protocols. This may increase the likelihood that maximal effort will be determined by a central cardiopulmonary function limitation rather than by a peripheral muscle limitation (40). Therefore, treadmill exercise may provide a more accurate reporting of cardiopulmonary function in the obese child. This may be particularly important for the obese child who has peripheral skeletal muscle deconditioning or a diagnosed musculoskeletal condition. Finally, a treadmill-based protocol may be more beneficial than cycle ergometry-based exercise testing because it involves a form of exercise that may be familiar to both obese and normal-weight children.

Nonetheless, cycle ergometry protocols naturally offer a non-weight-bearing exercise. This is advantageous when comparing cardiopulmonary exercise performance in obese and normal-weight children. Furthermore, cycle ergometry-based exercise protocols may be safer than treadmill exercise for the obese child because they feature improved balance, stability, and ease of blood pressure monitoring.

While these common exercise testing modes are safe and feasible when testing the obese child, the use of an arm crank ergometer may provide an effective alternative. An arm crank ergometer may also be useful for the obese child with balance or lower-body musculoskeletal limitations (including joint pain or skeletal muscle deconditioning). Arm crank ergometry is limited in that normative data for children are not widely reported in the literature.

Exercise Testing Equipment

Appropriate-sized exercise equipment should be used for obese children. A properly sized blood pressure cuff is needed to reduce measurement error and enable blood pressure monitoring during exercise. Electrocardiographic lead placement near the rib margins between the midclavicular and anterior axillary line may improve electrocardiographic signals in obese children.

Protocol and Exercise Modality Familiarization

Obese children and adolescents may lack confidence in a formal exercise testing environment due to a history of poor physical activity and sport performance. Their anxiety before and during testing may be reduced with a thorough introduction to the laboratory environment, a longer familiarization or warm-up period to build confidence, the use of field tests, or the use of a portable $\dot{V}O_2$ gas analyzer.

Conclusion

Childhood obesity has direct implications for the performance and interpretation of cardiopulmonary exercise testing. Physiological deconditioning may be present within the cardiovascular, pulmonary, and musculoskeletal systems of the obese child. The interpretation of exercise performance needs to be carefully evaluated in obese versus normal-weight children. Whole body mass–independent $\dot{V}O_{2peak}$ measures suggest that cardiopulmonary function during exercise may be similar in obese and normal-weight children. This has been reported in studies where an exponential function (allometric scaling) was applied to adjust for differences in whole body mass. Minor protocol modifications may be warranted depending on the severity of obesity and of any pathophysiological adaptations or conditions that may affect the cardiopulmonary exercise response. Nonetheless, most obese children and adolescents are physically able to perform cardiopulmonary exercise testing satisfactorily in a supervised laboratory environment.

Intellectual Disability

Bo Fernhall, PhD, and Tracy Baynard, PhD

Most people with intellectual disabilities (ID) live in the community, participate in community activities, and use community resources like anyone else. Thus, it is very likely that exercise professionals will encounter people with ID. It is important to understand the unique aspects of exercise testing with this population because very useful clinical information can be obtained from exercise testing of persons with ID. This population presents with several important concerns that affect our ability to obtain valid and reliable test results. They also exhibit some unique physiological responses that need to be considered when interpreting exercise test results. This chapter will review definitions of ID, special concerns of the population, the underlying physiology of the exercise response, procedures and exercise test protocol selection, and exercise test interpretation for people with ID.

Intellectual disability is the largest subcategory of developmental disabilities, and it is the most common developmental disorder in Western society, with an estimated prevalence of 3% of the population (7, 19). There are an estimated 10 million people with ID in the United States. The American Association on Intellectual and Developmental Disabilities (AAIDD) defines ID as a disability with limitations in both intellectual function and adaptive behavior that originates before age 18 (6).

ID is diagnosed based on the results of a combination of standardized tests, including tests of IQ, such as the Wechsler Adult Intelligence Scale or the Stanford-Binet scale, and tests of adaptive behavior, such as the Diagnostic Adaptive Behavior Scale (6). People with ID will usually need support services in one or more of the areas of adaptive behavior, and the intensity of support needed is used to plan services, but it is not part of the definition or diagnosis of ID. Most individuals with ID live either at home or in community-based group homes, and very few now live in state-supported institutions. Further, some people with

ID may also be "declassified" from an ID diagnosis following early interventions (7). ID is usually classified as mild, moderate, or severe and profound, based primarily on IQ. People with mild ID have higher IQ (typically above 50), can usually understand directions, communicate reasonably well, and can function intellectually at an elementary school level. People with moderate ID have lower levels of IQ (typically 30-50), and although they can communicate, the level of communication is limited. Individuals with severe and profound ID have IQs below 30 and have very limited ability to communicate. Consequently, everyone with ID will usually require simple and direct forms of communication, and the ability to understand and carry out directions declines with decreasing IQ (6, 7, 19, 20).

There is no single underlying cause of ID, and the etiology of ID can vary considerably. Often the specific cause is unknown. Important factors related to the development of ID include genetic and maternal disorders, birth trauma, infectious diseases, behavioral or societal factors such as poverty, malnutrition, maternal drug and alcohol use, and severe stimulus deprivation (7, 18).

Life expectancy is lower in people with ID than in the general population. Mortality rates are one and a half to four times higher than expected compared populations without ID (42, 59). Life expectancy has been increasing rapidly, however, and is now approaching that of the general population (51). Individuals with Down syndrome (DS) still have a much lower life expectancy than others with ID, approximately age 60 (51). Cardiovascular and pulmonary disorders are the most common medical problems in persons with ID (42, 54). In addition, in people with DS, leukemia, infections, and the early development of Alzheimer disease are the most frequent causes of both mortality and morbidity (16, 51).

Schooling for children with ID can take various forms, ranging from homeschooling to special-education classes to fully inclusive schools. For many students, schooling can be a combination of adapted and inclusive experiences. Children with ID have access to many in-school or after-school activities and programs. Many children also participate in Special Olympics or in sports through their school systems. It is reasonable to expect an increasing rate of participation in all aspects of recreation for children with ID. Thus, the need for exercise testing of children with ID is also expected to increase.

Physiological Implications

Some children with ID, especially children with DS, have unique physiological responses to exercise. These responses cannot be explained by any limitations in test performance, assuming that appropriate test procedures are followed. Recognizing these characteristics and physiological responses is important to the proper interpretation of test results.

Children With ID Without Down Syndrome

Early data on children aged 6 to 16 with ID showed that aerobic capacity ($\dot{V}O_{2max}$) was within normal limits (43-47 ml $\cdot$ kg^{-1} $\cdot$ min^{-1} for girls and 48-52 ml $\cdot$ kg^{-1} $\cdot$ min^{-1} for boys) (8, 13); however, many of those subjects were not technically classified with ID, although they had below-average intelligence. Several other studies have reported low but close to normal $\dot{V}O_{2max}$ values for both boys and girls with ID, obtained using cycle ergometry (48, 61). However, cycle ergometry is not a valid testing mode in this population, and it also yields lower $\dot{V}O_{2max}$ values than treadmill testing in general.

Several more recent studies have found normal or close to normal $\dot{V}O_{2max}$ values in children with ID, with mean values of 39 to 41 ml $\cdot$ kg^{-1} $\cdot$ min^{-1} in boys and girls combined, with several individual values above 50 ml $\cdot$ kg^{-1} $\cdot$ min^{-1} (29, 60). Another study reported similar $\dot{V}O_{2max}$ values between age-matched controls and boys with ID (46 ml $\cdot$ kg^{-1} $\cdot$ min^{-1}), but girls with ID exhibited lower levels of $\dot{V}O_{2max}$ (31 ml $\cdot$ kg^{-1} $\cdot$ min^{-1}) than their peers without ID (57). The lower $\dot{V}O_{2max}$ in girls with ID may have been a function of body size; the BMI for the girls with ID was 23.4 kg/m^2 compared to 19.9 kg/m^2 for the girls without disabilities. In contrast, the boys with ID had a BMI similar to that of their peers without disabilities (18.1 vs. 19.5 kg/m^2, respectively). However, several other studies have

reported lower-than-normal $\dot{V}O_{2max}$ values (26-35 ml $\cdot$ kg^{-1} $\cdot$ min^{-1}) for boys and girls with ID (31, 52). Unfortunately, most of these studies were small, and the limited number of subjects precludes generalization of the findings.

The largest study to date (over 600 subjects) evaluated $\dot{V}O_{2max}$ and BMI over a variety of ages in individuals with and without ID; 180 of the subjects with ID did not have DS (9). As shown in figure 19.1, children with ID in the younger age group exhibited lower $\dot{V}O_{2max}$ values than children without ID, but there was no difference in $\dot{V}O_{2max}$ in the 16- to 21-year-old group. The lower $\dot{V}O_{2max}$ in the younger group was explained by higher body weight in children with ID, as the absolute $\dot{V}O_{2max}$ values were similar between children with and without ID (2,318 vs. 2,442 ml/min, respectively). Therefore, in children with ID without DS, $\dot{V}O_{2max}$ appears to

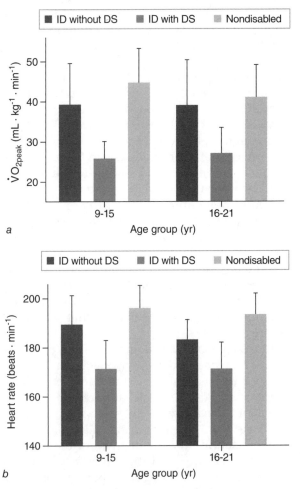

Figure 19.1 *(a)* Maximal aerobic capacity and *(b)* maximal heart rate in individuals with ID.

Data from Baynard et al. 2008.

be similar to their nondisabled peers. However, girls and some younger children with ID (boys and girls) may have lower relative $\dot{V}O_{2max}$ values than their nondisabled peers, primarily because of higher body weight. This is a concern because the rate of overweight and obesity in children with ID can be twice that of their nondisabled peers (15).

Even though $\dot{V}O_{2max}$ values may be similar between children with and without ID, run performance is substantially reduced in children with ID (29, 32). Children without ID nearly double the run performance (i.e., number of laps completed in a 20 m shuttle run) of children with ID, even if their $\dot{V}O_{2max}$ values are similar (13, 14, 29, 31, 32, 34). Part of the explanation for the reduced run performance in children with ID may be reduced leg strength (12, 27, 28, 56) because leg strength is related to both run performance and $\dot{V}O_{2max}$ in this population (12, 27). Run performance also does not change with maturation (ages 10-18) in children with ID, which is different from their nondisabled peers, who improve run performance considerably as they mature (58). The potential difference in BMI does not explain the difference in run performance (58). The reason for the different developmental path in run performance of children with ID compared to their peers without ID is not known at this time. Consequently, absolute levels of physical work or performance are reduced in children with ID, even if $\dot{V}O_{2max}$ is similar to that of their peers without disabilities.

Children With Down Syndrome

Children with DS typically have small stature, short limbs and digits, malformations of the feet and toes, small mouth and nasal cavities, and a large, protruding tongue, all of which can influence exercise performance (18). These characteristics are coupled with joint laxity and often skeletal muscle hypotonia, and the possibility of pulmonary hypoplasia in some individuals (17, 19, 28, 36, 59). There is also a potential for atlantoaxial instability, which may severely affect the safety of exercise in children with DS (18, 28). Many people with DS have impaired immune function, leading to increased risk of infections and also an increased risk for developing leukemia (59). Close to 50% of people with DS have congenital heart disease, which is now usually surgically corrected early in life (59).

Aerobic capacity is markedly reduced in children, adolescents, and adults with DS, and most studies show $\dot{V}O_{2peak}$ values of 18 to 25 ml $\cdot$ kg^{-1} $\cdot$ min^{-1},

with little influence of age (30, 35, 50, 55). These findings are markedly different from children and adolescents with ID without DS or without disabilities (see figure 19.1). Furthermore, aerobically trained adults with DS have higher aerobic capacity than their untrained counterparts, but their $\dot{V}O_{2peak}$ values were still below 40 ml $\cdot$ kg^{-1} $\cdot$ min^{-1}, even though many trained as much as 10 h per week (39). Thus, DS per se negatively affects aerobic capacity. The low levels of $\dot{V}O_{2peak}$ reported cannot be explained by the high incidence of congenital heart disease in this population because children with congenital heart disease were excluded from all of these studies.

Chronotropic incompetence has a major role in explaining the low work capacity of individuals with DS, with maximal heart rates about 30 bpm lower than those of peers without disabilities (23). Average maximal heart rates in children with DS are between 160 and 175 bpm when tested on a treadmill, and they do not typically go over 180 bpm, regardless of age (28). Both the chronotropic index and maximal heart rates indicate that these attenuated maximal heart rates to exercise constitute chronotropic incompetence. The low maximal heart rate accounts for the difference in $\dot{V}O_{2max}$ between persons with DS and their peers with ID without DS and children without disabilities (30). This is most likely due to a reduction in maximal cardiac output as a function of the low maximal heart rate, but no data are available on maximal cardiac output in children (or adults) with DS. One study showed similar stroke volume between individuals with and without DS during submaximal walking (55), but it is unknown if maximal stroke volume is altered in persons with DS. Although obesity rates are very high in children with DS, obesity does not affect maximal heart rate, and it minimally affects $\dot{V}O_{2max}$ in children with DS compared to children with ID without DS (19).

We have recently proposed that the reduced maximal heart rate and $\dot{V}O_{2max}$ in individuals with DS is a consequence of autonomic dysfunction (24). This concept is illustrated in figure 19.2. People with DS exhibit reduced heart rate and blood pressure responses to sympathetic stressors such as isometric handgrip exercise and cold pressor testing (26). The reduced responses are at least partially due to reduced vagal withdrawal during the perturbation (37) coupled with reduced sympathetic stimulation. Recent data also showed that people with DS had little or no change in epinephrine and norepinephrine in response to

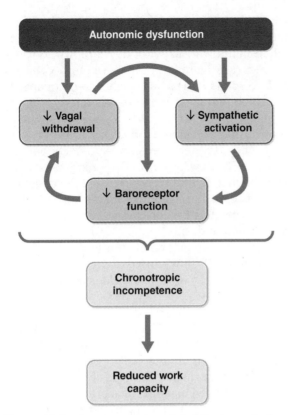

Figure 19.2 A conceptual model of the impact of autonomic dysfunction on chronotropic incompetence and work capacity on children with Down syndrome.
Based on Fernhall, Mendonca, and Baynard 2013.

maximal exercise testing, providing evidence for severely reduced sympathetic activation in this population (21). People with DS also exhibit reduced heart rate and blood pressure response to orthostatic challenges such as upright tilt or sit-to-stand tasks (1, 22, 46). It is likely that altered baroreceptor function contributes to the reduced heart rate and blood pressure changes to both upright tilt and isometric handgrip exercise in persons with DS (43, 46). Consequently, autonomic dysfunction appears to contribute to the substantial reduction in $\dot{V}O_{2max}$ in children with DS.

It is also possible that changes in vascular function may contribute to the reduced work capacity in persons with DS. Both circulating endothelial progenitor cells and flow-mediated dilation are reduced in people with DS (10, 11), and endothelial function is associated with work capacity in populations without disabilities (44). Recent data suggest that vascular reserve in response to exercise is reduced in individuals with DS, and this

was dependent on $\dot{V}O_{2max}$, suggesting that reduced vascular function affects $\dot{V}O_{2max}$ in persons with DS (45).

Run performance is also severely reduced in children with DS, even more so than in their peers with ID without DS (5). In contrast to children with ID without DS, whose run performance is reduced but still associated with $\dot{V}O_{2max}$ (32), there is no relationship between run performance and $\dot{V}O_{2max}$ in children with DS (5). In addition, both walking and general movement economy are reduced in persons with DS (2-4), and thus they spend more energy walking and running at the same speed as their peers without DS. Therefore, the $\dot{V}O_2$ or physical work reserve is much lower in children with DS. This is illustrated in figure 19.3.

Exercise Testing

As with the general population, patient safety is a top priority when conducting a cardiopulmonary exercise test in persons with ID. Congenital heart defects may pose a potential cause for concern because approximately 50% of people with DS are diagnosed with these early in life. However, with surgical corrections occurring within the first few years of life, there are generally few limitations imposed once the patient is fully recovered and deemed ready for normal physical activity. Also, a small number of persons with DS may have atlantoaxial instability, which could affect exercise safety. However, there are no data suggesting exercise testing is not safe among individuals with ID. A more relevant concern is the ability of the patient to understand test-related instructions. Despite this concern, standardized cardiopulmonary exercise testing is valid and reliable in persons with ID (17, 20, 25, 32, 33, 35).

Pretest Preparation

To ensure a reliable test, review the following points:

1. Follow the pre-screening recommendations in the ACSM. A detailed health history and potentially a pretest physical exam (or physician clearance) are warranted for patients with DS, given their propensity for congenital heart defects and possible atlantoaxial instability.

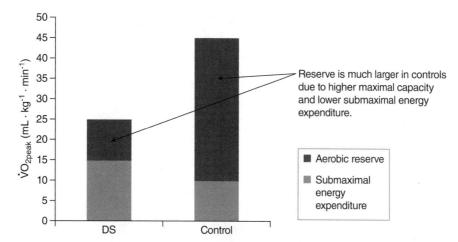

Figure 19.3 Both $\dot{V}O_{2peak}$ (highest value, top of bar) and aerobic reserve (difference between the highest value and submaximal energy expenditure) are lower in people with DS compared to nondisabled controls. The lower aerobic reserve is a function of lower $\dot{V}O_{2peak}$ and higher submaximal energy expenditure.

2. Familiarization is absolutely critical to ensure a maximal effort by the participant. Participants should be fully familiarized with the physical space of the lab, the laboratory and testing staff, testing equipment, and testing procedures. There is no set number of recommended familiarization sessions needed to guarantee the best outcome; this will depend on the participant. Therefore, be prepared to schedule a tentative test date with the understanding that no "real data" may be collected that day. Nevertheless, one to three familiarization sessions are often adequate to achieve the desired comfort level for the participant. When scheduling familiarization sessions, be sure to allow for adequate time for personnel to demonstrate a given expectation first, followed by some practice by the participant. To enhance the participant's experience, provide regular and positive feedback. With regard to familiarization and equipment, if measuring indirect calorimetry, it will be necessary to practice fitting the mouthpieces. Smaller mouthpieces (e.g., masks or "snorkel" mouthpieces) may need to be purchased in order to fit the smaller faces of people with DS. Masks may be more comfortable. Some children and adolescents with DS may have rela-

tively large tongues that may influence mouthpiece size selection. See *Suggested Sequence for Pretest Familiarization on a Treadmill* for more information.

3. Single-step instructions are critical for individuals with ID. A new environment, new personal interactions, and new expectations are often overwhelming. Unfamiliarity may affect the comprehension rate; therefore, give the participant more time to understand instructions. To enhance listening skills, explicit one-step instructions are important. Allow for the completion of one task before providing instructions for the next task (e.g., "Hold onto the handrails," followed by, "Start walking" after the subject holds onto the handrails). Lastly, minimizing distractions is important. For instance, do not set up testing near open doors because hallway activity will be distracting.

4. While two workers are commonly recommended for conducting exercise tests for the non-ID population, three or more may be useful for children with ID, especially for treadmill testing, when those with DS in particular may have balance issues. We have found it helpful to have one staff member in front of the treadmill and an additional staff member on each side of the person being tested.

Suggested Sequence for Pretest Familiarization on a Treadmill

Introduce participant to one or two lab staff at a time.

Explain necessary paperwork.

Provide a tour of space.

Ask if the participant has been on a treadmill before.

Have lab staff demonstrate treadmill (if needed).

Have participant practice walking on the treadmill until he or she is comfortable, with little to no handrail use.

Instruct participant of any changes in speed or grade prior to initiation.

After adequate practice, help the participant to get off the treadmill and to be seated.

If using indirect calorimetry, familiarize the participant with your mouthpiece and headgear. Do so seated or standing. If possible, have the participant practice walking with mouthpiece in the lab.

After successful familiarization, have the participant walk on the treadmill with the headgear and mouthpiece.

Reminder—the number of familiarization sessions and the pace of familiarization will depend on each participant's ability.

From T.W. Rowland, American College of Sports Medicine, and North American Society for Pediatric Exercise Medicine, 2018, *Cardiopulmonary exercise testing in children and adolescents* (Champaign, IL: Human Kinetics).

5. All personnel should be positive and calm.

6. Laboratory staff may want to ask the caregiver(s) and participant for useful reinforcement and communication suggestions. It may be useful to provide a small reward for completion of the task, such as a sticker. Do not provide a food-based reward. Seek the caregiver's approval for any award before testing.

7. Due to poor validity and reliability, one should not use pure running protocols, arm ergometry, or longer field runs (1 or 1.5 mi). While some cardiovascular field tests are reliable in this population, they are valid only in children with ID without DS. Field tests have been demonstrated not to be valid in predicting peak aerobic capacity in persons with DS. Cycle ergometry protocols should be avoided because of the difficulty in coordinating forward pedal movement. This is also usually an unfamiliar exercise modality for persons with ID.

8. Try not to allow handrails to be used during the test, if possible. This may be difficult depending on the subject's eyesight, balance, proprioception, and comfort level. If handrails are used, try to minimize their use as much as possible (e.g., touching with tips of fingers instead of full-handed gripping).

9. Children and youth with ID and DS are often more active when a social benefit is perceived (41, 49). For this reason, consider music and other social motivators to modify the environment. Other examples include introducing a game-like activity to correspond with each stage of the treadmill protocol.

10. Providing a timeline of activities organized as a small book (often referred to as a social story) is extremely effective in communicating expected tasks and alleviating anxiety in children with ID. This story should include pictures and brief,

child-friendly text explaining the tasks, the equipment, anticipated rewards, and that they will feel tired and out of breath but that the activity is short. Provide the story in advance so that caregiver(s) can read it multiple times before testing. This will greatly assist with familiarity and comprehension. This strategy has been used effectively in other field-based exercise and fitness-related testing (40, 47).

Graded Exercise Test Protocol

Individualized walking protocols are most commonly used for testing people with ID, with or without DS. The key is to provide a systematic approach in designing each person's protocol, with some consistency among different individuals. Following are recommendations for developing an individualized protocol for children with ID, with or without DS.

1. Start all participants at a comfortable walking speed, which will have been determined during their previous familiarization session(s).

2. Then increase speed in an equal-graded fashion (e.g., 0.5 mph/3 min stages).

3. Once a brisk walking speed is achieved that does not compromise the participant's balance, the percent incline can be increased in a graded manner as well (e.g., 2.5% increments every 2 min).

4. Finally, increase speed to a jog or a run for 1 to 2 min at the end of the test. This will require strong and positive encouragement giving to the participant. Handrail use is not uncommon during the steeper inclines and the jogging phase of this type of protocol. The key will be to allow as much for safety and comfort as possible for the participant.

5. At the completion of peak effort, reduce both the speed and the grade as quickly as possible. The extra personnel are important during this transition to ensure the participant's safety on the treadmill. Use caution in saying "Done" before the start of the recovery phase, which could prompt the participant to try to step off the treadmill while the belt is in motion. Remind the participant to continue moving until the belt stops.

6. Lower the speed to approximately 1.8 to 2.2 mph for the recovery phase. It is difficult to obtain blood pressure measurements during a graded cardiopulmonary exercise test in this population, due to the added complexity of the task and potential balance issues. An automated blood pressure cuff may be useful but again must be part of familiarization. Furthermore, obtaining accurate and reliable ratings of perceived exertion is difficult for most people with ID, although it may work with some. This difficulty may be due to the type of mouthpiece used if using indirect calorimetry, a lack of understanding of a rating of perceived exertion, or increased task complexity. See *Recommended Treadmill Protocol for Cardiopulmonary Exercise Testing in Children and Adolescents With Intellectual Disability* for a sample treadmill protocol with recommended treadmill settings in table 19.1.

Due to differences in walking and running economy, this type of protocol cannot be used to predict peak oxygen uptake; in addition, there are no formulas that exist for persons with ID. However, peak work capacity can be measured from this protocol provided handrail support is minimized.

Alternative Protocols

Dual action cycle ergometers (cycle ergometers that use both arms and legs simultaneously) have been reported to be valid for obtaining maximal values in this population (53), with recommended stage increases of 25 W. However, submaximal data are probably not reliable due to problems with maintaining a steady pedal cadence. Field tests, such as the 20 m shuttle run, the 600 yd run/walk, and the 1-mi Rockport walk test have been validated among children and adolescents with ID without DS. These tests are not valid in children with DS (29, 31, 32, 34, 60). Table 19.2 outlines population-specific prediction formulas for these field tests. Non-ID-specific formulas will not yield accurate data; therefore, pay careful attention to the formula used. Lastly, the prediction equations for fitness may not be ideal for estimating an individual's fitness level due to the wide variation associated with these tests; they are an acceptable alternative when examining group mean data (29, 31).

Recommended Treadmill Protocol for Cardiopulmonary Exercise Testing in Children and Adolescents With Intellectual Disability

Begin with a comfortable walking speed observed during familiarization, which may be between 2 and 3.5 mph. The warm-up is included in this initial stage. A common speed for use during a majority of the test is 2.5 mph, which is indicated in table 19.1.

Increase the grade by 2.5% every 4 min until 7.5% is achieved, after which continue to increase the grade 2.5% every 2 min until 12.5% is reached.

Keeping grade at 12.5%, increase speed in an incremental fashion, paying close attention to the participant. Try to have the participant jog for a minimum of 1 min. Peak effort is generally achieved when the heart rate plateaus with no further increase in work rate.

Recover for several minutes at a speed at or under 2.5 mph. Remove mouthpiece as soon as safely possible to increase comfort level of the participant.

From T.W. Rowland, American College of Sports Medicine, and North American Society for Pediatric Exercise Medicine, 2018, *Cardiopulmonary exercise testing in children and adolescents* (Champaign, IL: Human Kinetics).

Table 19.1   Recommended Treadmill Settings

Time (min)	Speed (mph)	Grade (%)
0–4	2.5	0
4–8	2.5	2.5
8–12	2.5	5
12–14	2.5	7.5
14–16	2.5	10
16–18	2.5	12.5
18–19	2.9	12.5
19–20	3.3	12.5
20–21	4.3*	12.5

*Jogging pace (usually greater than 3.5 mph).

Similar to a graded exercise test, certain parameters are recommended to help ensure the best possible test. With field tests, ensure the following:

1. Extra personnel to assist with the administration of the test; this includes a person walking or running next to the participant without touching them, for support and encouragement.
2. Clear, one-step instructions with example(s) by staff.
3. Practice and familiarization sessions (may need to consider the "real" test on a separate day).
4. Positive encouragement and reminder instructions throughout familiarization and "real" testing.
5. Small positive reward (e.g., sticker).

Criterion for Peak Effort

The traditional gold standard of a maximal effort is a plateau in oxygen uptake with an increase in workload. However, a plateau in oxygen uptake is difficult to observe in children with ID, as in children without disabilities. For example, it would not be unusual to observe a peak oxygen uptake value between 21 and 26 ml $\cdot$ kg^{-1} $\cdot$ min^{-1} for a young adult

Table 19.2 Intellectual Disability–Specific Prediction Formulas for Peak Aerobic Capacity From Field Tests

Reference	Field test	Formula
Teo-Koh et al. 1999 (60)	1-mile Rockport walk test	$\dot{V}O_{2peak}$ (L/min) = −0.18(walk time in min) + 0.03(body weight in kg) + 2.9
Fernhall et al. 1998 (32)	20 m shuttle run	$\dot{V}O_{2peak}$ (ml · kg^{-1} · min^{-1}) = 0.35(no. of 20 m laps) − 0.59(BMI) − 4.5(gender; 1 = boys, 2 = girls) + 50.8
Fernhall et al. 1998 (32)	600 yd run/walk	$\dot{V}O_{2peak}$ (ml · kg^{-1} · min^{-1}) = −5.24(600 yd run/walk time in min) − 0.37(BMI) − 4.61(gender; 1 = boys, 2 = girls) + 73.64

with DS. Their smaller reserve would make it even more difficult to observe a traditional plateau. Additional criteria of peak effort are often difficult to ascertain in people with ID. For reasons stated previously, obtaining a rating of perceived exertion is often not feasible. Using heart rate within 10 bpm of age-predicted maximal heart rate is also not accurate due to the lower-than-expected heart rates observed in this population, especially among those with DS. A validated prediction equation during exercise testing for peak heart rate among persons with ID, with and without DS, has been developed (23)

$$\text{Estimated peak heart rate} = 210 - (0.56 \times \text{age}) - (15.5 \times \text{ID})$$

where ID = 1 for non-DS and ID = 2 for DS.

It is often common for persons with DS to have a heart rate approximately 30 to 40 bpm below the predicted maximal rate using the common 220 − age formula; persons with ID without DS have a closer-to-expected response. It is often more practical to observe the heart rate response during the exercise test and to terminate the test when a plateau in heart rate has been achieved (e.g., less than a three-beat difference from the previous stage, or less than a three- to four-beat difference during the last 30 s of running).

In individuals without ID, it is common to use a respiratory exchange ratio over 1.1 or 1.15 to help determine if peak aerobic capacity has been achieved. This criterion is difficult to use for people with ID, in particular with DS. Instead, a respiratory exchange ratio over 1.0 (or 1.05) is recommended. Fernhall and Tymeson observed that nearly all subjects had a plateau in HR and $\dot{V}O_{2peak}$ with RER at or over 1.0 (33). The physiology of this response still needs to be elucidated, but it is likely due to altered anaerobic metabolism.

Obtaining a lactate measure would not be considered reasonable during the course of a normal exercise test in this population. For instance, obtaining a measure while they were on the treadmill would not be safe due to balance issues. Further, stopping and restarting a test (i.e., a discontinuous protocol) to obtain the sample has not been validated in this population. There is also the issue that multiple sticks may be necessary to obtain the requisite samples, which may create problems with subject cooperation. Often the best criterion of peak effort among individuals with ID is observed volitional exhaustion, as well as the plateau in heart rate during the last few minutes of the exercise test. Volitional exhaustion can be observed as substantial changes in gait pattern and handrail use (e.g., more pulling vs. using the handrail for balance) and difficulty in maintaining speed. If measuring with indirect calorimetry, verbal communication with participants can be quite difficult, but if they are trying to verbally communicate at high levels of exercise, it should be assumed they would like to stop.

Conclusion

The etiology of ID is varied, and often the true cause of ID is unknown. The most common genetic cause of ID is DS. Most people with ID live in the community either at home or in group homes. Exercise testing in persons with ID is feasible, valid, and reliable, provided appropriate testing procedures are followed. This includes the use of familiarization and the selection of appropriate protocols. For children with ID without DS, $\dot{V}O_{2max}$ can be similar to that of their peers without ID. However, children with DS have lower $\dot{V}O_{2max}$ than their peers both with and without ID. The low $\dot{V}O_{2max}$ of people with DS appears to be a function of autonomic dysfunction. In addition, people with ID have lower levels of leg strength than their peers without ID, and this affects run performance because leg strength is related to run performance in children with DS. Furthermore, run performance is considerably lower in children with ID than in those without ID, even when $\dot{V}O_{2max}$ is similar. It is not unusual to find lower maximal heart rates in children with ID, especially in those with DS. Importantly, cycle ergometry has not been validated as a testing modality for individuals with ID, and only three field tests have been validated in this population (1-mi Rockport walk test, 20 m shuttle run, and the 600 yd run/walk). Thus, the selection of appropriate test protocols and procedures is crucial to obtaining valid and reliable results in this population.

Neuromuscular Disease

Olaf Verschuren, PT, PhD, Janke de Groot, PT, PhD, and Tim Takken, PhD

Indications for exercise testing in children with neuromuscular disease (NMD) are broad and have as a general goal the evaluation of exercise performance, although the tests selected depend on the goal(s) of the intervention. In fitness- and exercise-related studies in youth with NMD, a variety of generic and disease-specific exercise tests have been used. The tests can be divided into the following categories: (a) submaximal exercise tests, (b) maximal exercise tests, and (c) anaerobic exercise tests. We are using the following definitions:

- Submaximal exercise tests are tests that predict maximum or peak oxygen uptake ($\dot{V}O_{2max}$, or $\dot{V}O_{2peak}$) from a submaximal exercise intensity (e.g., the test does not require a maximal effort from the participant) (20, 21), or a test that requires the patient to exercise to a certain end point (e.g., 6 min walking).
- Maximal exercise tests are tests to directly determine the maximal aerobic fitness of a patient using a graded exercise test to exhaustion (20).
- Anaerobic exercise tests are tests that measure the anaerobic performance of a patient. Usually these tests last between 5 and 30 s and require maximal effort from the participant (20). Because of their high intensity and short duration, the anaerobic energy pathways (adenosine triphosphate, creatine phosphate, and anaerobic glycolysis) are mainly used during anaerobic exercise (6).

Because of the physical limitations imposed by NMD, standard testing protocols should often be modified for the physical ability of the patients. This chapter will review those adaptations for two conditions, cerebral palsy and muscular dystrophy, and will highlight how information from testing patients with NMD can be used clinically.

Cerebral Palsy

As defined by Bax et al., cerebral palsy (CP) describes a group of disorders of the development of movement and posture that cause activity limitation and are attributed to nonprogressive disturbances that occurred in the developing fetal or infant brain (3). The motor disorders of CP are often accompanied by disturbances of sensation, cognition, communication, perception or behavior, and by a seizure disorder (3).

There are many different indications for exercise testing in children and adolescents with CP. The general aim for performing an exercise test on the child with CP is to evaluate the exercise performance of the child and to elucidate any limiting physiological factors. These factors might differ between patients based on the child's clinical issues (23). Some of the patients with CP might have a heart disease, for example.

In general, performing a maximal exercise test is regarded as safe for the pediatric age group (1), even for children with chronic conditions like CP. No studies investigating maximal exercise in children and adolescents with CP have reported adverse health or safety concerns (5, 12, 16, 24, 30, 32). Based to the available evidence, it is our opinion that maximal exercise testing for children and adolescents with CP is suitable and safe. However we recommend that the proper safety precautions, as described by Paridon et al. (23), be taken into account.

When conducting exercise tests in children with CP it is important that the patient understands exactly what is requested of him or her. The exercise test should also not require too much technical skill from the subject. Otherwise it will be more like a motor competence test than a fitness test.

The Gross Motor Function Classification System (GMFCS) is an often-used five-level classification system to describe the gross motor function of patients with CP. This description is made on the basis of their self-initiated movement with particular emphasis on sitting, walking, and wheeled mobility (22). For subjects classified at GMFCS levels I and II, the most effective way to test their fitness is with a walking- or running-based test. Often-used treadmill protocols in the clinic are the Bruce protocol and the Balke protocol. These general protocols are often not appropriate for subjects with CP because most subjects have problems with movement coordination and an equinus position of the foot. The increasing speed and high inclination of the treadmill are problematic and difficult for these subjects. For children who use canes or walkers or a wheelchair, the most functional way of testing should be walking or propelling a wheelchair.

Based on a recently performed Delphi study, a core set of measures in the three fitness categories was identified for various GMFCS levels (36). This core set was updated based on recent literature and divided into field-based tests (table 20.1) and laboratory tests (table 20.2).

Because almost all patients classified as GMFCS levels I or II can walk and cycle, we have listed both walking- and cycling-based exercise tests. Patients classified as GMFCS level III can walk and cycle. Some patients are wheelchair-bound for covering short or long distances. Therefore, we have identified tests for walking, cycling, arm cranking, and wheelchair riding for this group. Arm cranking mimics wheelchair propelling more closely than either cycling or walking (figure 20.1). For patients classified as GMFCS level IV, the proposed tests are cycling, wheelchair riding, and arm cranking.

Recently we have reported CP-specific norm values for two field exercise tests. These norm values are available for the 10 m shuttle run test (GMFCS I and II) (35) and the muscle power sprint test (GMFCS I and II) (34). For all other listed tests in tables 20.1 and 20.2, norm values should be established for individual testing facilities.

Duchenne and Becker Muscular Dystrophy

The Becker (BMD) and Duchenne (DMD) types of muscular dystrophy are inherited diseases with a male distribution pattern. BMD and DMD are caused by a deficiency or reduced expression of the muscle protein dystrophin due to deletions or point mutations in the dystrophin gene. BMD and DMD result in a progressive loss of muscle mass

and strength and a decline in physical activity and in the ability to perform activities of daily living during childhood (4, 9). DMD is an X-linked disease that affects 1 in 3,600 live male births (9). Mutations in the dystrophin gene lead to an absence of or defect in the protein dystrophin, which results in progressive proximal weakness of the upper and lower extremities with the loss of independent ambulation by age 13 or earlier (9). BMD is a milder variant of DMD, with patients showing an in-frame mutation but still producing, in varying amounts, reduced functional dystrophin (9). Becker dystrophy patients therefore generally have a better functional prognosis than DMD patients.

About 90% of BMD and DMD patients have cardiomyopathy. It is not surprising that cardiomyopathy is the cause of death in about 50% of BMD and 20% of DMD patients (8, 10). The involvement of the heart muscle in patients with BMD may be disproportionate to the skeletal muscle involvement, but in general it is less affected than in DMD (8). While cardiac problems are common, cardiomyopathy should not be regarded as an absolute contraindication for testing. Given the frequent cardiac events in this population, however, exercise testing should *always* be performed in consultation with a cardiologist.

Many parents have questions about the appropriate levels of physical activity for their children with DMD. Current research has not adequately clarified the importance and recommended amounts of exercise for these patients. Therefore exercise testing is mostly performed in the evaluation of physical functioning or to establish the effectiveness of medical interventions in these children. For people with BMD, recent studies have shown beneficial effects of low- to moderate-intensity exercise without obvious signs of cardiac events or rhabdomyolysis. Even so, in both groups exercise testing and training should be done in close collaboration with the child's physician.

The international TREAT-NMD guidelines recommend a number of functional tests such as the North Star Ambulatory Assessment to assess motor function, quantitative muscle testing, and the six-minute walk test to evaluate functional capacity. More information on functional outcome measures can be found at the TREAT-NMD, www.treat-nmd.eu/dmd/overview. No disease-specific norm values are available for the laboratory and field exercise tests for children with muscle disease described next. These tests can only be used

Table 20.1   Field Exercise Tests

	Mode of testing	GMFCS I and II	GMFCS III	GMFCS IV	Norm values
SUBMAXIMAL EXERCISE TEST					
6-min walk test (15, 28)	Walking	+	+	Not applicable	Healthy (11)
6-min push test (37)	Wheelchair propelling	Not applicable	+	+	Not available
MAXIMAL EXERCISE TEST					
10 m shuttle run test (SRT-I and SRT-II) (32)	Walking	+	Not applicable	Not applicable	GMFCS I and II
7.5-m shuttle run test (SRT-II protocol) (7)	Walking	Not applicable	+	Not applicable	Not available
10 m shuttle ride test (SRiT) (38)	Wheelchair propelling	Not applicable	+	+	Not available
ANAEROBIC EXERCISE TEST					
Muscle power sprint test (33)	Walking	+	Not applicable	Not applicable	Healthy, GMFCS I and II
Muscle power sprint test (39)	Wheelchair propelling	Not applicable	+	+	Not available

+ = use of test is recommended for this population.

Table 20.2   Laboratory Exercise Tests

	Mode of testing	GMFCS I and II	GMFCS III	GMFCS IV	Norm values
SUBMAXIMAL EXERCISE TEST					
Arm cranking ergometer protocol (30)	Arm cranking	Not applicable	+	+	Not available
MAXIMAL EXERCISE TEST					
Graded cycle ergometry test (5)	Cycling	+	+	+	Healthy only
Graded arm ergometry test (38)	Arm cranking	Not applicable	+	+	Not available
ANAEROBIC EXERCISE TEST					
20 and 30 s Wingate cycle test (5)	Cycling	+	+	Not applicable	Not available
30 s Wingate arm cranking test (29, 39)	Arm cranking	Not applicable	+	+	Not available

+ = use of test is recommended for this population.

for evaluator purposes or for providing exercise training recommendations.

North Star Ambulatory Assessment

The North Star Ambulatory Assessment (NSAA) scale consists of 17 functional items related to ambulatory function, including standing, rising from the floor, hopping, standing on heels, and running. Additional information on how to score individual items is available from the TREAT-NMD website. Each item can be scored on a 3-point scale ranging from 0 (unable to achieve independently) to 2 (normal—achieves goal without any assistance).

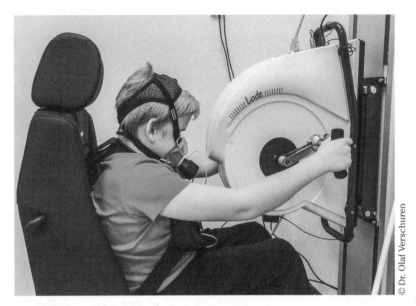

Figure 20.1 Child performing graded arm ergometry test (GAET).

Six-Minute Walk Test

The six-minute walk test (6-MWT) is often used to assess functional ambulation in children with DMD (18). While it has shown good reproducibility and good correlation with the NSAA, it has not been validated to predict $\dot{V}O_{2peak}$ and has only moderate correlation to self-reported health-related quality of life in children with DMD (18). One study looked at the relationship between the 6-MWT and energy expenditure index (EEI, measured by heart rate during 6-MWT minus resting heart rate divided by walking speed) and found a negative correlation between the two outcome measures (19). Children with a lower EEI covered more distance on the 6-MWT. Therefore, EEI might be a useful physiological marker for predicting ambulatory function in children with muscular dystrophy.

Six-Minute Assisted Bike Test

Recently a submaximal exercise test was developed for nonambulatory children with DMD for the legs and arms (13). This test uses a motor-assisted cycle ergometer device on which children are asked to pedal for 6 min at zero-load resistance. Cumulative number of revolutions as well as heart rate and rating of perceived exertion are recorded every minute. A detailed description of the test procedures can be found in the appendix of the original publication (13).

This test showed good correlation with the Motor Function Measure, a measure of disease-related disability, in boys with DMD and a significant correlation with the 6 min walking distance in healthy boys (13).

Graded Exercise Testing

There is increasing interest in assessing maximal exercise capacity as an outcome measure for interventions in patients with BMD and DMD (17, 26), but the use of cardiopulmonary exercise testing (CPET) in children with muscular dystrophies has been limited by the fear of exercise-induced muscle damage or cardiac side effects (29). However, there are no well-controlled studies available that are suggesting safety concerns in this population (14). Sockolov et al. investigated 40 years ago the maximal exercise capacity of ambulatory boys with DMD, and they did not describe any adverse events (25). In a more recent study, Andersen et al. reported the safety of different exercise intensities in adult BMD patients. They showed that the serum level of creatine kinase (CK), a biological marker for muscle damage, was still elevated 24 h after a single bout of high-intensity exercise during which patients cycled five times for 4 min at 95% of $\dot{V}O_{2peak}$, alternating with 3 min of unloaded rest (2). This increase was not observed in patients with other muscle diseases (limb-girdle dystrophy and facioscapulohumeral muscular dystrophy) (2).

This indicates that care should be taken when maximal exercise testing is performed on patients with BMD (2). However, Sveen et al. (26) observed no increase in CK levels after a 12 wk training program in adults with BMD. More exercise testing studies investigating CPET in children with BMD or DMD are needed because evidence about its safety and its importance for clinical decision making is still scarce.

The advantage of CPET includes registration of ECG and respiratory gas exchange, by which it is possible to distinguish between limiting factors that could reduce exercise capacity in patients with MD (i.e., muscular or cardiac vs. pulmonary limitations) (31). Both Sockolov et al. (25) and Sveen et al. (26) reported lower values in peak heart rate (HR_{peak}) with a mean of 136 ($\pm$13) in DMD and 176 ($\pm$9) in adults with BMD. Takken et al. reported on CPET response in a 14-year-old boy (relative $\dot{V}O_{2peak}$ [ml $\cdot$ kg^{-1} $\cdot$ min^{-1}]: +2.7 SD, HR_{peak}: 202), showing that mildly affected patients with BMD may show normal exercise capacity with adequate peak heart rate values (27).

Limited knowledge about exercise testing in this pediatric population warrants the need to further explore the feasibility of CPET, especially given upcoming new therapeutic strategies for patients with DMD and BMD, including exercise as a possible therapeutic intervention to prevent early functional decline.

Conclusion

Exercise testing in children with neuromuscular diseases poses challenges for the clinician because standard testing protocols are often not suitable for these patients. Therefore, modified protocols with lower increments per minute are needed. Wheelchair-using patients need a different approach and should be tested in their main mode of activity. For both exercise modalities, suggestions for exercise testing are provided for laboratory exercise tests as well as field exercise tests. There still is only a limited scientific basis for understanding exercise responses in patients with NMD and the role of exercise as a therapeutic modality for these patients. Information we have to date suggests that greater insights gained about exercise capacities in youth with NMD may provide important diagnostic, prognostic, and therapeutic benefits.

References

Preface

1. Paridon SM, Alpert BS, Boas SR, et al. Clinical stress testing in the pediatric age group. Circulation. 2006;113:1905-20.

2. Rhodes J, Tikkanen AU, Jenkins KJ. Exercise testing and training in children with congenital heart disease. Circulation. 2010;122:1957-67.

3. Rowland TW (ed). Pediatric laboratory exercise testing: clinical guidelines. Champaign, IL: Human Kinetics; 1993.

4. Stephens P, Paridon SM. Exercise testing in pediatrics. Pediatr Clin N Am. 2004;51:1569-87.

Chapter 1

1. Ahmad F, Kavey RE, Kveselis DA, et al. Responses of non-obese white children to treadmill exercise. J Pediatr. 2001;139:284-90.

2. Alpert BS, Verrill DE, Flood NL, et al. Complications of ergometer exercise in children. Pediatr Cardiol. 1983;4:91-96.

3. Armstrong N, Williams J, Balding J, et al. The peak oxygen uptake of British children with reference to age, sex, and sexual maturity. Eur J Appl Physiol. 1991;62:369-75.

4. Bongers BC, van Brussel M, Blank AC, et al. The oxygen efficiency slope in children with congenital heart disease: construct and group validity. Eur J Cardiovasc Prev Rehabil. 2011;18:384-92.

5. Cooper DM. Rethinking exercise testing in children: a challenge. Am J Respir Crit Care Med. 1995;152:1154-57.

6. Cumming GR, Everatt D, Hastman L. Bruce treadmill test in children: normal values in a clinic population. Am J Cardiol. 1978;41:69-75.

7. Freed MD. Exercise testing in children: a survey of techniques and safety. Circulation. 1981;64 (suppl IV): IV-278.

8. Harrell JS, McMurray RG, Baggett CD, et al. Energy costs of physical activities in children and adolescents. Med Sci Sports Exerc. 2005;37:329-36.

9. Lenk MK, Alehan D, Celiker A, et al. Bruce treadmill test in healthy Turkish children: endurance time, heart rate, blood pressure and electrocardiographic changes. Turk J Pediatr. 1998;40:167-75.

10. Noakes TD. Is it time to retire the A.V. Hill model? A rebuttal to the article by Professor Roy Shepard. Sports Med. 2011;41:263-77.

11. Olivotto I, Montereggi A, Mazzuoli F, et al. Clinical utility and safety of exercise testing in patients with hypertrophic cardiomyopathy. G. Ital Cardiol. 1999;29:11-19.

12. Rehan TA, Munkvik M, Lunde PK, et al. Intrinsic skeletal muscle alteration in chronic heart failure patients: a disease-specific myopathy or a result of deconditioning? Heart Fail Rev. 2012;17:421-36.

13. Rhodes J, Tikkanen AU, Jenkins KJ. Exercise testing and training in children with congenital heart disease. Circulation. 2010;122:1957-67.

14. Riopel DA, Taylor AB, Hohen RA. Blood pressure, heart rate, pressure-pulse product, and electrocardiographic changes in healthy children during treadmill exercise. Am J Cardiol. 1979;44:697-704.

15. Ruf K, Winkler B, Hebestreit A, et al. Risks associated with exercise testing and sports participation in cystic fibrosis. J Cyst Fibros. 2010;9:339-45.

16. Smith G, Reyes JT, Russell JL, et al. Safety of maximal cardiopulmonary exercise testing in pediatric patients with pulmonary hypertension. Chest. 2009;135:1209-14.

17. Ten Harkel AD, Takken T, Osch-Gevers V, et al. Normal values for cardiopulmonary exercise testing in children. Eur J Cardiovasc Prev Rehabil. 2011;18:48-54.

18. van der Cammen-van Zijp MM, Ijsseltijn H, Takken T, et al. Exercise testing of pre-school children using the Bruce treadmill protocol: new reference values. Eur J Appl Physiol. 2010;108:393-99.

19. Washington RL, van Gundy JC, Cohen C, et al. Normal aerobic and anaerobic exercise data for North American school-age children. J Pediatr. 1988;112:223-33.

20. Wasserman K, Hansen JE, Sue DY, et al. Principles of exercise testing and interpretation. 4th ed. Philadelphia: Lippincott Williams & Wilkins; 2005. p. 612.

21. Welsman JR, Armstrong N. Interpreting exercise performance data in relation to body size. In: Armstrong N, van Mechelen W, editors. Paediatric exercise science and medicine. 2nd ed. Oxford: Oxford University Press; 2008. p. 13-21.

Chapter 2

1. Bar-Or, O. Exercise in pediatric assessment and diagnosis. Scand J Sports Sci. 1985;7:35-39.

2. Barber G. Pediatric exercise testing: methodology, equipment, and normal values. Prog Pediatr Cardiol. 1993;2:4-10.

3. Chang R-KR, Gurvitz M, Rodriguez S, et al. Current practice of exercise stress testing among

pediatric cardiology and pulmonology centers in the United States. Pediatr Cardiol. 2006;27:110-16.

4. Connuck DM. The role of exercise stress testing in pediatric patients with heart disease. Prog Pediatr Cardiol. 2005;20:45-52.

5. Cooper DM, Weiler-Ravell D, Whipp BJ, et al. Aerobic parameters of exercise as a function of body size during growth in children. J Appl Physiol Respir Environ Exerc Physiol. 1984;56:628-34.

6. Cumming GR, Everatt D, Hastman L. Bruce treadmill test in children: normal values in a clinic population. Am J Cardiol. 1978;41:69-75.

7. Freed MD. Exercise testing in children: a survey of techniques and safety. Circulation. 1981;64(suppl IV):IV-278.

8. Gibbons RJ, Balady GJ, Beasley JW, et al. ACC/AHA guidelines for exercise testing: a report of the American College of Cardiology/American Heart Association task force on practice guidelines (Committee on Exercise Testing). J Am Coll Cardiol. 1997;30:260-315.

9. Godfrey S, Davies CT, Wozniak E, et al. Cardiorespiratory response to exercise in normal children. Clin Sci. 1971;40:419-31.

10. Hermansen L, Saltin B. Oxygen uptake during maximal treadmill and bicycle exercise. J Appl Physiol. 1969;26:31-37.

11. James FW, Blomquist CG, Freed MD, et al. Standards for exercise testing in the pediatric age group. Council on Cardiovascular Disease in the Young Ad Hoc Committee on Exercise Testing. Circulation. 1982;66:1377A-97A.

12. James FW, Kaplan S, Glueck CJ, et al. Responses of normal children and young adults to controlled bicycle exercise. Circulation. 1980;61:902-12.

13. McArdle WD, Katch FI, Pechar GS. Comparison of continuous and discontinuous treadmill and bicycle tests for max $\dot{V}O_2$. Med Sci Sports. 1973;5:156-60.

14. Myers J, Buchanan N, Walsh D, et al. Comparison of ramp versus standard exercise protocols. J Am Coll Cardiol. 1991;17:1334-42.

15. Paridon SM, Alpert BS, Boas SR, et al. Clinical stress testing in the pediatric age group: a statement from the American Heart Association Council on Cardiovascular Disease in the Young, Committee on Atherosclerosis, Hypertension, and Obesity in Youth. Circulation. 2006;113:1905-20.

16. Rodgers GP, Ayanian JZ, Balady G, et al. American College of Cardiology/American Heart Association clinical competence statement on stress testing. J Am Coll Cardiol. 2000;36:1441-53.

17. Smith G, Reyes JT, Russell JL, et al. Safety of maximal cardiopulmonary exercise testing in pediatric patients with pulmonary hypertension. Chest. 2009;135:1209-14.

18. Stephens P, Paridon SM. Exercise testing in pediatrics. Pediatr Clin North Am. 2004;51:1569-87.

19. Tanner CS, Heise CT, Barber G. Correlation of the physiologic parameters of a continuous ramp versus an incremental James exercise protocol in normal children. Am J Cardiol. 1991;67:309-12.

20. Tomassoni TL. Conducting the pediatric exercise test. In: Rowland TW, editor. Pediatric laboratory exercise testing: clinical guidelines. Champaign IL: Human Kinetics Publishers; 1993. p. 1-17.

21. Turley KR, Wilmore JH. Cardiovascular responses to treadmill and cycle ergometer exercise in children and adults. J Appl Physiol. 1997;83:948-57.

22. Washington RL, van Gundy JC, Cohen C, et al. Normal aerobic and anaerobic exercise data for North American school-age children. J Pediatr. 1988;112:223-33.

23. Washington RL, Bricker JT, Alpert BS et al. Guidelines for testing in the pediatric age group: from the Committee on Atherosclerosis and Hypertension in Children, Council on Cardiovascular Disease in the Young, the American Heart Association. Circulation. 1994;90:2166-78.

24. Yetman AT, Taylor AL, Doran A, et al. Utility of cardiopulmonary stress testing in assessing disease severity in children with pulmonary arterial hypertension. Am J Cardiol. 2005;95:697-99.

25. Zhang YY, Johnson MC II, Chow N, et al. Effect of exercise testing protocol on parameters of aerobic function. Med Sci Sports Exerc. 1991;23:625-30.

Chapter 3

1. Alpert BS, Flood NL, Strong WB, et al. Responses to ergometer exercise in a healthy biracial population of children. J Pediatr. 1982;101(4):538-45.

2. American Thoracic Society. American Thoracic Society Statement: Guidelines for the six-minute walk test. Am J Respir Crit Care Med. 2002;166:111-17.

3. Armstrong N, Balding J, Gentle P, et al. Peak oxygen uptake and physical activity in 11- to 16-year-olds. Pediatr Exer Sci. 1990;2(4):349-58.

4. Armstrong N, Williams J, Balding J, et al. The peak oxygen uptake of British children with reference to age, sex and sexual maturity. Eur J Appl Physiol Occup Physiol. 1991;62(5):369-75.

5. Åstrand P-O. Experimental studies of physical working capacity in relation to sex and age. Copenhagen: Ejnar Munksgaard; 1952.

6. Aurora P, Prasad S, Balfour-Lynn I, et al. Exercise tolerance in children with cystic fibrosis undergoing lung transplantation assessment. Eur Respir J. 2001;18(2):293-97.

7. Balke B, Ware RW. An experimental study of "physical fitness" of air force personnel. US Arm Forc Med J. 1959;10:675-88.

8. Bar-Or O, Rowland TW. Pediatric exercise medicine: from physiologic principles to health care application. Champaign, IL: Human Kinetics; 2004.

9. Bar-Or O. Pediatric sports medicine for the practitioner: from physiologic principles to clinical applications. New York: Springer-Verlag; 1983.

10. Ben Saad H, Prefaut C, Tabka Z, et al. 6-minute walk distance in healthy North Africans older than 40 years: influence of parity. Respir Med. 2009;103(1):74-84.

11. Binkhorst R, van't Hof M, Saris W. Maximal exercise in children; reference values girls and boys, 6–18 years of age. Nederlandse Hartstichting, Den Haag. 1992.

12. Boileau RA, Bonen A, Heyward VH, et al. Maximal aerobic capacity on the treadmill and bicycle ergometer of boys 11-14 years of age. J Sports Med. 1977;17:153-62.

13. Boone J, Bourgois J. The oxygen uptake responses to incremental ramp exercise: methological and physiological issues. Sports Med. 2012;42(6):511-26.

14. Bruce RA, Blackmon JR, Jones JW, et al. Exercise testing in adult normal subjects and cardiac patients. Pediatrics. 1963;32(4):742-56.

15. Bruce RA, Fisher LD, Hossack KF. Validation of exercise-enhanced risk assessment of coronary heart disease events: longitudinal changes in incidence in Seattle community practice. J Am Coll Cardiol. 1985;5(4):875-81.

16. Bruce RA, Gey GO Jr, Cooper MN, et al. Seattle Heart Watch: initial clinical, circulatory and electrocardiographic responses to maximal exercise. Am J Cardiol. 1974;33(4):459-69.

17. Bruce RA, Hossack KF, DeRouen TA, et al. Enhanced risk assessment for primary coronary heart disease events by maximal exercise testing: 10 years' experience of Seattle Heart Watch. J Am Coll Cardiol. 1983;2(3):565-73.

18. Bruce RA, Lovejoy FW Jr, Pearson R, et al. Normal respiratory and circulatory pathways of adaptation in exercise. J Clin Invest. 1949;28:1423-30.

19. Buchfuhrer MJ, Hansen JE, Robinson TE, et al. Optimizing the exercise protocol for cardiopulmonary assessment. J Appl Physiol. 1983;55:1558-64.

20. Cabrera M, DiBella J, Johnson E. Can a ramped Bruce yield similar exercise parameters in children as the standard Bruce? Pediatr Exer Sci. 2002;13(3):285.

21. Chang RK, Gurvitz M, Rodriguez S, et al. Current practice of exercise stress testing among pediatric cardiology and pulmonology centers in the United States. Pediatr Cardiol. 2007;27:110-16.

22. Chatrath R, Shenoy R, Serratto M, et al. Physical fitness of urban American children. Pediatr Cardiol. 2002;23(6):608-12.

23. Cooper D, Weiler-Ravell D, Whipp BJ, et al. Aerobic parameters of exercise as a function of body size during growth in children. J Appl Physiol. 1984;56(3):628-34.

24. Cumming G, Langford S. Comparison of nine exercise tests used in pediatric cardiology. In: Binkhorst RA, Kemper Hg, Saris WM, editors. Children and exercise XI. Champaign, IL: Human Kinetics; 1985. p. 58-68.

25. Cunha MT, Rozov T, de Oliveira RC, et al. Six-minute walk test in children and adolescents with cystic fibrosis. Pediatr Pulmonol. 2006;41(7):618-22.

26. DiBella, JA, Johnson EM, Cabrera ME. Ramped vs. standard Bruce protocol in children: A comparison of exercise responses. Pediatr Exer Sci. 2002;14(4):391-400.

27. Dubowy KO, Baden W, Bernitzki S, et al. A practical and transferable new protocol for treadmill testing of children and adults. Cardiol Young. 2008;18:615-23.

28. DuRant RH, Linder CW. An evaluation of five indexes of relative body weight for use with children. J Am Diet Assoc. 1981;78(1):35.

29. Felsing N, Brasel J, Cooper D. Effect of low and high intensity exercise on circulating growth hormone in men. J Clin Endocrinol Metab. 1992;75(1):157-62.

30. Geiger R, Strasak A, Treml B, et al. Six-minute walk test in children and adolescents. J Pediatr. 2007;150(4):395-99.

31. Godfrey S, Davies C, Wozniak E, et al. Cardiorespiratory response to exercise in normal children. Clin Sci. 1971;40(Pt 5):419-31.

32. Godfrey S. Exercise testing in children: applications in health and disease. London: W.B. Saunders Company Ltd; 1974.

33. Gratz A, Hess J, Harger A. Self-estimated physical functioning poorly predicts actual exercise capacity in adolescents and adults with congenital heart disease. Eur Heart J. 2009;30:497-504.

34. Green M, Foster C. Effect of magnitude of handrail support on prediction of oxygen uptake during treadmill testing. Med Sci Sports Exer. 1991;23:166.

35. Gruber W, Orenstein D, Braumann K, et al. Health-related fitness and trainability in children with cystic fibrosis. Pediatr Pulmonol. 2008;43(10):953-64.

36. Gulmans V, Van Veldhoven N, De Meer K, et al. The six-minute walking test in children with cystic

fibrosis: reliability and validity. Pediatr Pulmonol. 1996;22(2):85-89.

37. Gumming GR, Everatt D, Hastman L. Bruce treadmill test in children: normal values in a clinic population. Am J Cardiol. 1978;41(1):69-75.

38. Haile L, Gallagher M Jr, Robertson RJ. Perceived exertion laboratory manual. New York: Springer; 2015.

39. Hassan J, van der Net J, Helders PJ, et al. Six-minute walk test in children with chronic conditions. Brit J Sports Med. 2010;44(4):270-74.

40. Hebestreit H. Exercise testing in children—what works, what doesn't, and where to go? Paediatr Respir Rev. 2004;5(Supplement 1):S11-14.

41. James F, Blomqvist C, Freed M, et al. Standards for exercise testing in the pediatric age group. American Heart Association Council on Cardiovascular Disease in the Young. Ad hoc Committee on Exercise Testing. Circulation. 1982;66(6):1377A.

42. James FW, Kaplan S, Glueck CJ, et al. Responses of normal children and young adults to controlled bicycle exercise. Circulation. 1980;61:902-12.

43. Kaminsky LA, Whaley MH. Evaluation of a new standardized ramp protocol: the BSU/Bruce Ramp protocol. J Cardiopulm Rehabil. 1998;18:438-44.

44. Klepper SE, Muir N. Reference values on the 6-minute walk test for children living in the United States. Pediatr Phys Ther. 2011;23(1):32-40.

45. Kotte EM, De Groot JF, Bongers BC, et al. Validity and reproducibility of a new treadmill protocol: the Fitkids treadmill test. Med Sci Sports Exerc. 2016;47(10): 2241-47.

46. Krahenbuhl G, Pangrazi R, Chomokos E. Aerobic responses of young boys to submaximal running. Res Q Exerc Sport 1979;50(3):413-21.

47. Krahenbuhl GS, Pangrazi RP. Characteristics associated with running performance in young boys. Med Sci Sports Exerc. 1983;15(6):486-90.

48. Lammers AE, Hislop AA, Flynn Y, et al. The 6-minute walk test: normal values for children of 4–11 years of age. Arch Dis Child 2008;93(6):464-68.

49. Lelieveld OT, Takken T, van der Net J, et al. Validity of the 6-minute walking test in juvenile idiopathic arthritis. Arthritis Care Res. 2005;53(2):304-7.

50. Li AM, Yin J, Au JT et al. Standard reference for the six-minute-walk test in healthy children aged 7 to 16 years. Am J Respir Crit Care Med. 2007;176(2):174-80.

51. Limsuwan A, Wongwandee R, Khowsathit P. Correlation between 6-min walk test and exercise stress test in healthy children. Acta Paediatr. 2010;99(3):438-41.

52. Macek M, Vavra J, Novosadova J. Prolonged exercise in prepubertal boys. Eur J Appl Physiol. 1976;35:291-98.

53. Maher CA, Williams MT, Olds TS. The six-minute walk test for children with cerebral palsy. Int J Rehabil Res. 2008;31(2):185-88.

54. Maksud MG, Coutts KD. Application of the Cooper twelve-minute run-walk test to young males. Res Q Exerc Sport. 1971;42(1):54-59.

55. Moalla W, Gauthier R, Maingourd Y, et al. Six-minute walking test to assess exercise tolerance and cardiorespiratory responses during training program in children with congenital heart disease. Int J Sports Med. 2005;26(09):756-62.

56. Morinder G, Mattsson E, Sollander C, et al. Six-minute walk test in obese children and adolescents: reproducibility and validity. Physiother Res Int. 2009;14(2):91-104.

57. Myers J, Bellin D. Ramp exercise protocols for clinical and cardiopulmonary exercise testing. Sports Med. 2000;30(1):23-29.

58. Myers J, Buchanan N, Smith D, et al. Individualized ramp treadmill: observations on a new protocol. Chest. 1992;101(5 Suppl):236S-41S.

59. Myers J, Buchanan N, Walsh D et al. Comparison of the ramp versus standard exercise protocols. J Am Coll Cardiol. 1991;17(6):1334-42.

60. Nixon PA, Joswiak ML, Fricker FJ. A six-minute walk test for assessing exercise tolerance in severely ill children. J Pediatr. 1996;129(3):362-66.

61. Olsson LG, Swedberg K, Clark AL, et al. Six minute corridor walk test as an outcome measure for the assessment of treatment in randomized, blinded intervention trials of chronic heart failure: a systematic review. Eur Heart J. 2005;26:778-93.

62. Paap E, Net Jvd, Helders P, et al. Physiologic response of the six-minute walk test in children with juvenile idiopathic arthritis. Arthritis Care Res. 2005;53(3):351-56.

63. Paridon SM, Alpert BS, Boas SR et al. Clinical stress testing in the pediatric age group: A statement from the American Heart Association Council on Cardiovascular Disease in the Young, Committee on Atherosclerosis, Hypertension, and Obesity in Youth. Circulation. 2006;113:1905-20.

64. Pescatello LS, ed. ACSM's guidelines for exercise testing and prescription. 9th ed. Philadelphia: Lippincott Williams & Wilkins; 2014.

65. Priesnitz CV, Rodrigues GH, da Silva Stumpf C, et al. Reference values for the 6-min walk test in healthy children aged 6–12 years. Pediatr Pulmonol. 2009;44(12):1174-79.

66. Ragg KE, Murray TF, Karbonit LM, et al. Errors in predicting functional capacity from a treadmill

exercise stress test. Am Heart J. 1980;100(4):581-83.

67. Rhodes J, Tikkanen AU, Jenkins KJ. Exercise testing and training in children with congenital heart disease. Circulation. 2010;122:1957-67.

68. Riopel DA, Taylor AB, Hohn AR. Blood pressure, heart rate, pressure-rate product and electrocardiographic changes in healthy children during treadmill exercise. Am J Cardiol. 1979;44(4):697-704.

69. Rivera-Brown AM, Rivera MA, Frontera WR. Achievement of $\dot{V}O_{2max}$ criteria in adolescent runners: effects of testing protocol. Pediatr Exerc Sci. 1994;6:236-45.

70. Rivera-Brown AM, Rivera MA, Frontera WR. Applicability of criteria for $\dot{V}O_{2max}$ in active adolescents. Pediatr Exerc Sci. 1992;4(4):331-39.

71. Rowland T, Auchinachie J, Keenan T, et al. Physiologic responses to treadmill running in adult and prepubertal males. Int J Sports Med. 1987;8(04):292-97.

72. Rowland TW. Aerobic exercise testing protocols. In: TW Rowland, editor. Pediatric laboratory exercise testing: clinical guidelines. Champaign, IL: Human Kinetics; 1993, p. 19-41.

73. Rowland TW. Developmental exercise physiology. Champaign, IL: Human Kinetics; 1996.

74. Shah BN. On the 50th anniversary of the first description of a multistage exercise treadmill test: re-visiting the birth of the 'Bruce protocol.' Heart. 2013;99(24):1793-94.

75. Sheehan J, Rowland T, Burke E. A comparison of four treadmill protocols for determination of maximum oxygen uptake in 10- to 12-year-old boys. Int J Sports Med. 1987;8(01):31-34.

76. Sheth KK. Activity-induced asthma. Pediatr Clin North Am. 2003;50(3):697-716.

77. Skinner J, Bar-Or O, Bergsteinova V, et al. Comparison of continuous and intermittent tests for determining maximal oxygen intake in children. Acta Paediatr. 1971;60(S217):24-28.

78. Stephens P Jr, Paridon SM. Exercise testing in pediatrics. Pediatr Clin North Am. 2004;51(6):1569-87.

79. Swain DP, Brawner CA (eds). ACSM's resource manual for guidelines for exercise testing and prescription. 6th ed. Philadelphia: Lippincott Williams & Wilkins; 2010.

80. Takken T, Engelbert R, van Bergen M, et al. Six-minute walking test in children with ESRD: discrimination validity and construct validity. Pediatr Nephrol. 2009;24(11):2217-23.

81. Tanner CS, Heise CT, Barber G. Correlation of the physiologic parameters of a continuous ramp versus an incremental James exercise protocol in normal children. Am J Cardiol. 1991;67:309-12.

82. Thompson P, Beath T, Bell J, et al. Test–retest reliability of the 10-metre fast walk test and 6-minute walk test in ambulatory school-aged children with cerebral palsy. Dev Med Child Neurol. 2008;50(5):370-76.

83. van der Cammen-van MH, IJsselstijn H, Takken T et al. Exercise testing of pre-school children using the Bruce treadmill protocol: new reference values. Eur J Appl Physiol. 2010;108(2):393-99.

84. Washington R, Bricker J, Alpert B, et al. Guidelines for exercise testing in the pediatric age group. From the Committee on Atherosclerosis and Hypertension in Children, Council on Cardiovascular Disease in the Young, the American Heart Association. Circulation. 1994;90(4):2166-79.

85. Washington RL, van Gundy JC, Cohen C, et al. Normal aerobic and anaerobic exercise data for North American school-age children. J Pediatr. 1988;112(2):223-33.

86. Wasserman K, Hansen JE, Sue DY, et al. Principles of exercise testing and interpretation: including pathophysiology and clinical applications. Philadelphia: Lippincott Williams & Wilkins; 2005.

87. Wessel HU, Strasburger JF, Mitchell BM. New standards for the Bruce treadmill protocol in children and adolescents. Pediatr Exerc Sci. 2001;13(4):392-401.

88. Whipp BJ, Davis JA, Torres F, et al. A test to determine parameters of aerobic function during exercise. J Appl Physiol. 1981;50(1):217-21.

89. Zhang Y, Johnson M 2nd, Chow N, et al. Effect of exercise testing protocol on parameters of aerobic function. Med Sci Sports Exerc. 1991;23(5):625.

Chapter 4

1. Atkov OY, Bednenko VS, Formina GA. Ultrasound techniques in space medicine. Aviat Environ Med. 1987;58 Suppl 9:A69-A73.

2. Bainbridge FA. The influence of venous filling upon the rate of the heart. J Physiol. 1915;50:65-84.

3. Bar-Or O, Shephard RJ, Allen CL. Cardiac output of 10-13 year old boys and girls during submaximal exercise. J Appl Physiol. 1971;30:219-23.

4. Bevegard S, Holmgren A, Johnsson B. The effect of body position on the circulation at rest and during exercise, with special reference to the influence on the stroke volume. Acta Physiol Scand. 1960;49:279-98.

5. Bevegard BS, Shepherd JT. Regulation of the circulation during exercise in man. Physiol Rev. 1967;47:178-213.

6. Bogaard HJ, Woltjer HH, Dekker BM, et al. Haemodynamic response to exercise in healthy young and elderly subjects. Eur J Appl Physiol. 1997;75:435-42.

7. Christie JL, Sheldahl LMN, Tristahi FE. Cardiovascular regulation during head out water immersion. J Appl Physiol. 1990;69:657-64.

8. Clausen JP. Circulatory adjustments to dynamic exercise and effect of physical training in normal subjects and in patients with coronary artery disease. Prog Cardiovasc Dis. 1976;18:459-95.

9. Eriksson BO, Grimby G, Saltin B. Cardiac output and arterial blood gases during exercise in pubertal boys. J Appl Physiol. 1971;31:348-52.

10. Guyton AC. Regulation of cardiac output. N Engl J Med. 1967;277:805-12.

11. Higginbotham M, Morris KG, Williams RS, et al. Regulation of stroke volume during submaximal and maximal upright exercise in normal man. Circ Res. 1986;58:281-91.

12. LaGerche A, Heidbuchel H, Burns AT, et al. Disproportionate exercise load and remodeling of the athlete's heart. Med Sci Sports Exerc. 2011;43:974-81.

13. LaGerche A, MacIsaac AI, Burns AT, et al. Pulmonary transit of agitated contrast is associated with enhanced pulmonary reserve and right ventricular function during exercise. J Appl Physiol. 2010;109:1307-17.

14. Linden RJ. The size of the heart. Cardioscience. 1994;12:5225-33.

15. Magel JR, Mcardle WD, Toner M, et al. Metabolic and cardiovascular adjustment to arm training. J Appl Physiol. 1978;45:75-79.

16. Nottin S, Vinet A, Stecken F, et al. Central and peripheral cardiovascular adaptations to exercise in endurance-trained children. Acta Physiol Scand. 2002;175:85-92.

17. Nottin S, Vinet A, Stecken F, et al. Central and peripheral cardiovascular adaptations during maximal cycle exercise in boys and men. Med Sci Sports Exerc. 2002;33:456-63.

18. Obert P, Mandigout S, Nottin S, et al. Cardiovascular responses to endurance training in children: effect of gender. Eur J Clin Invest. 2003;33:199-208.

19. Oxborough D, Shave R, Warburton D, et al. Dilation and dysfunction of the right ventricle immediately after ultraendurance exercise. Circ Cardiovasc Imaging. 2011;4:253-63.

20. Patterson SW, Starling EH. On the mechanical factors which determine the output of the ventricles. J Physiol. 1914;48:357-59.

21. Perrault HM, Turcotte RA. Do athletes have the "athlete's heart"? Progr Pediatr Cardiol. 1993;2:40-50.

22. Pokan R, von Duvillard SP, Hoffman P. Change in left atrial and ventricular dimensions during and immediately after exercise. Med Sci Sports Exerc. 2000;32:1713-18.

23. Poliner LR, Dhemer GJ, Lewis SE, et al. Left ventricular performance in normal subjects: a comparison of the responses to exercise in the upright and supine positions. Circulation. 1980;62:528-34.

24. Ross J, Linhart JW, Braunwald E. Effects of changing heart rate in man by electrical stimulation of the right atrium. Circulation. 1965;32:549-58.

25. Rowland T. Echocardiography and circulatory response to progressive endurance exercise. Sports Med. 2008;38:541-51.

26. Rowland T, Blum JW. Cardiac dynamics during upright cycle exercise in boys. Am J Hum Biol. 2000;12:749-57.

27. Rowland T, Bougault V, Waslter G, et al. Cardiac responses to swim bench exercise in age-group swimmer and non-athletic children. J Sci Med Sport. 2009;12:266-72.

28. Rowland T, Garrard M, Marwood S, et al. Myocardial performance during progressive exercise in athletic adolescent males. Med Sci Sports Exerc. 2009;41:1721-28.

29. Rowland T, Garrison A, Delulio A. Circulatory responses to progressive exercise: insights from positional differences. Int J Sports Med. 2003;24:512-17.

30. Rowland T, Kline G, Goff D, et al. One-mile run performance and cardiovascular fitness in children. Arch Pediatr Adolesc Med. 1999;153:845-49.

31. Rowland T, Kline G, Goff D, et al. Physiological determinants of maximal aerobic power in healthy 12-year-old boys. Pediatr Exerc Sci. 1999;11:317-26.

32. Rowland T, Mannie E, Gawle L. Dynamics of left ventricular diastolic filling during exercise: a Doppler echocardiographic study of 10-14 year old boys. Chest. 2001;120:145-50.

33. Rowland T, Unnithan V. Myocardial inotropic response to progressive exercise in health subjects: a review. Curr Sports Med Rep. 2013;12:93-100.

34. Rowland T, Unnithan V. Stroke volume dynamics during progressive exercise in healthy adolescents. Pediatr Exerc Sci. 2013;25:173-85.

35. Rowland T, Unnithan V, Fernhall BO, et al. Left ventricular response to dynamic exercise in young cyclists. Med Sci Sports Exerc. 2002;34:637-42.

36. Rowland T, Unnithan V, Garrard M, et al. Sex influence on myocardial function with exercise in adolescents. Am J Hum Biol. 2010;22:680-82.

37. Rowland T, Unnithan V, Roche D. Myocardial function and aerobic fitness in adolescent females. Eur J Appl Physiol. 2011;111:1991-97.

38. Rowland T, Wehnert M, Miller K. Cardiac responses to exercise in competitive child cyclists. Med Sci Sports Exerc. 2000;32:747-52.

39. Rowland T, Wharton M, Masters T, et al. Right ventricular myocardial responses to progressive exercise in young adult males. Int J Sports Med, in press.

40. Rushmer RF, Smith OA. Cardiac control. Physiol Rev. 1959;39:41-68.

41. Sullivan MJ, Cobb FR, Higginbotham MB. Stroke volume increases by similar mechanisms during upright exercise in normal men and women. Am J Cardiol. 1991;67:1405-12.

42. Tschakovsky ME, Pike KE. Cardiovascular responses to exercise and limitations to human performance. In Taylor NAS, Groeller H, editors. Physiological bases of human performance during work and exercise. London: Elsevier; 2008, p. 5-28.

43. Tschakovsky ME, Shoemaker JK, Hughson RL. Vasodilation and muscle pump contribution to immediate exercise hyperemia. Am J Physiol. 1996;271:H1697-H1701.

Chapter 5

1. Allen SW, Shaffer EM, Harrigan LA, et al. Maximal voluntary work and cardiorespiratory fitness in patients who have had Kawasaki syndrome. J Pediatr. 1992;121:221-25.

2. Aziz PF, Wieand TS, Ganley J, et al. Genotype- and mutation site-specific adaptation during exercise, recovery, and postural changes in children with long QT syndrome. Circ Arrhythm Electrophysiol. 2011;4:867-73.

3. Bailey DA, Ross WD, Mirwald RI, et al. Size dissociation of maximal aerobic power during growth in boys. Med Sport. 1978;11:140-51.

4. Basso C, Maron BJ, Corrado D, et al. Clinical profile of congenital coronary artery anomalies with origin from the wrong aortic sinus leading to sudden death in young competitive athletes. J Am Coll Cardiol. 2000;35:1493-1501.

5. Bengtsson E. The exercise electrocardiogram in healthy children and in comparison with adults. Acta Med Scand. 1956;154:225-44.

6. Benson DW, Wang DW, Dyment M, et al. Congenital sick sinus syndrome caused by recessive mutations in the cardiac sodium channel gene (SCN5A). J Clin Invest. 2003;112:1019-28.

7. Berger WR, Gow RM, Kamberi S, et al. The QT and corrected QT interval in recovery after exercise in children. Circ Arrhythm Electrophysiol. 2011;4:448-55.

8. Bodner ME, Rhodes EC. A review of the concept of the heart rate deflection point. Sports Med. 2000;30:31-36.

9. Bricker JT, Porter CJ, Garson A, et al. Exercise testing in children with Wolff-Parkinson-White syndrome. Am J Cardiol. 1985;55:1001-4.

10. Bricker JT, Traweek MS, Danford DA, et al. Exercise related bundle branch block in children. Am J Cardiol. 1985;56:796-97.

11. Bricker JT, Traweek MS, Smith RT, et al. Exercise-related ventricular tachycardia in children. Am Heart J. 1986;122:186-88.

12. Bricker JT, Traweek MS, Smith RT, et al. SVT during exercise testing in children [abstract]. In: Abstracts of the 2nd World Congress of Pediatric Cardiology. New York: Springer Verlag; 1985. p. 107.

13. Bricker JT, Vargo TA. Advances in exercise testing. In: Bricker JT, McNamara DG, editors. Pediatric cardiology: its current practice. London: Edward Arnold; 1988. p. 99-111.

14. Bricker JT, Ward KA, Zinner A, et al. Decrease in canine endocardial and epicardial electrogram voltages with exercise: implications for pacemaker sensing. Pacing Clin Electrophysiol. 1988;11:460-64.

15. Cagdas D, Celiker A, Ozer S. Premature ventricular contractions in normal children. Turk J Pediatr. 2008;50:260-64.

16. Chandramouli B, Ehmke DA, Lauer RM. Exercise-induced electrocardiographic changes in children with congenital aortic stenosis. J Pediatr. 1975;87:725-30.

17. Coelho A, Palileo E, Ashley W. Tachyarrhythmias in young athletes. J Am Coll Cardiol. 1986;7:237-43.

18. Cole CR, Foody JM, Blackstone EH, et al. Heart recovery after submaximal exercise testing as a predictor of mortality in a cardiovascularly healthy cohort. Ann Intern Med. 2000;132:552-55.

19. Committee on Rheumatic Fever, Endocarditis, and Kawasaki Disease. Guidelines for long-term management of patients with Kawasaki disease. Circulation. 1994;89:916-22.

20. Cumming GR. Hemodynamics of supine bicycle exercise in 'normal' children. Am Heart J. 1977;93:617-22.

21. Cumming GR, Langford S. Comparison of nine exercise tests used in pediatric cardiology. In: Binkhorst RA, Kemper HCG, Saris WHM, editors. Children and exercise XI. Champaign, IL: Human Kinetics; 1985. p. 58-68.

22. Draper DE, Giddins NG, McCort J, et al. Diagnostic usefulness of graded exercise testing in pediatric supraventricular tachycardia. Can J Cardiol. 2009;25:407-10.

23. Halloran KH. The telemetered exercise electrocardiogram in congenital aortic stenosis. Pediatrics. 1971;47:31-39.

24. Hijazki ZM, Udelson JE, Snapper H, et al. Physiologic significance of chronic coronary aneurysms in patients with Kawasaki disease. J Am Coll Cardiol. 1994;24:1633-38.

25. Jacobsen JR, Garson A, Gillette PC, et al. Premature ventricular contractions in normal children. J Pediatr. 1978;92:36-38.

26. James FW. Exercise testing in children and young adults: an overview. Cardiovasc Clin. 1978;9:187-203.

27. Karpawich PP, Gillette PC, Garson A, et al. Congenital complete heart block: clinical and electrophysiologic predictors of need for pacemaker insertion. Am J Cardiol. 1981;48:1098-1102.

28. Kato H, Ichinoise E, Yoshioka F. Fate of coronary aneurysms in Kawasaki disease: serial coronary angiography and long-term follow-up study. Am J Cardiol. 1982;49:1768-66.

29. Kennedy HL, Underhill SJ. Frequent or complex ventricular ectopy in apparently healthy subjects. A clinical study of 25 cases. Am J Cardiol. 1976;38:141-48.

30. Lin LY, Kuo HK, Lai LP, et al. Inverse correlation between heart rate recovery and metabolic risks in healthy children and adolescents: insight from the National Health and Nutrition Examination Survey 1999-2002. Diabetes Care. 2008;31:1015-20.

31. Marcus B, Gillette PC, Garson A. Intrinsic heart rate in children and young adults: an index of sinus node function isolated from autonomic control. Am Heart J. 1990;119:911-16.

32. Mohler PJ, Schott JJ, Gramolini AO, et al. Ankyrin-B mutation causes type 4 long QT cardiac arrhythmia and sudden death. Nature. 2003;421:634-39.

33. Moltedo JM, Iyer RV, Forman H, et al. Is exercise stress testing a cost-saving strategy for risk assessment of pediatric Wolff-Parkinson-White syndrome patients? Ochsn J. 2006;6:64-67.

34. Nilsson G, Hedberg P, Jonason T, et al. Heart rate recovery is more strongly associated with the metabolic syndrome, waist circumference, and insulin sensitivity in women than in men among the elderly in the general population. Am Heart J. 2007;154(3):460.e1-e7.

35. Obeyesekere MN, Klein GJ, Modi S, et al. How to perform and interpret provocative testing for the diagnosis of Brugada syndrome, long QT syndrome, and catecholamine polymorphic ventricular tachycardia. Circ Arrhythm Electrophysiol. 2011;4:958-64.

36. Osaki M, McCrindle BW, Van Arsdell G, et al. Anomalous origin of a coronary artery from the opposite sinus of Valsvalva with an intra-arterial course: clinical profile and approach to management in the pediatric population. Pediatr Cardiol. 2008;29:24-30.

37. Paridon SM, Bricker JT. Quantitative QRS changes with exercise in children and adolescents. Med Sci Sports Exerc. 1990;22:159-64.

38. Paridon SM, Galioto FM, Vincent JA, et al. Exercise capacity and incident of myocardial perfusion defects after Kawasaki disease in children and adolescents. J Am Coll Cardiol. 1991;17:729-32.

39. Paridon SM, Ross RD, Kuhns LR, et al. Myocardial performance and perfusion during exercise in patients with coronary artery disease caused by Kawasaki disease. J Pediatr. 1990;116:52-56.

40. Reilly T, Brooks GA. Selective persistence of circadian rhythms in physiological responses to exercise. Chronobiol Int. 1990;7:59-67.

41. Riopel DA, Taylor AB, Hohn AR. Blood pressure, heart rate, and electrocardiographic changes in healthy children during treadmill exercise. Am J Cardiol. 1979;44:697-704.

42. Rocchini AP, Chun PO, Dick M. Ventricular tachycardia in children. Am J Cardiol. 1981;47:1091-97.

43. Rozanski JJ, Dimich I, Steinfeld L, et al. Maximal exercise stress testing in evaluation of arrhythmias in children: results and reproducibility. Am J Cardiol. 1979;43:951-56.

44. Rowland T, Cunningham L. Heart rate deceleration during treadmill exercise in children [abstract]. Pediatr Exerc Sci. 1993;5:463.

45. Rowland T, Unnithan V, Barker P, et al. Time-of-day effect on cardiac responses to progressive exercise. Chronobiol Int. 2011;28:1-6.

46. Rowland T, Unnithan V, Fernhall B, et al. Left ventricular response to dynamic exercise in young cyclists. Med Sci Sports Exerc. 2002;34:637-42.

47. Schmidt-Nielsen K. Scaling. Why is animal size so important? Cambridge: Cambridge University Press; 1984.

48. Sheehan JM, Rowland T, Burke EJ. A comparison of four treadmill protocols for determination of maximal oxygen uptake in 10-12 year old boys. Int J Sports Med. 1987;8:31-34.

49. Sholler GF, Walsh EP. Congenital complete heart block in patients without anatomic cardiac defects. Am Heart J. 1989;118:1193-98.

50. Simhaee D, Corriveau N, Gurm R, et al. Recovery heart rate: an indicator of cardiovascular risk among middle school children. Pediatr Cardiol. 2013;34:1431-37.

51. Spar DS, Silver ES, Hordof AJ, et al. Relation of the utility of exercise testing for risk assessment in pediatric patients with ventricular preexcitation to pathway location. Am J Cardiol. 2012;109:1011-14.

52. Strasberg B, Ashley WW, Wyndham CRC, et al. Treadmill testing in the Wolff-Parkinson-White syndrome. Am J Cardiol. 1980;45:742-48.

53. Swan H, Toivonen L, Viitasalo M. Rate adaptation of QT intervals during and after exercise in children with congenital long QT syndrome. Eur Heart J. 1998;19:508-13.

54. Takenaka K, Ai T, Shimizu W, et al. Exercise stress amplifies genotype-phenotype correlation in the LQT1 and LQT2 forms of the long QT syndrome. Circulation. 2003;107:838-44.

55. Thapar MK, Strong WB, Miller MD, et al. Exercise electrocardiography in healthy black children. Am J Dis Child. 1978;132:592-595.

56. Thompson WM, Nida JR, Riley HD. Electrocardiographic findings in adolescents at rest and at maximal exertion. Pediatrics. 1969;43:438-42.

57. Viitasalo M, Rovamo L, Toivonen L, et al. Dynamics of the QT interval during and after exercise in healthy children. Eur Heart J. 1996;17:1723-28.

58. Walsh EP, Breul CI, Triedman JK. Cardiac arrhythmias. In: Keane JF, Lock JE, Fyler DC. Nadas' pediatric cardiology. 2nd ed. Philadelphia: Saunders; 2006. p. 477-524.

59. Washington RL, van Gundy JC, Cohen C, et al. Normal aerobic and anaerobic exercise data for North American school age children. J Pediatr. 1988;112:223-33.

60. Weintraub RG, Gow RM, Wilkinson JL. The congenital long QT syndromes in childhood. J Am Coll Cardiol. 1990;16:674-80.

61. Wiles HB. Exercise testing for arrhythmia. Prog Pediatr Cardiol. 1993;2:51-60.

62. Wilson BA, Chisholm D. Total body maximal aerobic power in children measured by the Concept II rowing ergometer [abstract]. Pediatr Exerc Sci. 1993;5:487.

63. Winkler RB, Freed MD, Nadas AS. Exercise-induced ventricular ectopy in children and young adults with complete heart block. Am Heart J. 1980;99:87-92.

64. Wren C. Arrhythmias in children: the influence of exercise and the role of exercise testing. Eur Heart J. 1987;8 suppl D:25-28.

65. Yabek SM. Ventricular arrhythmias in children with an apparently normal heart. J Pediatr. 1991;119:1-11.

Chapter 6

1. Ahmad F, Kavey RE, Kveselis DA, et al. Responses of non-obese white children to treadmill exercise. J Pediatr. 2001;139:284-90.

2. Alpert BS, Dover EV, Booker DL, et al. Blood pressure response to dynamic exercise in healthy children —black vs white. J Pediatr. 1981;99:556-60.

3. Alpert BS, Flood NL, Balfour IC, et al. Automated blood pressure measurement during ergometer exercise in children. Cathet Cardiovasc Diagn. 1982;8:525-33.

4. Alpert BS, Flood NL, Strong WB, et al. Responses to ergometer exercise in a healthy biracial population of children. J Pediatr. 1982;101:538-45.

5. Alpert BS, Gilman PA, Strong WB, et al. Hemodynamic and ECG responses to exercise in children with sickle cell anemia. Am J Dis Child. 1981;135:362-66.

6. Alpert BS, Moes DM, Durant RH, et al. Hemodynamic responses to ergometer exercise in children and young adults with left ventricular pressure or volume overload. Am J Cardiol. 1983;52:563-67.

7. Alpert BS, Verrill DE, Flood NL, et al. Complications of ergometer exercise in children. Pediatr Cardiol. 1983;4:91-96.

8. Amoore JN, Scott DH. Can simulators evaluate systematic differences between oscillometric non-invasive blood-pressure monitors? Blood Press Monit. 2000;5:81-89.

9. Arensman FW, Treiber FA, Gruber MP, et al. Exercise-induced differences in cardiac output, blood pressure, and systemic vascular resistance in a healthy biracial population of 10-year-old boys. Am J Dis Child. 1989;143:212-16.

10. Barber G, Danielson GK, Heise CT, et al. Cardiorespiratory response to exercise in Ebstein's anomaly. Am J Cardiol. 1985;56:509-14.

11. Canzanello VJ, Jensen PL, Schwartz GL. Are aneroid sphygmomanometers accurate in hospital and clinic settings? Arch Intern Med. 2001;161:729-31.

12. Carletti L, Rodrigues AN, Perez AJ, et al. Blood pressure response to physical exertion in adolescents: influence of overweight and obesity. Arq Bras Cardiol. 2008;91:24-30.

13. Caru B, Colombo E, Santoro F, et al. Regional flow responses to exercise. Chest. 1992;101:223S-25S.

14. Casiglia E, Palatini P, Bongiovi S, et al. Haemodynamics of recovery after strenuous exercise in physically trained hypertensive and normotensive subjects. Clin Sci (Lond). 1994;86:27-34.

15. Connor TM. Evaluation of persistent coarctation of aorta after surgery with blood pressure measurement and exercise testing. Am J Cardiol. 1979;43:74-78.

16. Connor TM, Baker WP. A comparison of coarctation resection and patch angioplasty using postexercise blood pressure measurements. Circulation. 1981;64:567-72.

17. Dipla K, Zafeiridis A, Koidou I, et al. Altered hemodynamic regulation and reflex control during exercise and recovery in obese boys. Am J Physiol Heart Circ Physiol. 2010;299:H2090-96.

18. Dlin R. Blood pressure response to dynamic exercise in healthy and hypertensive youths. Pediatrician. 1986;13:34-43.

19. Dlin RA, Hanne N, Silverberg DS, et al. Follow-up of normotensive men with exaggerated blood pressure response to exercise. Am Heart J. 1983;106:316-20.

20. Driscoll DJ, Danielson GK, Puga FJ, et al. Exercise tolerance and cardiorespiratory response to exercise after the Fontan operation for tricuspid atresia or functional single ventricle. J Am Coll Cardiol. 1986;7:1087-94.

21. Ehrman J, Keteyian S, Fedel F, et al. Cardiovascular responses of heart transplant recipients to graded exercise testing. J Appl Physiol (1985). 1992;73:260-64.

22. Engvall J, Nylander E, Wranne B. Arm and ankle blood pressure response to treadmill exercise in normal people. Clin Physiol. 1989;9:517-24.

23. Fletcher GF, Ades PA, Kligfield P, et al. Exercise standards for testing and training: a scientific statement from the American Heart Association. Circulation. 2013;128:873-934.

24. Freed MD, Rocchini A, Rosenthal A, et al. Exercise-induced hypertension after surgical repair of coarctation of the aorta. Am J Cardiol. 1979;43:253-58.

25. Gibbons RJ, Balady GJ, Bricker JT, et al. ACC/AHA 2002 guideline update for exercise testing: summary article: a report of the American College of Cardiology/American Heart Association Task Force on Practice Guidelines (Committee to Update the 1997 Exercise Testing Guidelines). Circulation. 2002;106:1883-92.

26. Giordano U, Calzolari A, Matteucci MC, et al. Exercise tolerance and blood pressure response to exercise testing in children and adolescents after renal transplantation. Pediatr Cardiol. 1998;19:471-73.

27. Godfrey S. Exercise testing in children. London: Saunders; 1974.

28. Guenthard J, Wyler F. Exercise-induced hypertension in the arms due to impaired arterial reactivity after successful coarctation resection. Am J Cardiol. 1995;75:814-17.

29. Guenthard J, Zumsteg U, Wyler F. Arm–leg pressure gradients on late follow-up after coarctation repair. Possible causes and implications. Eur Heart J. 1996;17:1572-75.

30. Guimarães GV, Bellotti G, Mocelin AO, et al. Cardiopulmonary exercise testing in children with heart failure secondary to idiopathic dilated cardiomyopathy. Chest. 2001;120:816-24.

31. Guo Y, Zhou AQ, Gao W, et al. [Evaluation of physiological index on treadmill exercise testing of 294 healthy children in Shanghai area]. Zhonghua Er Ke Za Zhi. 2003;41:338-43.

32. Hauser M. Exercise blood pressure in congenital heart disease and in patients after coarctation repair. Heart. 2003;89:125-26.

33. Hauser M, Kuehn A, Wilson N. Abnormal responses for blood pressure in children and adults with surgically corrected aortic coarctation. Cardiol Young. 2000;10:353-57.

34. Hirschfeld S, Tuboku-Metzger AJ, Borkat G, et al. Comparison of exercise and catheterization results following total surgical correction of tetralogy of Fallot. J Thorac Cardiovasc Surg. 1978;75:446-51.

35. Jackson AS, Squires WG, Grimes G, et al. Prediction of future resting hypertension from exercise blood pressure. J Cardiopulm Rehab. 1983;3:263-68.

36. James FW, Blomqvist CG, Freed MD, et al. Standards for exercise testing in the pediatric age group. American Heart Association Council on Cardiovascular Disease in the Young. Ad hoc committee on exercise testing. Circulation. 1982;66:1377A-97A.

37. James FW, Kaplan S, Glueck CJ, et al. Responses of normal children and young adults to controlled bicycle exercise. Circulation. 1980;61:902-12.

38. James FW, Schwartz DC, Kaplan S, et al. Exercise electrocardiogram, blood pressure, and working capacity in young patients with valvular or discrete subvalvular aortic stenosis. Am J Cardiol. 1982;50:769-75.

39. Kappetein PA, Guit GL, Bogers AJ, et al. Noninvasive long-term follow-up after coarctation repair. Ann Thorac Surg. 1993;55:1153-59.

40. Lauer RM, Burns TL, Clarke WR, et al. Childhood predictors of future blood pressure. Hypertension. 1991;18:I74-81.

41. Lunt D, Briffa T, Briffa NK, et al. Physical activity levels of adolescents with congenital heart disease. Aust J Physiother. 2003;49:43-50.

42. MacRae HS, Allen PJ. Automated blood pressure measurement at rest and during exercise: evaluation of the motion tolerant CardioDyne NBP 2000. Med Sci Sports Exerc. 1998;30:328-31.

43. Mahoney LT, Schieken RM, Clarke WR, et al. Left ventricular mass and exercise responses predict future blood pressure. The Muscatine Study. Hypertension. 1988;12:206-13.

44. Markel H, Rocchini AP, Beekman RH, et al. Exercise-induced hypertension after repair of coarctation of the aorta: arm versus leg exercise. J Am Coll Cardiol. 1986;8:165-71.

45. Maron BJ, Humphries JO, Rowe RD, et al. Prognosis of surgically corrected coarctation of the

aorta. A 20-year postoperative appraisal. Circulation. 1973;47:119-26.

46. Maron BJ, Zipes DP. Introduction: eligibility recommendations for competitive athletes with cardiovascular abnormalities—general considerations. J Am Coll Cardiol. 2005;45:1318-21.

47. Mathews RA, Fricker FJ, Beerman LB, et al. Exercise studies after the Mustard operation in transposition of the great arteries. Am J Cardiol. 1983;51:1526-1529.

48. McConnell ME, Daniels SR, Lobel J, et al. Hemodynamic response to exercise in patients with sickle cell anemia. Pediatr Cardiol. 1989;10:141-44.

49. McHam SA, Marwick TH, Pashkow FJ, et al. Delayed systolic blood pressure recovery after graded exercise: an independent correlate of angiographic coronary disease. J Am Coll Cardiol. 1999;34:754-59.

50. Morise AP. Exercise testing in nonatherosclerotic heart disease: hypertrophic cardiomyopathy, valvular heart disease, and arrhythmias. Circulation. 2011;123:216-25.

51. Paridon SM, Alpert BS, Boas SR, et al. Clinical stress testing in the pediatric age group: a statement from the American Heart Association Council on Cardiovascular Disease in the Young, Committee on Atherosclerosis, Hypertension, and Obesity in Youth. Circulation. 2006;113:1905-20.

52. Rhodes J, Ubeda Tikkanen A, Jenkins KJ. Exercise testing and training in children with congenital heart disease. Circulation. 2010;122:1957-67.

53. Riopel DA, Taylor AB, Hohn AR. Blood pressure, heart rate, pressure-rate product and electrocardiographic changes in healthy children during treadmill exercise. Am J Cardiol. 1979;44:697-704.

54. Ross RD, Clapp SK, Gunther S, et al. Augmented norepinephrine and renin output in response to maximal exercise in hypertensive coarctectomy patients. Am Heart J. 1992;123:1293-99.

55. Rowell LB. Reflex control of the circulation during exercise. Int J Sports Med. 1992;13 Suppl 1:S25-27.

56. Safar ME, Smulyan H. The blood pressure measurement—revisited. Am Heart J. 2006;152:417-19.

57. Sarubbi B, Pacileo G, Pisacane C, et al. Exercise capacity in young patients after total repair of tetralogy of Fallot. Pediatr Cardiol. 2000;21:211-15.

58. Sehested J, Baandrup U, Mikkelsen E. Different reactivity and structure of the prestenotic and poststenotic aorta in human coarctation. Implications for baroreceptor function. Circulation. 1982;65:1060-65.

59. Smith RT Jr, Sade RM, Riopel DA, et al. Stress testing for comparison of synthetic patch aorto-plasty with resection and end to end anastomosis for repair of coarctation in childhood. J Am Coll Cardiol. 1984;4:765-70.

60. Takken T, Blank AC, Hulzebos EH, et al. Cardiopulmonary exercise testing in congenital heart disease: (contra)indications and interpretation. Neth Heart J. 2009;17:385-92.

61. Tanaka H, Bassett DR Jr, Turner MJ. Exaggerated blood pressure response to maximal exercise in endurance-trained individuals. Am J Hypertens. 1996;9:1099-1103.

62. The fourth report on the diagnosis, evaluation, and treatment of high blood pressure in children and adolescents. National High Blood Pressure Education Program Working Group on High Blood Pressure in Children and Adolescents. Pediatrics. 2004;114:555-76.

63. Treiber FA, Musante L, Strong WB, et al. Racial differences in young children's blood pressure. Responses to dynamic exercise. Am J Dis Child. 1989;143:720-23.

64. Turmel J, Bougault V, Boulet LP, et al. Exaggerated blood pressure response to exercise in athletes: dysmetabolism or altered autonomic nervous system modulation? Blood Press Monit. 2012;17:184-92.

65. Vibarel N, Hayot M, Pellenc PM, et al. Non-invasive assessment of inspiratory muscle performance during exercise in patients with chronic heart failure. Eur Heart J. 1998;19:766-73.

66. Wanne OP, Haapoja E. Blood pressure during exercise in healthy children. Eur J Appl Physiol Occup Physiol. 1988;58:62-67.

67. Washington RL, van Gundy JC, Cohen C, et al. Normal aerobic and anaerobic exercise data for North American school-age children. J Pediatr. 1988;112:223-33.

68. Weber KT, Janicki JS. Cardiopulmonary exercise testing for evaluation of chronic cardiac failure. Am J Cardiol. 1985;55:22A-31A.

69. Wessel HU, Cunningham WJ, Paul MH, et al. Exercise performance in tetralogy of Fallot after intracardiac repair. J Thorac Cardiovasc Surg. 1980;80:582-93.

70. Whitmer JT, James FW, Kaplan S, et al. Exercise testing in children before and after surgical treatment of aortic stenosis. Circulation. 1981;63:254-63.

71. Wilson NV, Meyer BM. Early prediction of hypertension using exercise blood pressure. Prev Med. 1981;10:62-68.

72. Zellers TM, Driscoll DJ, Mottram CD, et al. Exercise tolerance and cardiorespiratory response to

exercise before and after the Fontan operation. Mayo Clin Proc. 1989;64:1489-97.

Chapter 7

1. Alghaeed Z, Reilly JJ, Chastin SF, et al. The influence of minimum sitting periods of the ActivPAL on the measurement of breaks in sitting in young children. PLoS One. 2013; 8:e71854.

2. Armstrong N, Balding J, Gentle P, et al. Peak oxygen uptake and physical activity in 11 to 16 year olds. Pediatr Exerc Sci. 1990;2:349-58.

3. Armstrong N, Williams J, Balding J, et al. The peak oxygen uptake of British children with reference to age, sex and sexual maturity. Eur J Appl Physiol. 1991;62:369-75.

4. Armstrong N, Welsman JR. Assessment and interpretation of aerobic fitness in children and adolescents. Exerc Sport Sci Rev. 1994;22:435-76.

5. Armstrong N, Kirby BJ, McManus AM, et al. Aerobic fitness of pre-pubescent children. Ann Hum Biol. 1995;22:427-41.

6. Armstrong N, McManus AM, Welsman JR, et al. Physical activity patterns and aerobic fitness among pre-pubescents. Eur Phy Educ Rev. 1996;2:7-18.

7. Armstrong N, Welsman J, Winsley R. Is peak $\dot{V}O_2$ a maximal index of children's aerobic fitness? Int J Sports Med. 1996;17:356-59.

8. Armstrong N, Welsman JR. Young people and physical activity. Oxford: Oxford University Press; 1997.

9. Armstrong N, Welsman JR, Kirby BJ. Peak oxygen uptake and maturation in 12-yr olds. Med Sci Sports Exerc. 1998;30:165-69.

10. Armstrong N, Welsman JR, Nevill AM, et al. Modeling changes in peak $\dot{V}O_2$ in 11-13 year olds. J Appl Physiol. 1999;87:2230-36.

11. Armstrong N, Welsman JR, Kirby BJ. Longitudinal changes in 11-13-year-olds' physical activity. Acta Paediatr 2000;89:775-80.

12. Armstrong N, Welsman JR. Peak oxygen uptake in relation to growth and maturation. Eur J Appl Physiol. 2001;28:259-65.

13. Armstrong N, Welsman JR. Aerobic fitness. In: Armstrong N, van Mechelen W, editors. Paediatric exercise science and medicine. Oxford: Oxford University Press; 2008. p. 97-108.

14. Armstrong N, McManus AM, Welsman JR. Aerobic fitness. In: Armstrong N, van Mechelen W, editors. Paediatric exercise science and medicine. Oxford: Oxford University Press; 2008. p. 269-82.

15. Armstrong N, Barker AR. Oxygen uptake kinetics in children and adolescents: a review. Pediatr Exerc Sci. 2009;2:130-47.

16. Armstrong N, Barker AR. Endurance training and elite young athletes. Med Sport Sci. 2011;56:59-83.

17. Armstrong N, McManus AM. Physiology of the elite young male athlete. Med Sport Sci. 2011;56:1-22.

18. Armstrong N, Barker AR. New insights in paediatric exercise metabolism. J Sport Health Sci. 2012;1:18-26.

19. Armstrong N. Aerobic fitness and physical activity in children. Pediatr Exerc Sci. 2013;25:548-60.

20. Åstrand PO. Experimental studies of physical working capacity in relation to sex and age. Copenhagen: Munksgaard; 1952.

21. Åstrand PO, Rodahl K. Textbook of work physiology. New York: McGraw-Hill; 1986.

22. Balemans ACJ, van Wely L, De Heer SJA, et al. Maximal aerobic and anaerobic exercise responses in children with cerebral palsy. Med Sci Sports Exerc. 2013;45:561-68.

23. Barker AR, Williams CA, Jones AM, et al. Establishing maximal oxygen uptake in young people during a ramp cycle test to exhaustion. Br J Sports Med. 2011;45:498-503.

24. Bar-Or O, Rowland TW. Pediatric exercise medicine. From physiologic principles to health care applications. Champaign, IL: Human Kinetics; 2004.

25. Berling J, Foster C, Gibson M, et al. The effect of handrail support on oxygen uptake during steady-state treadmill exercise. J Cardiopulm Rehabil. 2006;26:391-94.

26. Berman N, Bailey R, Barstow TJ, et al. Spectral and bout detection analysis of physical activity patterns in healthy, prepubertal boys and girls. Am J Hum Biol. 1998;10:289-97.

27. Blank AC, Hakim S, Strengers JL, et al. Exercise capacity in children with isolated congenital complete atrioventricular block: does pacing make a difference? Pediatr Cardiol. 2012;33:576-85.

28. Bloxham SR, Welsman JR, Armstrong N. Ergometer-specific relationships between peak oxygen uptake and short-term power output in children. Pediatr Exerc Sci. 2005;17:136-48.

29. Boileau RA, Bonen A, Heyward VH, et al. Maximal aerobic capacity on the treadmill and bicycle ergometer of boys 11-14 years of age. J Sports Med Phys Fitness. 1977;17:153-62.

30. Booth FW, Lees SJ. Physically active subjects should be the control group. Med Sci Sports Exerc. 2006;38:405-6.

31. Brehm MA, Balemans ACJ, Becher JG, et al. Reliability of a progressive maximal cycle ergometer test to assess peak oxygen uptake in children with mild to moderate cerebral palsy. Phys Ther. 2014;94:121-8.

32. Buys R, Budts W, Reybrouck T, et al. Serial exercise testing in children, adolescents and young adults with Senning repair for transposition of the great arteries. BMC Cardiovasc Disord. 2012;12:88.

33. Cumming GR. Hemodynamics of supine bicycle exercise in "normal" children. Am Heart J 1977;93:617-22.

34. Cumming GR, Everatt D, Hastman L. Bruce treadmill test in children: normal values in a clinic population. Am J Cardiol. 1978;41:69-75.

35. Dallman PR, Siimes MA. Percentile curves for hemoglobin and red cell volume in infancy and childhood. Pediatrics. 1979;94:26-31.

36. de Groot JF, Takken T, de Graaff S, et al. Treadmill testing of children who have spina bifida and are ambulatory: does peak oxygen uptake reflect maximum oxygen uptake? Phys Ther. 2009;89:679-87.

37. Dencker M, Thorsson O, Karlsson MK, et al. Daily physical activity and its relation to aerobic fitness in children aged 8-11 years. Eur J Appl Physiol. 2006; 96:587-92.

38. Dencker M, Bugge A, Hermansen B, et al. Aerobic fitness in prepubertal children according to level of body fat. Acta Pædiatrica. 2010;99:1854-60.

39. Di Bonito P, Capaldo B, Forziato C, et al. Central adiposity and left ventricular mass in obese children. Nutr Metab Cardiovasc Dis. 2008;18:613-17.

40. Driscoll DJ, Offord KP, Feldt RH, et al. Five- to 15-year follow-up after Fontan procedure. Circulation. 1992;85:469-96.

41. Eiberg S, Hasselstrom H, Gronfeldt V, et al. Maximum oxygen uptake and objectively measured physical activity in Danish children 6-7 years of age: the Copenhagen school child intervention study. Br J Sports Med. 2005;39:725-30.

42. Fawkner SG, Armstrong N, Potter CR, et al. Oxygen uptake kinetics in children and adults after the onset of moderate-intensity exercise. J Sports Sci. 2002;20:319-26.

43. Fawkner SG, Armstrong N. Sex differences in the oxygen uptake kinetic response to heavy-intensity exercise in prepubertal children. Eur J Appl Physiol. 2004;93:210-16.

44. Friedberg MK, Fernandes FP, Roche SL, et al. Relation of right ventricular mechanics to exercise tolerance in children after tetralogy of Fallot repair. Am Heart J. 2013;165:551-57.

45. Giardini A, Fenton M, Andrews RE, et al. Peak oxygen uptake correlates with survival without clinical deterioration in ambulatory children with dilated cardiomyopathy. Circulation. 2011;124:1713-18.

46. Green DJ, O'Driscoll G, Blanksby BA, et al. Control of skeletal muscle blood flow during dynamic exercise: contribution of endothelium-derived nitric oxide. Sports Med. 1996;21:119-46.

47. Guimarães GV, d'Avila VM, Camargo PR, et al. Prognostic value of cardiopulmonary exercise testing in children with heart failure secondary to idiopathic dilated cardiomyopathy in a non-β-blocker therapy setting. Eur J Heart Fail. 2008;10:560-65.

48. He ZH, Ma LH. The aerobic fitness (VO_2 peak) and β–fibrinogen genetic polymorphism in obese and non-obese Chinese boys. Int J Sports Med. 2005;26:253-57.

49. Hebestreit H. Exercise testing in children—what works, what doesn't and where to go? Paediatr Respir Rev. 2004;5:S11-14.

50. Higgins LW, Robertson RJ, Kelsey SF, et al. Exercise intensity self-regulation using the OMNI scale in children with cystic fibrosis. Pediatr Pulmonol. 2013;48:497-505.

51. Howley ET, Bassett DR, Welch HG. Criteria for maximal oxygen uptake: review and commentary. Med Sci Sports Exerc. 1995;27:1292-1301.

52. Kemper HCG. Amsterdam growth and health longitudinal study. Med Sport. 2004;47:1-198.

53. Kemper HCG, Koppes LLJ. Is physical activity important for aerobic power in young males and females? Med Sports Sci. 2004;47:153-66.

54. Koopman LP, McCrindle BW, Slorach C, et al. Interaction between myocardial and vascular changes in obese children: a pilot study. J Am Soc Echocardiogr. 2012;25:401-10.

55. Krahenbuhl GS, Skinner JS, Kohrt WM. Developmental aspects of maximal aerobic power in children. Exerc Sport Sci Rev. 1985;13:503-38.

56. Krovetz LJ, McLoughlin TG, Mitchell MB, et al. Hemodynamic findings in normal children. Pediatr Res. 1967;1:22-30.

57. Lambrick DM, Rowlands AV, Eston RG. The perceptual response to treadmill exercise using the Eston-Parfitt scale and marble dropping task, in children age 7 to 8 years. Pediatr Exerc Sci. 2001;23:36-48.

58. Lambrick D, Faulkner J, Westrupp N, et al. The influence of body weight on the pulmonary oxygen uptake kinetics in pre-pubertal children during moderate- and heavy intensity treadmill exercise. Eur J Appl Physiol. 2013;113:1947-55.

59. Loftin M, Sothern M, Trosclair L, et al. Scaling VO_2 peak in obese and non-obese girls. Obes Res. 2001;9:290-96.

60. Madsen A, Green K, Buchvald F, et al. Aerobic fitness in children and young adults with primary ciliary dyskinesia. PLoS ONE 2013;8:e71409.

61. Marelli AJ, Alejos JC. Exercise response in atrial septal defect. Prog Pediatr Cardiol. 1993;2:20-23.

62. May LJ, Punn R, Olson I, et al. Supine cycling in pediatric exercise testing: disparity in performance measures. Pediatr Cardiol. 2014;35:705-110.

63. McManus AM, Armstrong N. Physiology of the elite young female athlete. Med Sport Sci. 2011;56:23-46.

64. McManus AM, Chu EYW, Yu CCW, et al. How children move—activity pattern characteristics in lean and obese Chinese children. J Obes. 2011;2011:679328. DOI: 10.1155/2011/679328.

65. Mellecker RR, McManus AM. Measurement of resting energy expenditure in healthy children. JPEN J Parenter Enteral Nutr. 2009;33:640-45.

66. Mirwald RL, Bailey DA. Maximal aerobic power. London, Ontario: Sports Dynamics; 1986.

67. Mirwald RL, Baxter-Jones ADG, Bailey DA, et al. An assessment of maturity from anthropometric measurements. Med Sci Sports Exerc. 2002;34:689-94.

68. Myers J, Walsh D, Sullivan M, et al. Effect of sampling on variability and plateau in oxygen uptake. J Appl Physiol. 1990;68:404-10.

69. Myers JN. Essentials of cardiopulmonary exercise testing. Champaign, IL: Human Kinetics; 1996.

70. Myers J, Gullestad L, Vagelos R, et al. Cardiopulmonary exercise testing and prognosis in severe heart failure: 14 mL/kg/min revisited. Am Heart J. 2000;139:78-84.

71. Nevill A, Rowland T, Goff D, et al. Scaling or normalising maximum oxygen uptake to predict 1-mile run time in boys. Eur J Appl Physiol. 2004;92:285-88.

72. Nixon PA, Orenstein DM, Kelsey SF, et al. The prognostic value of exercise testing in patients with cystic fibrosis. N Engl J Med. 1992;327:1785-88.

73. Noble BJ, Robertson RJ. Perceived exertion. Champaign, IL: Human Kinetics; 1996.

74. Nottin S, Vinet A, Lecoq A-M, et al. Test-retest reproducibility of submaximal and maximal cardiac output by Doppler echocardiography and CO_2 rebreathing in prepubertal children. Pediatr Exerc Sci. 2001;13:214-24.

75. Obert P, Mandigout S, Nottin S, et al. Cardiovascular responses to endurance training in children: effect of gender. Eur J Clin Invest. 2003;33:199-208.

76. Obert P, Gueugnon C, Nottin S, et al. Two-dimensional strain and twist by vector velocity imaging in adolescents with severe obesity. Obesity. 2012;20:2397-2405.

77. Opocher F, Varnier M, Sanders SP, et al. Effects of aerobic exercise training in children after the Fontan operation. Am J Cardiol. 2005;95:150-52.

78. Paridon SM. Congenital heart disease: cardiac performance and adaptation to exercise. Pediatr Exerc Sci. 1997;9:308-23.

79. Paridon SM, Alpert BS, Boas SR, et al. Clinical stress testing in the pediatric age group: a statement from the American Heart Association Council on Cardiovascular Disease in the Young, Committee on Atherosclerosis, Hypertension, and Obesity in Youth. Circulation. 2006;113:1905-20.

80. Pérez M, Groeneveld IF, Santana-Sosa E, et al. Aerobic fitness is associated with lower risk of hospitalization in children with cystic fibrosis. Pediatr Pulmonol. 2014;49:641-9.

81. Pianosi P, LeBlanc J, Almudevar A. Peak oxygen uptake and mortality in children with cystic fibrosis. Thorax. 2005;60:50-54.

82. Porapakkham P, Porapakkham P, Krum H. Is target dose of beta-blocker more important than achieved heart rate or heart rate change in patients with systolic chronic heart failure? Cardiovasc Ther. 2010;28:93-100.

83. Potter CR, Childs DJ, Houghton W, et al. Breath-to-breath "noise" in the ventilatory and gas exchange responses of children to exercise. Eur J Appl Physiol Occup Physiol. 1999;80:118-24.

84. Robben KE, Poole DC, Harms CA. Maximal oxygen uptake validation in children with expiratory flow limitation. Pediatr Exerc Sci. 2013;25:84-100.

85. Robertson R. Perceived exertion for practitioners rating effort with the OMNI Picture System. Champaign, IL: Human Kinetics; 2004.

86. Robinson S. Experimental studies of physical fitness in relation to age. Arbeitsphysiologie. 1938;10:251-323.

87. Rowland TW, Cunningham LN. Oxygen uptake plateau during maximal treadmill exercise in children. Chest. 1992;101:485-89.

88. Rowland TW. Aerobic exercise testing protocols. In: Rowland TW, editor. Pediatric laboratory exercise testing. Champaign, IL: Human Kinetics; 1993. p. 19-41.

89. Rowland TW. Does peak $\dot{V}O_2$ reflect $\dot{V}O_{2max}$ in children? Evidence from supramaximal testing. Med Sci Sports Exerc. 1993;25:689-93.

90. Rowland T, Popowski B, Ferrone L. Cardiac responses to maximal upright cycle exercise in healthy boys and men. Med Sci Sports Exerc. 1997;29:1146-51.

91. Rowland TW, Melanson EL, Popowski BE, et al. Test-retest reproducibility of maximum cardiac output by Doppler echocardiography. Am J Cardiol. 1998;81:1228-29.

92. Rowland TW, Blum JW. Cardiac dynamics during upright cycle exercise in boys. Am J Hum Biol. 2000;12:749-57.

93. Rowland TW, Goff D, Martel L, et al. Influence of cardiac functional capacity on gender differences in maximal oxygen uptake in children. Chest 2000;117:629-35.

94. Rowland TW, Obert P. Doppler echocardiography for the estimation of cardiac output with exercise. Sports Med. 2002;32:973-86.

95. Rowland TW, Bhargava R, Parslow D, et al. Cardiac response to progressive cycle exercise in moderately obese adolescent females. J Adolesc Health. 2003;32:422-27.

96. Rowland TW, Garrison A, Delulio A. Circulatory responses to progressive exercise: insights from positional differences. Int. J Sports Med. 2003;24:512-17.

97. Rowland TW. Children's exercise physiology. 2nd ed. Champaign, IL: Human Kinetics; 2005.

98. Rowland TW, Garrard M, Marwood S, et al. Myocardial performance during progressive exercise in athletic adolescent males. Med Sci Sports Exerc. 2009;41:1721-28.

99. Rowland TW, Unnithan VB, Roche D, et al. Myocardial function and aerobic fitness in adolescent females. Eur J Appl Physiol. 2011;111:1991-97.

100. Rowland T. Oxygen uptake and endurance fitness in children, revisited. Pediatr Exerc Sci. 2013; 25:508-14.

101. Rutenfranz J, Andersen KL, Seliger V, et al. Maximum aerobic power and body composition during the puberty growth period: similarities and differences between children of two European countries. Eur J Pediatr. 1981;136:123-33.

102. Singh TP, Curran TJ, Rhodes J. Cardiac rehabilitation improves heart rate recovery following peak exercise in children with repaired congenital heart disease. Pediatr Cardiol. 2007;28:276-79.

103. Staats BA, Grinton SF, Mottram CD, et al. Quality control in exercise testing. *Prog Pediatr Cardiol.* 1993;2:11-17.

104. Stevens D, Oades PJ, Armstrong N, et al. Early oxygen uptake recovery following exercise testing in children with chronic chest diseases. Pediatr Pulmonol. 2009;44:480-88.

105. Vinet A, Mandigout S, Nottin S, et al. Influence of body composition, hemoglobin concentration, and cardiac size and function on gender differences in maximal oxygen uptake in prepubertal children. Chest. 2003;124:1494-99.

106. Wasserman K. Diagnosing cardiovascular and lung pathophysiology from exercise gas exchange. Chest. 1997;112:1091-1101.

107. Wells JCK. Sexual dimorphism of body composition. Best Pract Res Clin Endocrinol Metab. 2007;21:415-30.

108. Welsman JR, Armstrong N, Kirby BJ, et al. Scaling peak $\dot{V}O_2$ for differences in body size. Med Sci Sports Exerc. 1996;28:259-65.

109. Welsman JR, Bywater K, Farr C, et al. Reliability of peak $\dot{V}O_2$ and cardiac output assessed using thoracic impedance in children. Eur J Appl Physiol. 2005;94:228-34.

110. Werkman MS, Hulzebos HJ, Arets HGM, et al. Is static hyperinflation a limiting factor during exercise in adolescents with cystic fibrosis? Pediatr Pulmonol. 2001;46:119-24.

111. Werkman MS, Hulzebos HJ, van de Weert-van Leeuwen PB, et al. Supramaximal verification of peak oxygen uptake in adolescents with cystic fibrosis. Pediatr Phys Ther. 2011;23:15-21.

112. Wharburton DER, Nettlefold L, McGuire KA, et al. Cardiovascular function. In: Armstrong N, van Mechelen W, editors. Paediatric exercise science and medicine. Oxford: Oxford University Press; 2008. p. 77-95.

113. Winsley RJ, Fulford J, Roberts AC, et al. Sex differences in peak oxygen uptake in prepubertal children. J Sci Med Sport. 2009;12:647-51.

114. Wolfe RR, Bartle L, Daberkow E, et al. Exercise responses in ventricular septal defect. Prog Pediatr Cardiol. 1993;2:24-29.

115. Yeung JP, Human DG, Sandor GGS, et al. Serial measurements of exercise performance in pediatric heart transplant patients using stress echocardiography. Pediatr Transplant. 2011;15:265-71.

116. Yoshizawa S. A comparative study of aerobic work capacity in urban and rural adolescents. J Hum Erg. 1972;1:45-65.

117. Yu CCW, McManus AM, Li MA, et al. Cardiopulmonary exercise testing in children. Hong Kong J Paediatr. 2010;15:35-47.

Chapter 8

1. Ainsworth B, Haskell W, Herrmann S, et al. 2011 compendium of physical activities: a second update of codes and MET values. Med Sci Sports Exerc. 2011;43:1575-81.

2. Akkerman M, van Brussel M, Bongers B, et al. Oxygen uptake efficiency slope in healthy children. Pediatr Exerc Sci. 2010;22:431-41.

3. Antoine-Jonville S, Pichon A, Vazir A, et al. Oxygen uptake efficiency slope, aerobic fitness, and V_E -$\dot{V}CO_2$ slope in heart failure. Med Sci Sports Exerc. 2012;44:428-34.

4. Armstrong L, Balady G, Berry M, et al. ACSM's guidelines for exercise testing and prescription. Philadelphia: Lippincott Williams & Wilkins; 2006.

5. Armstrong N, Welsman J. Cardiovascular responses to submaximal treadmill running in 11 to 13 year olds. Acta Paediatr. 2002;91:125-31.

6. Åstrand P, Rodahl K, Dahl H, et al. Textbook of work physiology. 4th ed. Champaign, IL: Human Kinetics; 2003.

7. ATS Statement: Guidelines for the six-minute walk test. Am J Respir Crit Care Med. 2002;166:111-17.

8. Baba R, Nagashima M, Goto M, et al. Oxygen uptake efficiency slope: A new index of cardiorespiratory functional reserve derived from the relation between oxygen uptake and minute

ventilation during incremental exercise. J Am Coll Cardiol. 1996;28:1567-72.

9. Baba R, Nagashima M, Nagano Y, et al. Role of the oxygen uptake efficiency slope in evaluating exercise tolerance. Arch Dis Child. 1999;81:73-75.

10. Baba R, Tsuyuki K, Yano H, et al. Robustness of the oxygen uptake efficiency slope to exercise intensity in patients with coronary artery disease. Nagoya J Med Sci. 2010;72:83-89.

11. Barboza de Andrade L, Silva D, Salgado T, et al. Comparison of six-minute walk test in children with moderate/severe asthma with reference values for healthy children. J Pediatr. 2014;90(3):250-57

12. Barkley J, Roemmich J. Validity of a pediatric RPE scale when different exercise intensities are completed on separate days. J Exerc Sci Fit. 2011;9:52-57.

13. Barron AJ, Medlow KI, Giannoni A, et al. Reduced confounding by impaired ventilatory function with oxygen uptake efficiency slope and VE/$\dot{V}CO_2$ slope rather than peak oxygen consumption to assess exercise physiology in suspected heart failure. Congest Heart Fail. 2010;16:259-64.

14. Beaver W, Wasserman K, Whipp B. A new method for detecting anaerobic threshold by gas exchange. J Appl Physiol. 1986;60: 2020-27.

15. Bland J, Pfeiffer K, Eisenmann J. The PWC170: comparison of different stage lengths in 11-16 year olds. 2012;112:1955-61.

16. Bongers B, Hulzbos E, Arets B, et al. Validity of the oxygen uptake efficiency slope in children with cystic fibrosis and mild-to-moderate airflow obstruction. Pediatr Exerc Sci. 2012;24:129-41.

17. Bongers B, Hulzebos H, Blank A, et al. The oxygen uptake efficiency slope in children with congenital heart disease: construct and group validity. Eur J Cardiovasc Prev Rehabil. 2011;18:384-92.

18. Breithaupt P, Adamo K, Colley R. The HALO submaximal treadmill protocol to measure cardiorespiratory fitness in obese children and youth: a proof of principle study. Appl Physiol Nutr Metab. 2012;37:308-14.

19. Breithaupt PG, Colley RC, Adamo KB. Using the oxygen uptake efficiency slope as an indicator of cardiorespiratory fitness in the obese pediatric population. Pediatr Exerc Sci. 2012;24:357-68.

20. Burri P. Structural aspects of postnatal lung development. Alveolar formation and growth. Biol Neonate. 2006;89:313-22.

21. Cahalin, L.P., Mathier, M.A., Semigran, M.J., et al. The six minute walk predicts peak oxygen consumption in patients with advanced heart failure. Chest. Aug;110 (2): 325-332.

22. Canter C, Shaddy R, Berstein D, et al. Indications for heart transplantation in pediatric heart disease. Circulation. 2007;115:658-76.

23. Casanova C, Celli B, Barria P, et al. The 6-min walk distance in healthy subjects: reference standards from seven countries. Eur Respir J. 2011;37:150-56.

24. Chen C, Chen S, Chiu H, et al. Prognostic value of submaximal exercise data for cardiac morbidity in Fontan patients. Med Sci Sports Exerc. 2013;46:10-15.

25. Cink R, Thomas T. Validity of the Åstrand-Ryhming nomogram for predicting maximal oxygen intake. Br J Sports Med. 1981;15:182-85.

26. Cooper D, Weiler-Ravell D, Whipp B, et al. Aerobic parameters of exercise as a function of body size during growth in children. J Appl Physiol. 1984;56:628-34.

27. Cumming G, Everatt D, Hastman L. Bruce treadmill test in children: normal values in a clinical population. Am J Cardiol. 1978;41:69-75.

28. Czaprowski D, Kotwicki T, Biernat R, et al. Physical capacity of girls with mild and moderate idiopathic scoliosis: influence of the size, length and number of curvatures. Eur Spine J. 2012;21:1099-1105.

29. D'Silva C, Vaishali K, Venkatesan P. Six-minute walk test. Normal values of school children aged 7-12 y in India: across-sectional study. Indian J Pediatr. 2011;79(5):597-601.

30. Das B, Taylor A, Boucek M, et al. Exercise capacity in pediatric heart transplant candidates: is there any role for the 14 ml/kg/min guideline? Pediatr Cardiol. 2006;27:226-29.

31. Davies L, Wensel R, Georgiadou P, et al. Enhanced prognostic value from cardiopulmonary exercise testing in chronic heart failure by non-linear analysis: oxygen uptake efficiency slope. Eur Heart J. 2006; 27:684-90.

32. De Backer I, Singh-Grewal D, Helders P, et al. Can peak work rate predict peak oxygen uptake in children with juvenile idiopathic arthritis? Arthritis Care Res. 2010;62:960-64.

33. Defoor J, Shepers D, Reybrouck T, et al. Oxygen-uptake efficiency slope in coronary artery disease, clinical use and response to training. Int J Sports Med. 2006;27:730-37.

34. Dencker M, Thorsson O, Karlsson M, et al. Gender differences and determinants of aerobic fitness in children aged 8-11 years. Eur J Appl Physiol. 2007;99:19-26.

35. Dencker M, Thorsson O, Karlsson M, et al. Maximal oxygen uptake versus maximal power output in children. J Sports Sci. 2008;13:1397-1402.

36. Diller G, Giardini A, Dimopoulos K, et al. Predictors of morbidity and mortality in contemporary

Fontan patients: results from a multicenter study including cardiopulmonary exercise testing in 321 patients. Eur Heart J. 2010;31:3073-83.

37. Doutreleau S, Di Marco P, Talha S, et al. Can the six-minute walk test predict peak oxygen uptake in men with heart transplant? Arch Phys Med Rehabil. 2009;90:51-7.

38. Drinkard B, Roberts M, Ranzenhofer L, et al. Oxygen uptake efficiency slope as a determinant of fitness in overweight adolescents. Med Sci Sports Exerc. 2007;39:1811-16.

39. Eston R. Perceived exertion: recent advances and novel applications in children and adults. J Exerc Sci Fit. 2009;7:S11-S17.

40. Fawkner S, Armstrong N, Childs D, et al. Reliability of the visually identified ventilatory threshold and V-slope in children. Pediatr Exerc Sci. 2002;14:181-92.

41. Gademan M, Swenne C, Verwey H, et al. Exercise training increases oxygen uptake efficiency slope in chronic heart failure. Eur J Cardiovasc Prev Rehabil. 2008;15:140-44.

42. Garofano R, Barst R. Exercise testing in children with primary pulmonary hypertension. Pediatr Cardiol. 1999;20:61-64.

43. Gaskill S, Ruby B, Walker A, et al. Validity and reliability of combining three methods to determine ventilator threshold. Med Sci Sports Exerc. 2001;33:1841-48.

44. Gaskill S, Walker A, Serfass R, et al. Changes in ventilatory threshold with exercise training in a sedentary population: the Heritage Family Study. Int J Sports Med. 2001;22:586-92.

45. Geiger R, Strasak A, Treml B, et al. Six-minute walk test in children and adolescents. J Pediatr. 2007;150:395-99.

46. Giardini A, Odendaal D, Khambadkone S, et al. Physiologic decrease of ventilatory response to exercise in the second decade of life in healthy children. Am Heart J. 2011;161:1214-19.

47. Giardini A, Specchia S, Gargiulo G, et al. Accuracy of oxygen uptake efficiency slope in adults with congenital heart disease. Int J Cardiol. 2009;133:74-79.

48. Groen W, Hulzebos H, Helders P, et al. Oxygen uptake to work rate slope in children with a heart, lung or muscle disease. Int J Sports Med. 2010;31:202-6.

49. Gruet M, Brisswalter J, Mely L, et al. Clinical utility of the oxygen uptake efficiency slope in cystic fibrosis patients. J Cyst Fibros. 2010;9:307-13.

50. Hansen H, Frober K, Rokkedal J, et al. A new approach to assessing maximal aerobic power in children: the Odense School Child Study. Eur J Appl Physiol. 1989;58:618-24.

51. Hebestreit H, Staschen B, Hebestreit A. Ventilatory threshold: a useful method to determine aerobic fitness in children? Med Sci Sports Exerc. 2000;32:1964-69.

52. Heyman E, Briard D, Dekerdanet M, et al. Accuracy of physical working capacity 170 to estimate aerobic fitness in prepubertal diabetic boys and in 2 insulin dose conditions. J Sports Med Phys Fitness. 2006;46:315-21.

53. Hollenberg M, Tager I. Oxygen uptake efficiency slope: an index of exercise performance and cardiopulmonary reserve requiring only submaximal exercise. J Am Coll Cardiol. 2000;36:194-201.

54. Houghton K, Tucker L, Potts J, et al. Fitness, fatigue, disease activity, and quality of life in pediatric lupus. Arthritis Rheum. 2008;59:537-545.

55. Ilarraza-Lomelí H, Castañeda-López J, Myers J, et al. Cardiopulmonary exercise testing in healthy children and adolescents at moderately high altitude. Arch Cardiol Mex. 2013;83:176-82.

56. Jung A, Nieman D, Kernodle M. Prediction of maximal aerobic power in adolescents from cycle ergometry. Pediatr Exerc Sci. 2001;13:167-72.

57. Laitinen P, Räsänen J. Measured versus predicted oxygen consumption in children with congenital heart disease. Heart. 1998;80:601-5.

58. Lammers A, Hislop A, Flynn Y, et al. The 6-minute walk test: normal values for children of 4-11 years of age. Arch Dis Child. 2008;93:464-68.

59. Li A, Yin J, Yu C, et al. The six-minute walk test in healthy children: reliability and validity. Eur Respir J. 2005;25:1057-60.

60. Limsuwan A, Wongwandee R, Khowsathit P. Correlation between 6-min walk test and exercise stress test in healthy children. Acta Paediatr. 2010;99:438-41.

61. Lucas C, Stevenson L, Johnson W, et al. The 6-min walk and peak oxygen consumption in advanced heart failure: Aerobic capacity and survival. Am Heart J. 1999;138:618-24.

62. Machado F, Denadai B. Validity of maximum heart rate prediction equations for children and adolescents. Arq Bras Cardiol. 2011;97:136-40.

63. Mahon A, Cheatham C. Ventilatory threshold in children: a review. Pediatr Exerc Sci. 2002;14:16-29.

64. Mancini D, Lietz K. Selection of cardiac transplantation candidates in 2010. Circulation. 2010;122:173-83.

65. Marinov B, Kostianev S. Exercise performance and oxygen uptake efficiency slope in obese children performing standardized exercise. Acta Physiol Pharmacol Bulg. 2003;27:59-64.

66. Mimura K, Maeda K. Heart response to treadmill exercise in children of ages 4-6 years. Ann Physiol Anthropol. 1989;8:143-50.

67. Morinder G, Mattsson E, Sollander C, et al. Six-minute walk test in obese children and adolescents: reproducibility and validity. Physiother Res Int. 2009;14:91-104.

68. Mourot L, Perrey A, Tordi N, et al. Evaluation of fitness level by the oxygen uptake efficiency slope after a short-term intermittent endurance training. Int J Sports Med. 2004;25:85-91.

69. Mucci P, Baquet G, Nourry C, et al. Exercise testing in children: Comparison in ventilatory threshold changes with interval training. Pediatr Pulmonol. 2013;48;809-16.

70. Müller J, Böhm B, Semsch S, et al. Currently, children with congenital heart disease are not limited in their submaximal performance. Eur J Cardiothorac Surg. 2013;43(6):1096-1100.

71. Nemeth B, Carrel A, Eickhoff J, et al. Submaximal treadmill test predicts $\dot{V}O_{2max}$ in overweight children. J Pediatr. 2009;154:677-81.

72. Nielson D, George J, Vehrs P, et al. Predicting $\dot{V}O_{2max}$ in college-aged participants using cycle ergometry and perceived functional ability. Meas Phys Educ Exerc Sci. 2010;14:252-64.

73. Niemeijer V, van't Veer M, Schep G, et al. Causes of nonlinearity of the oxygen uptake efficiency slope: a prospective study in patients with chronic heart failure. Eur J Prev Cardiol. 2012;10:1-7.

74. Ohuchi H, Nakajima T, Kawade M, et al. Measurement and validity of the ventilatory threshold in patients with congenital heart disease. Pediatr Cardiol. 1996;17:7-14.

75. Petzl D, Haber P, Echuster E, et al. Reliability of estimation of maximum performance capacity on the basis of submaximum ergometric stress tests in children 10-14 years old. Eur J Pediatr. 1988;147:174-78.

76. Potter C, Childs D, Houghton W, et al. Breath-to-breath "noise" in the ventilator and gas exchange responses of children to exercise. Eur J Appl Physiol. 1999;80:118-24.

77. Reybrouck T, Weymans M, Stijns H, et al. Ventilatory anaerobic threshold in healthy children. Eur J Appl Physiol. 1985;54:278-84.

78. Ridley K, Olds T. Assigning energy costs to activities in children: a review and synthesis. Med Sci Sports Exerc. 2008;40:1439-46.

79. Ross R, Murthy J, Wollak I, et al. The six minute walk test accurately estimates mean peak oxygen uptake. BMC Pulm Med. 2010;10:31.

80. Rowland T, Rambusch J, Staab J, et al. Accuracy of physical working capacity (PWC170) in estimating aerobic fitness in children. J Sports Med Phys Fitness. 1993;33:184-88.

81. Schulze-Neick I, Wessel H, Paul M. Heart rate and oxygen uptake response to exercise in children with low peak exercise heart rate. Eur J Pediatr. 1992;151:160-66.

82. Storer T, Davis J, Caiozzo V. Accurate prediction of $\dot{V}O_{2max}$ in cycle ergometry. Med Sci Sports Exerc. 1989;22:704-12.

83. Svedahl K, MacIntosh B. Anaerobic threshold: the concept and methods of measurement. Can J Appl Physiol. 2003;28:299-323.

84. Tanaka H, Monahan K, Seals D. Age-predicted maximal heart rate revisited. J Am Coll Cardiol. 2001;37:153-56.

85. Ten Harkel AD, Takken T, Van Osch-Gevers M, et al. Normal values for cardiopulmonary exercise testing in children. Eur J Cardiovasc Prev Rehabil. 2011;18(1):48-54.

86. Ten Harkel A, Takken T. Exercise testing and prescription in patients with congenital heart disease. Int J Pediatr. 2010;2010:pii:791980. Available from: www.hindawi.com/journals/ijpedi/ DOI: 10.1155/2010/791980.

87. Terziyski K, Marinov B, Aliman O, et al. Oxygen uptake efficiency slope and chronotropic incompetence in chronic heart failure and chronic obstructive pulmonary disease. Folia Medica. 2009;51(4):18-24.

88. Ulrich S, Hildenbrand F, Treder U, et al. Reference values for the 6-minute walk test in healthy children and adolescents in Switzerland. BMC Pulm Med. 2013;13:49.

89. Van der Cammen-van Zijp M, van den Berg-Emons R, Willemsen S, et al. Exercise capacity in Dutch children: new reference values for the Bruce treadmill protocol. Scand J Med Sci Sports. 2010;20:e130-e136.

90. Van Laethem C, De Sutter J, Peersman W, et al. Intratest reliability and test-retest reproducibility of the oxygen uptake efficiency slope in healthy participants. Eur J Cardiovasc Prev Rehabil. 2009;16:493-98.

91. Van Laethem C, Van De Veire N, Backer G, et al. Response of the oxygen uptake efficiency slope to exercise training in patients with chronic heart failure. Eur J Heart Fail. 2007;9:625-29.

92. Verschuren O, Maltais D, Takken T. The 220 – age equation does not predict maximum heart rate in children and adolescents. Dev Med Child Neurol. 2011;53:861-64.

93. Washington R, van Gundy J, Cohen C, et al. Normal aerobic and anaerobic exercise data for North American school-age children. J Pediatr. 1988;112:223-33.

94. Washington R. Anaerobic threshold in children. Pediatr Exerc Sci. 1989;1:244-56.

95. Wasserman K, Hansen JE, Sue DY, et al. Principles of exercise testing and interpretation. 3rd ed. Philadelphia: Lippincott Williams & Wilkins; 1999.

96. Wasserman K, McIlroy M. Detecting the threshold of anaerobic metabolism in cardiac patients during exercise. Am J Cardiol. 1964;14:844-52.

97. Werkman M, Hulzebos E, Helders P, et al. Estimating peak oxygen uptake in adolescents with cytic fibrosis. Arch Dis Child. 2014;99(1):21-25.

98. Williamson W, Fuld J, Westgate K, et al. Validity of reporting oxygen uptake efficiency slope from submaximal exercise using respiratory exchange ratio as secondary criterion. Pulm Med. 2012;2012:1-8. Available from: http://dx.doi.org/10.1155/2012/874020

99. Wirth A, Träger E, Scheele K, Mayer D, et al. Cardiopulmonary adjustment and metabolic response to maximal and submaximal physical exercise of boys and girls at different stages of maturity. Eur J Appl Physiol. 1978;39:229-40.

100. Zijp M, IJsselstijn H, Takken T, et al. Exercise testing of pre-school children using the Bruce treadmill protocol: new reference values. Eur J Appl Physiol. 2010;108:393-99.

Chapter 9

1. Aladangady N, Leung T, Costeloe K, et al. Measuring circulating blood volume in newborn infants using pulse dye densitometry and indocyanine green. Paediatr Anaesth. 2008;18(9):865-71.

2. Aoyagi T. Pulse oximetry: its invention, theory, and future. J Anesth. 2003;17:259-66.

3. Armstrong N, Welsman JR. Assessment and interpretation of aerobic fitness in children and adolescents. Exerc Sport Sci Rev. 1994;22:435-76.

4. Barker RC, Hopkins SR, Kellogg N, et al. Measurement of cardiac output during exercise by open-circuit acetylene uptake. J Appl Physiol. 1999;87(4):1506-12.

5. Baulig W, Bernhard EO, Bettex D, et al. Cardiac output measurement by pulse dye densitometry in cardiac surgery. Anaesthesia. 2005;60(10):968-73.

6. Beekman RH, Katch V, Marks C, et al. Validity of CO_2-rebreathing cardiac output during rest and exercise in young adults. Med Sci Sports Exerc. 1984;16(3):306-10.

7. Bell C, Monahan KD, Donato AJ, et al. Use of acetylene breathing to determine cardiac output in young and older adults. Med Sci Sports Exerc. 2003;35(1):58-64.

8. Bowdle TA, Freund PR, Rooke GA. Biophysical measurement series: cardiac output. Issaquah, WA: SpaceLabs Inc.; 1991.

9. Bremer F, Schiele A, Tschaikowsky K. Cardiac output measurement by pulse dye densitometry: a comparison with the Fick's principle and thermodilution method. Intensive Care Med. 2002;28(4):399-405.

10. Calbet JA, Boushel R. Assessment of cardiac output with transpulmonary thermodilution during exercise in humans. J Appl Physiol (1985). 2015;118(1):1-10.

11. Calbet JA, Mortensen SP, Munch GD, et al. Constant infusion transpulmonary thermodilution for the assessment of cardiac output in exercising humans. Scand J Med Sci Sports. 2016;26(5):518-27.

12. Card N, Gledhill N, Thomas S. A non-rebreathe technique for measuring cardiac output throughout progressive to max exercise. Med Sci Sports Exerc. 1996;28(5 Suppl):S120.

13. Chantler PD, Clements RE, Sharp L, et al. The influence of body size on measurements of overall cardiac function. Am J Physiol Heart Circ Physiol. 2005;289(5):H2059-65.

14. Charloux A, Lonsdorfer-Wolf E, Richard R, et al. A new impedance cardiograph device for the non-invasive evaluation of cardiac output at rest and during exercise: comparison with the "direct" Fick method. Eur J Appl Physiol. 2000;82(4):313-20.

15. Chew MS, Poelaert J. Accuracy and repeatability of pediatric cardiac output measurement using Doppler: 20-year review of the literature. Intensive Care Med. 2003;29(11):1889-94.

16. Christie J, Sheldahl LM, Tristani FE, et al. Determination of stroke volume and cardiac output during exercise: comparison of two-dimensional and Doppler echocardiography, Fick oximetry, and thermodilution. Circulation. 1987;76(3):539-47.

17. Collier CR. Determination of mixed venous CO_2 tensions by rebreathing. J Appl Physiol. 1956;9:25-29.

18. Cournand A, Riley RL, Breed ES, et al. Measurement of cardiac output in man using the technique of catheterization of the right auricle or ventricle. J Clin Invest. 1945;24(1):106-16.

19. Critoph CH, Patel V, Mist B, et al. Non-invasive assessment of cardiac output at rest and during exercise by finger plethysmography. Clin Physiol Funct Imaging. 2013;33(5):338-43.

20. Darovic GO. Hemodynamic monitoring: invasive and noninvasive clinical application. Philadelphia: W.B. Saunders; 1995.

21. Defares JG. Determination of $PvCO_2$ from the exponential CO_2 rise during rebreathing. J Appl Physiol. 1958;13:159-64.

22. Dewey FE, Rosenthal D, Murphy DJ Jr, et al. Does size matter? Clinical applications of scaling cardiac size and function for body size. Circulation. 2008;117(17):2279-87.

23. Dibski DW, Smith DJ, Jensen R, et al. Comparison and reliability of two non-invasive acetylene

uptake techniques for the measurement of cardiac output. Eur J Appl Physiol. 2005;94(5-6):670-80.

24. Driscoll DJ, Staats BA, Beck KC. Measurement of cardiac output in children during exercise: a review. Pediatr Exerc Sci. 1989;1:102-15.

25. Durando MM, Corley KT, Boston RC, et al. Cardiac output determination by use of lithium dilution during exercise in horses. Am J Vet Res. 2008;69(8):1054-60.

26. Ekblom B, Hermansen L. Cardiac output in athletes. J Appl Physiol. 1968;25(5):619-25.

27. Elliott AD, Skowno J, Prabhu M, et al. Measurement of cardiac output during exercise in healthy, trained humans using lithium dilution and pulse contour analysis. Physiol Meas. 2012;33(10):1691-1701.

28. Ferguson RJ, Faulkner JA, Julius S, et al. Comparison of cardiac output determined by CO_2 rebreathing and dye-dilution methods. J Appl Physiol. 1968;25(4):450-54.

29. Gan K, Nishi I, Chin I, et al. On-line determination of pulmonary blood flow using respiratory inert gas analysis. IEEE Trans Biomed Eng. 1993;40(12):1250-59.

30. Giraud R, Siegenthaler N, Park C, et al. Transpulmonary thermodilution curves for detection of shunt. Intensive Care Med. 2010;36(6):1083-86.

31. Gledhill N, Warburton D, Jamnik V. Haemoglobin, blood volume, cardiac function, and aerobic power. Can. J Appl Physiol. 1999;24(1):54-65.

32. Gledhill N, Warburton DER. Hemoglobin, blood volume and endurance. In: Shephard RJ, Astrand PO, editors. Endurance in sport. Oxford: Blackwell Scientific Publications; 2000, p. 301-15.

33. Habazettl H, Athanasopoulos D, Kuebler WM, et al. Near-infrared spectroscopy and indocyanine green derived blood flow index for noninvasive measurement of muscle perfusion during exercise. J Appl Physiol (1985). 2010;108(4):962-67.

34. Holmgren A, Pernow B. The reproducibility of cardiac output determination by the direct Fick method during muscular work. Scand J Clin Invest. 1960;12:224-27.

35. Iijima T, Aoyagi T, Iwao Y, et al. Cardiac output and circulating blood volume analysis by pulse dye-densitometry. J Clin Monit. 1997;13(2):81-9.

36. Imai T, Takahashi K, Fukura H, et al. Measurement of cardiac output by pulse dye densitometry using indocyanine green: a comparison with the thermodilution method. Anesthesiology. 1997;87(4):816-22.

37. Imai T, Takahashi K, Goto F, et al. Measurement of blood concentration of indocyanine green by pulse dye densitometry—comparison with the conventional spectrophotometric method. J Clin Monit Comput. 1998;14(7-8):477-84.

38. Jacob SV, Hornby L, Lands LC. Estimation of mixed venous PCO_2 for determination of cardiac output in children. Chest. 1997;111(2):474-80.

39. Johnson BD, Beck KC, Proctor DN, et al. Cardiac output during exercise by the open circuit acetylene washin method: comparison with direct Fick. J Appl Physiol. 2000;88(5):1650-58.

40. Jonas MM, Tanser SJ. Lithium dilution measurement of cardiac output and arterial pulse waveform analysis: an indicator dilution calibrated beat-by-beat system for continuous estimation of cardiac output. Curr Opin Crit Care. 2002;8(3):257-61.

41. Kemps HM, Thijssen EJ, Schep G, et al. Evaluation of two methods for continuous cardiac output assessment during exercise in chronic heart failure patients. J Appl Physiol (1985). 2008;105(6):1822-29.

42. Kurita T, Morita K, Kato S, et al. Comparison of the accuracy of the lithium dilution technique with the thermodilution technique for measurement of cardiac output. Br J Anaesth. 1997;79(6):770-75.

43. Kusaka T, Okubo K, Nagano K, et al. Cerebral distribution of cardiac output in newborn infants. Arch Dis Child Fetal Neonatal Ed. 2005;90(1):F77-78.

44. Lepretre PM, Koralsztein JP, Billat VL. Effect of exercise intensity on relationship between VO_{2max} and cardiac output. Med Sci Sports Exerc. 2004;36(8):1357-63.

45. Levy RJ, Chiavacci RM, Nicolson SC, et al. An evaluation of a noninvasive cardiac output measurement using partial carbon dioxide rebreathing in children. Anesth Analg. 2004;99(6):1642-47, table of contents.

46. Linton NW, Linton RA. Estimation of changes in cardiac output from the arterial blood pressure waveform in the upper limb. Br J Anaesth. 2001;86(4):486-96.

47. Linton RA, Band DM, Haire KM. A new method of measuring cardiac output in man using lithium dilution. Br J Anaesth. 1993;71(2):262-66.

48. Linton RA, Jonas MM, Tibby SM, et al. Cardiac output measured by lithium dilution and transpulmonary thermodilution in patients in a paediatric intensive care unit. Intensive Care Med. 2000;26(10):1507-11.

49. Mathews L, Singh RK. Cardiac output monitoring. Ann Card Anaesth. 2008;11(1):56-68.

50. Matsukawa K, Kobayashi T, Nakamoto T, et al. Noninvasive evaluation of cardiac output during postural change and exercise in humans: comparison between the Modelflow and pulse dye-densitometry. Jpn J Physiol. 2004;54(2):153-60.

51. McLuckie A, Murdoch IA, Marsh MJ, et al. A comparison of pulmonary and femoral artery thermodilution cardiac indices in paediatric intensive care patients. Acta Paediatr. 1996;85(3):336-38.

52. Nagano K, Kusaka T, Okubo K, et al. Estimation of circulating blood volume in infants using the pulse dye densitometry method. Paediatr Anaesth. 2005;15(2):125-30.

53. Nassis GP, Sidossis LS. Methods for assessing body composition, cardiovascular and metabolic function in children and adolescents: implications for exercise studies. Curr Opin Clin Nutr Metab Care. 2006;9(5):560-67.

54. Neilan TG, Pradhan AD, King ME, et al. Derivation of a size-independent variable for scaling of cardiac dimensions in a normal paediatric population. Eur J Echocardiogr. 2009;10(1):50-55.

55. Nottin S, Vinet A, Lecoq AM, et al. Study of the reproducibility of cardiac output measurement during exercise in pre-pubertal children by Doppler echocardiography and CO_2 inhalation. Arch Mal Coeur Vaiss. 2000;93(11):1297-1303.

56. Paterson DH, Cunningham DA. Comparison of methods to calculate cardiac output using the CO_2 rebreathing method. Eur J Appl Physiol. 1976;35(3):223-30.

57. Paterson DH, Cunningham DA, Plyley MJ, et al. The consistency of cardiac output measurement (CO_2 rebreathe) in children during exercise. Eur J Appl Physiol. 1982;49(1):37-44.

58. Pauli C, Fakler U, Genz T, et al. Cardiac output determination in children: equivalence of the transpulmonary thermodilution method to the direct Fick principle. Intensive Care Med. 2002;28(7):947-52.

59. Pianosi P, Hochman J. End-tidal estimates of arterial PCO_2 for cardiac output measurement by CO_2 rebreathing: a study in patients with cystic fibrosis and healthy controls. Pediatr Pulmonol. 1996;22(3):154-60.

60. Pianosi PT. Measurement of exercise cardiac output by thoracic impedance in healthy children. Eur J Appl Physiol. 2004;92(4-5):425-30.

61. Rang S, de Pablo Lapiedra B, van Montfrans GA, et al. Modelflow: a new method for noninvasive assessment of cardiac output in pregnant women. Am J Obstet Gynecol. 2007;196(3):235 e1-8.

62. Reddy PS, Curtiss EI, Bell B, et al. Determinants of variation between Fick and indicator dilution estimates of cardiac output during diagnostic catheterization. Fick vs. dye cardiac outputs. J Lab Clin Med. 1976;87(4):568-76.

63. Reekers M, Simon MJ, Boer F, et al. Cardiovascular monitoring by pulse dye densitometry or arterial indocyanine green dilution. Anesth Analg. 2009;109(2):441-46.

64. Richard R, Lonsdorfer-Wolf E, Charloux A, et al. Non-invasive cardiac output evaluation during a maximal progressive exercise test, using a new impedance cardiograph device. Eur J Appl Physiol. 2001;85(3-4):202-7.

65. Rosenthal M, Bush A. Haemodynamics in children during rest and exercise: methods and normal values. Eur Respir J. 1998;11(4):854-65.

66. Rowland T, Goff D, Martel L, et al. Normalization of maximal cardiovascular variables for body size in premenarcheal girls. Pediatr Cardiol. 2000;21(5):429-32.

67. Rowland T, Obert P. Doppler echocardiography for the estimation of cardiac output with exercise. Sports Med. 2002;32(15):973-86.

68. Rowland T, Whatley Blum J. Cardiac dynamics during upright cycle exercise in boys. Am J Hum Biol. 2000;12(6):749-57.

69. Rowland TW, Melanson EL, Popowski BE, et al. Test-retest reproducibility of maximum cardiac output by Doppler echocardiography. Am J Cardiol. 1998;81(10):1228-30.

70. Sady SP, Freedson PS, Gilliam TB. Calculation of submaximal and maximal cardiac output in children using the CO_2 rebreathing technique. J Sports Med Phys Fitness. 1981;21(3):245-52.

71. Shaw JG, Johnson EC, Voyles WF, et al. Noninvasive Doppler determination of cardiac output during submaximal and peak exercise. J Appl Physiol. 1985;59(3):722-31.

72. Stok WJ, Stringer RC, Karemaker JM. Noninvasive cardiac output measurement in orthostasis: pulse contour analysis compared with acetylene rebreathing. J Appl Physiol (1985). 1999;87(6):2266-73.

73. Sugawara J, Tanabe T, Miyachi M, et al. Non-invasive assessment of cardiac output during exercise in healthy young humans: comparison between Modelflow method and Doppler echocardiography method. Acta Physiol. Scand. 2003;179(4):361-66.

74. Taguchi N, Nakagawa S, Miyasaka K, et al. Cardiac output measurement by pulse dye densitometry using three wavelengths. Pediatr Crit Care Med. 2004;5(4):343-50.

75. Thomas SG, Gledhill N, Jamnik V. Single breath measurement of cardiac output: technique and reliability. Can. J Appl Physiol. 1997;22:52p.

76. Tibby SM, Hatherill M, Marsh MJ, et al. Clinical validation of cardiac output measurements using femoral artery thermodilution with direct Fick in ventilated children and infants. Intensive Care Med. 1997;23(9):987-91.

77. Turley KR, Stanforth PR, Rankinen T, et al. Scaling submaximal exercise cardiac output and stroke

volume: the HERITAGE Family Study. Int J Sports Med. 2006;27(12):993-99.

78. Vinet A, Mandigout S, Nottin S, et al. Influence of body composition, hemoglobin concentration, and cardiac size and function of gender differences in maximal oxygen uptake in prepubertal children. Chest. 2003;124(4):1494-99.

79. Vinet A, Nottin S, Lecoq AM, et al. Reproducibility of cardiac output measurements by Doppler echocardiography in prepubertal children and adults. Int J Sports Med. 2001;22(6):437-41.

80. Warburton DE, Gledhill N, Quinney HA. Blood volume, aerobic power, and endurance performance: potential ergogenic effect of volume loading. Clin J Sport Med. 2000;10(1):59-66.

81. Warburton DE, Haykowsky MJ. The evaluation of cardiac function across the health spectrum under diverse conditions of physiological stress: introduction. Appl. Physiol. Nutr. Metab. 2007;32(2):309-10.

82. Warburton DE, Haykowsky MJ, Quinney HA, et al. Reliability and validity of measures of cardiac output during incremental to maximal aerobic exercise. Part I: Conventional techniques. Sports Med. 1999;27(1):23-41.

83. Warburton DE, Haykowsky MJ, Quinney HA, et al. Reliability and validity of measures of cardiac output during incremental to maximal aerobic exercise. Part II: Novel techniques and new advances. Sports Med. 1999;27(4):241-60.

84. Warburton DE, Nicol C, Bredin SS. Health benefits of physical activity: the evidence. Can Med Assoc J. 2006;174(6):801-9.

85. Warburton DE, Taylor A, Bredin SS, et al. Central haemodynamics and peripheral muscle function during exercise in patients with chronic heart failure. Appl. Physiol. Nutr. Metab. 2007;32(2):318-31.

86. Warburton DER. Effect of alterations in blood volume on cardiac functioning during maximal exercise [letter: Reprint]. Med Sci Sports Exerc. 1998;30(8):1339-41.

87. Warburton DER, Nettlefold L, McGuire A, et al. Cardiovascular function. In: Armstrong N, van Mechelen, W, editors. Paediatric exercise science and medicine. 2nd ed. London, UK: Oxford University Press; 2008. p. 77-96.

88. Welsman J, Bywater K, Farr C, et al. Reliability of peak VO(2) and maximal cardiac output assessed using thoracic bioimpedance in children. Eur J Appl Physiol. 2005;94(3):228-34.

89. Welsman JR, Armstrong N. The measurement and interpretation of aerobic fitness in children: current issues. J R Soc Med. 1996;89(5):281P-85P.

90. Welsman JR, Armstrong N, Nevill AM, et al. Scaling peak VO$_2$ for differences in body size. Med Sci Sports Exerc. 1996;28(2):259-65.

91. Wesseling KH, Jansen JR, Settels JJ, et al. Computation of aortic flow from pressure in humans using a nonlinear, three-element model. J Appl Physiol (1985). 1993;74(5):2566-73.

92. Wesseling KH, Purschke R, Smith NT, et al. A computer module for the continuous monitoring of cardiac output in the operating theatre and the ICU. Acta Anaesthesiol Belg. 1976;27 suppl:327-41.

93. Wiegand G, Binder W, Ulmer H, et al. Noninvasive cardiac output measurement at rest and during exercise in pediatric patients after interventional or surgical atrial septal defect closure. Pediatr Cardiol. 2012;33(7):1109-14.

94. Wiegand G, Kerst G, Baden W, et al. Noninvasive cardiac output determination for children by the inert gas-rebreathing method. Pediatr Cardiol. 2010;31(8):1214-18.

95. Wyse SD, Pfitzner J, Rees A, et al. Measurement of cardiac output by thermal dilution in infants and children. Thorax. 1975;30:262-65.

96. Zeidifard E, Godfrey S, Davies EE. Estimation of cardiac output by an N_2O rebreathing method in adults and children. J Appl Physiol. 1976;41(3):433-38.

97. Zenger MR, Brenner M, Haruno M, et al. Measurement of cardiac output by automated single-breath technique, and comparison with thermodilution and Fick methods in patients with cardiac disease. Am J Cardiol. 1993;71(1):105-9.

Chapter 10

1. Akagi T, Benson LN, Green M, et al. Ventricular function during supine bicycle exercise in univentricular connection with absent right atrioventricular connection. Am J Cardiol. 1991;67:1273-78.

2. Borer JS, Kent KM, Bacharach SL. Sensitivity, specificity, and predictive accuracy of radionuclide cineangiography during exercise in patients with coronary artery disease. Circulation. 1979;60:572-79.

3. Bougault V, Nottin S, Doucende G, et al. Tissue Doppler reproducibility during exercise. Int J Sports Med. 2008;29:395-400.

4. Braunwald E, Sarnoff SJ, Stainsby SN. Determinants of duration and mean rate of ventricular ejection. Circ Res. 1958;6:319-25.

5. Chen RHS, Wong SJ, Wong WHS, et al. Left ventricular contractile reserve after arterial switch operation for complete transposition of the great arteries: an exercise echocardiographic study. Eur Heart J. 2013;14:480-86.

6. Cooper DM, Weiler-Ravell D, Whipp BJ, et al. Growth-related changes in oxygen uptake and heart rate during progressive exercise in children. Ped Res. 1984;18:845-51.

7. DeLuca PS, Arena R. Cardiopulmonary exercise testing abnormalities: a universal indicator of declining left ventricular function during exertion. Int J Cardiol. 2011;153:e59-e61.

8. DeSouza AM, Potts JE, Potts MT, et al. A stress echocardiography study of function during progressive exercise in pediatric oncology patients treated with anthracyclines. Pediatr Blood Cancer. 2007;49:56-64.

9. DeSouza M, Schaffer MS, Gilday DL, et al. Exercise radionucilde angiography in hyperlipidemic children with apparently normal hearts. Nucl Med Comm. 1984;5:13-17.

10. Froelicher VF. Exercise testing & training. New York: Le Jacq Publishing. 1983. p. 137-78.

11. Ha J-W, Ahn J-A, Kim J-M, et al. Abnormal longitudinal myocardial functional reserve assessed by exercise tissue Doppler echocardiography in patients with hypertrophic cardiomyopathy. J Am Soc Echocardiogr. 2006;19:1314-19.

12. Harada K, Toyono M, Yamamoto F. Assessment of right ventricular function during exercise with quantitative Doppler tissue imaging in children later after repair of tetralogy of Fallot. J Am Soc Echocardiogr. 2004;17:303-9.

13. Hijazi ZM, Udelson JE, Snapper H, et al. Physiologic significance of chronic coronary aneurysms in patients with Kawasaki disease. J Am Coll Cardiol. 1994;24:1633-38.

14. Kang SJ, Lim HS, Hwang J, et al. Impact of changes in myocardial velocity assessed by tissue Doppler imaging during exercise on dynamic mitral regurgitation in patients with nonischemic cardiomyopathy. Echocardiography. 2008;25:394-400.

15. Kondo C. Myocardial perfusion imaging in pediatric cardiology. Ann Nucl Med. 2004;18:551-61.

16. Lewis RP. The use of systolic time intervals for evaluation of left ventricular function. In: Fowler NO, editor. Noninvasive diagnostic methods in cardiology. Philadelphia: F.A. Davis Company; 1983. p. 335-53.

17. McIntosh RA, Silberbauer J, Veasey RA, et al. Tissue Doppler-derived contractile reserve is a simple and strong predictor of cardiopulmonary exercise performance across a range of cardiac diseases. Echocardiography. 2013;29:527-33.

18. Moon J, Hong YJ, Kim Y-J, et al. Extent of late gadolinium enhancement on cardiovascular magnetic resonance imaging and its relation to left ventricular longitudinal functional reserve during exercise in patients with hypertrophic cardiomyopathy. Circ J. 2013;77:1742-49.

19. Munhoz EC, Hollanda R, Vargas JP, et al. Flattening of oxygen pulse during exercise may detect extensive myocardial ischemia. Med Sci Sports Exerc. 2007;39:1221-26.

20. Norris SR, Bell GJ, Bhambani YN. Oxygen pulse as a predictor of stroke volume during cycle ergometer exercise [abstr]. Med Sci Sports Exerc. 1991;23 Suppl:S158.

21. Pahl E, Duffy CE, Chaudhry FA. The role of stress echocardiography in children. Echocardiography. 2000;17:507-12.

22. Parrish MD, Boucek RJ, Burger J, et al. Exercise radionuclide ventriculography in children: normal values for exercise variables and right and left ventricular function. Br Heart J. 1985;54:509-16.

23. Paul MH, Wessel HU. Exercise studies in patients with transposition of the great arteries after atrial repair operations (Mustard/Senning): a review. Pediatr Cardiol. 1999;20:49-55.

24. Poerner TC, Goebel B, Figulla HR, et al. Diastolic biventricular impairment at long-term follow-up after atrial switch operation for complete transposition of the great arteries: an exercise tissue Doppler echocardiographic study. J Am Soc Echocardiogr. 2007;20:1285-93.

25. Robbers-Visser D, Luijnenburg SE, van den Berg J, et al. Stress imaging in congenital heart disease. Cardiol Young. 2009;19:552-62.

26. Roberson DA, Cui W. Tissue Doppler imaging parameters of left ventricular function in children: mitral annulus displacement is superior to peak velocity. J Am Soc Echocardiogr. 2009;22:376-82.

27. Rowland T. Cardiovascular function. In: Armstrong N, van Michelen W, editors. Paediatric exercise science and medicine. Oxford: Oxford University Press; 2008. p.255-68.

28. Rowland T, Potts J, Potts T, et al. Cardiovascular responses to exercise in children and adolescents with myocardial dysfunction. Am Heart J. 1999;137:126-33.

29. Rowland T, Unnithan V. Myocardial inotropic response to progressive exercise in healthy subjects: a review. Curr Sports Med Rep. 2013;12:93-100.

30. Rowland T, Willers M. Reproducibility of Doppler measures of ventricular function during maximal upright cycling. Cardiol Young. 2010;20:676-79.

31. Rubis P, Drabik L, Kopec G, et al. The prognostic role of exercise echocardiography in heart failure. Kardiol Pol. 2011;69:656-63.

32. Sekiguchi M, Adachi H, Oshima S, et al. Effect of changes in left ventricular diastolic function during exercise on exercise tolerance assessed by exercise-stress tissue Doppler echocardiography. Int Heart J. 2009;50:763-71.

33. Suda K, Iwatani H, Mori C, et al. Radionuclide assessment of left ventricular performance on

exercise after external conduit operation. Acta Paediatr Jpn. 1993:35:283-88.

34. Unnithan V, Rowland TW. Use of oxygen pulse in predicting Doppler-derived maximal stroke volume in adolescents. Pediatr Exerc Sci. 2015;27:412-18.

35. Vavra J, Sova J, Macek M. Effect of age on systolic time intervals at rest and during exercise on a bicycle ergometer. Eur J Appl Physiol. 1982;50:71-78.

36. Washington RL, van Gundy JC, Cohen C, et al. Normal aerobic and anaerobic exercise data for North American school-age children. J Pediatr. 1988;112:223-33.

37. Wasserman K, Hansen JE, Sue DY, et al. Principles of exercise testing and interpretation. 4th ed. Philadelphia: Lippincott Williams & Wilkins; 2005.

Chapter 11

1. Adams RJ, Fuhlbrigge A, Finkelstein JA, et al. Impact of inhaled antiinflammatory therapy on hospitalization and emergency department visits for children with asthma. Pediatrics. 2001;107(4):706-11.

2. Akkerman M, van Brussel M, Bongers BC, et al. Oxygen uptake efficiency slope in healthy children. Pediatr Exerc Sci. 2010;22(3):431-41.

3. American Thoracic Society, American College of Chest Physicians. ATS/ACCP statement on cardiopulmonary exercise testing. Am J Respir Crit Care Med. 2003;167(2):211-77.

4. Anderson SD, Godfrey S. Transfer factor for CO during exercise in children. Thorax. 1971;26(1):51-54.

5. Armstrong N, Kirby BJ, McManus AM, et al. Prepubescents' ventilatory responses to exercise with reference to sex and body size. Chest. 1997;112(6):1554-60.

6. ATS Committee on Proficiency Standards for Clinical Pulmonary Function Laboratories. ATS statement: guidelines for the six-minute walk test. Am J Respir Crit Care Med. 2002;166(1):111-17.

7. Baba R, Nagashima M, Goto M, et al. Oxygen uptake efficiency slope: a new index of cardiorespiratory functional reserve derived from the relation between oxygen uptake and minute ventilation during incremental exercise. J Am Coll Cardiol. 1996;28(6):1567-72.

8. Baraldi E, Zanconato S, Santuz PA, et al. A comparison of two noninvasive methods in the determination of the anaerobic threshold in children. Int J Sports Med. 1989;10(2):132-34.

9. Berntsen S, Lødrup Carlsen KC, Anderssen SA, et al. Factors associated with aerobic fitness in adolescents with asthma. Respir Med. 2013;107(8):1164-71.

10. Beydon N, Davis SD, Lombardi E, et al. An official American Thoracic Society/European Respiratory Society statement: pulmonary function testing in preschool children. Am J Respir Crit Care Med. 2007;175(12):1304-45.

11. Bongers BC, Hulzebos EHJ, Arets BGM, et al. Validity of the oxygen uptake efficiency slope in children with cystic fibrosis and mild-to-moderate airflow obstruction. Pediatr Exerc Sci. 2012;24(1):129-41.

12. Bongers BC, Hulzebos HJ, Blank AC, et al. The oxygen uptake efficiency slope in children with congenital heart disease: construct and group validity. Eur J Cardiovasc Prev Rehabil. 2011;18(3):384-92.

13. Boule M, Gaultier C, Girard F. Breathing pattern during exercise in untrained children. Respir Physiol. 1989;75(2):225-33.

14. Buchfuhrer MJ, Hansen JE, Robinson TE, et al. Optimizing the exercise protocol for cardiopulmonary assessment. J Appl Physiol. 1983;55(5):1558-64.

15. Cahalin L, Pappagianopoulos P, Prevost S, et al. The relationship of the 6-min walk test to maximal oxygen consumption in transplant candidates with end-stage lung disease. Chest. 1995;108(2):452-59.

16. Carlsen KH, Engh G, Mørk M, et al. Cold air inhalation and exercise-induced bronchoconstriction in relationship to metacholine bronchial responsiveness: different patterns in asthmatic children and children with other chronic lung diseases. Respir Med. 1998;92(2):308-15.

17. Centers for Disease Control and Prevention. Asthma Table 3-1: Current asthma population estimates—in thousands by age, NHIS, 2012 [Internet]. Atlanta, GA: CDC; 2012. [cited 2014 May 11]. Available from: www.cdc.gov/asthma/nhis/2012/table3-1.htm

18. Clark JS, Votteri B, Ariagno RL, et al. Noninvasive assessment of blood gases. Am Rev Respir Dis. 1992;145(1):220-32.

19. Coates AL, Canny G, Zinman R, et al. The effects of chronic airflow limitation, increased dead space, and the pattern of ventilation on gas exchange during maximal exercise in advanced cystic fibrosis. Am Rev Respir Dis. 1988;138(6):1524-31.

20. Cooper DM, Kaplan MR, Baumgarten L, et al. Coupling of ventilation and CO_2 production during exercise in children. Pediatr Res. 1987;21(6):568-72.

21. Cooper DM, Leu S-Y, Galassetti P, et al. Dynamic interactions of gas exchange, body mass, and progressive exercise in children. Med Sci Sports Exerc 2014;46(5):877-86.

22. Crapo RO, Casaburi R, Coates AL, et al. Guidelines for methacholine and exercise challenge

testing—1999. Am J Respir Crit Care Med. 2000;161(1):309-29.

23. Cropp GJ, Pullano TP, Cerny FJ, et al. Exercise tolerance and cardiorespiratory adjustments at peak work capacity in cystic fibrosis. Am Rev Respir Dis. 1982;126(2):211-16.

24. Dempsey JA, Hanson PG, Henderson KS. Exercise-induced arterial hypoxaemia in healthy human subjects at sea level. J Physiol (Lond.). 1984;355:161-75.

25. Driessen JM, van der Palen J, van Aalderen WM, et al. Inspiratory airflow limitation after exercise challenge in cold air in asthmatic children. Respir Med. 2012;106(10):1362-68.

26. ERS Task Force, Palange P, Ward SA, et al. Recommendations on the use of exercise testing in clinical practice. Eur Respir J. 2007;29(1):185-209.

27. National Asthma Education and Prevention Program Coordinating Committee of NHLBI. Expert panel report 3 (EPR3): Guidelines for the diagnosis and management of asthma [Internet]. Washington DC: NIH; 2007 [cited 2014 May 11]. Available from: www.nhlbi.nih.gov/files/docs/guidelines/asthsumm.pdf

28. Feisal KA, Fuleihan FJ. Pulmonary gas exchange during exercise in young asthmatic patients. Thorax. 1979;34(3):393-96.

29. Ferrazza AM, Martolini D, Valli G, et al. Cardiopulmonary exercise testing in the functional and prognostic evaluation of patients with pulmonary diseases. Respiration. 2009;77(1):3-17.

30. Fink G, Kaye C, Blau H, et al. Assessment of exercise capacity in asthmatic children with various degrees of activity. Pediatr Pulmonol. 1993;15(1):41-43.

31. Fitzgerald NM, Kennedy B, Fitzgerald DA, et al. Diffusion capacity of carbon monoxide (DLCO) pre- and post-exercise in children in health and disease. Pediatr Pulmonol. 2014 Aug;49(8):782-89.

32. Fulton JE, Pivarnik JM, Taylor WC, et al. Prediction of maximum voluntary ventilation (MVV) in African-American adolescent girls. Pediatr Pulmonol. 1995;20(4):225-33.

33. Giardini A, Odendaal D, Khambadkone S, et al. Physiologic decrease of ventilatory response to exercise in the second decade of life in healthy children. Am Heart J. 2011;161(6):1214-19.

34. Gitt AK, Wasserman K, Kilkowski C, et al. Exercise anaerobic threshold and ventilatory efficiency identify heart failure patients for high risk of early death. Circulation. 2002;106(24):3079-84.

35. Godfrey S. Exercise testing in children. London: W.B. Saunders Company Ltd.; 1974, p. 69.

36. Godfrey S. Exercise Testing in Children. London: W.B. Saunders Company Ltd.; 1974, p. 81.

37. Graff-Lonnevig V, Bevegård S, Eriksson BO. Ventilation and pulmonary gas exchange at rest and during exercise in boys with bronchial asthma. Eur J Respir Dis. 1980;61(6):357-66.

38. Grant GP, Mansell AL, Garofano RP, et al. Cardiorespiratory response to exercise after the Fontan procedure for tricuspid atresia. Pediatr Res. 1988;24(1):1-5.

39. Hallstrand TS, Altemeier WA, Aitken ML, et al. Role of cells and mediators in exercise-induced bronchoconstriction. Immunol Allergy Clin North Am. 2013;33(3):313-28, vii.

40. Hanel B, Teunissen I, Rabol A, et al. Restricted postexercise pulmonary diffusion capacity and central blood volume depletion. J Appl Physiol. 1997;83(1):11-17.

41. Hankinson JL, Odencrantz JR, Fedan KB. Spirometric reference values from a sample of the general U.S. population. Am J Respir Crit Care Med. 1999;159(1):179-87.

42. Harms CA, McClaran SR, Nickele GA, et al. Exercise-induced arterial hypoxaemia in healthy young women. J Physiol (Lond.). 1998;507(Pt 2):619-28.

43. Haverkamp HC, Dempsey JA, Miller JD, et al. Gas exchange during exercise in habitually active asthmatic subjects. J Appl Physiol. 2005;99(5):1938-50.

44. Hebestreit H, Hebestreit A, Trusen A, et al. Oxygen uptake kinetics are slowed in cystic fibrosis. Med Sci Sports Exerc 2005;37(1):10-17.

45. Hebestreit H, Staschen B, Hebestreit A. Ventilatory threshold: a useful method to determine aerobic fitness in children? Med Sci Sports Exerc. 2000;32(11):1964-69.

46. Hedlin G, Graff-Lonnevig V, Freyschuss U. Working capacity and pulmonary gas exchange in children with exercise-induced asthma. Acta Paediatr Scand. 1986;75(6):947-54.

47. Henke KG, Orenstein DM. Oxygen saturation during exercise in cystic fibrosis. Am Rev Respir Dis. 1984;129(5):708-11.

48. Hsia CCW. Recruitment of lung diffusing capacity: update of concept and application. Chest. 2002;122(5):1774-83.

49. Javadpour S, Selvadurai H, Wilkes D, et al. Does carbon dioxide retention during exercise predict a more rapid decline in FEV1 in cystic fibrosis? Arch Dis Child. 2005;90(8):792-95.

50. West JB. Respiratory physiology: the essentials. 9th ed. Philadelphia: Lippincott Williams & Wilkins; 2011.

51. Johnson BD, Beck KC, Zeballos RJ, et al. Advances in pulmonary laboratory testing. Chest. 1999;116(5):1377-87.

52. Johnson BD, Scanlon PD, Beck KC. Regulation of ventilatory capacity during exercise in asthmatics. J Appl Physiol. 1995;79(3):892-901.

53. Johnson BD, Weisman IM, Zeballos RJ, et al. Emerging concepts in the evaluation of ventilatory limitation during exercise: the exercise tidal flow-volume loop. Chest. 1999;116(2):488-503.

54. Jones NL, Robertson DG, Kane JW. Difference between end-tidal and arterial PCO_2 in exercise. J Appl Physiol Respir Environ Exerc Physiol. 1979;47(5):954-60.

55. Kim Y-J, Hall GL, Christoph K, et al. Pulmonary diffusing capacity in healthy Caucasian children. Pediatr Pulmonol. 2012;47(5):469-75.

56. Kusenbach G, Wieching R, Barker M, et al. Effects of hyperoxia on oxygen uptake kinetics in cystic fibrosis patients as determined by pseudorandom binary sequence exercise. Eur J Appl Physiol Occup Physiol. 1999;79(2):192-96.

57. Laursen PB, Tsang GCK, Smith GJ, et al. Incidence of exercise-induced arterial hypoxemia in prepubescent females. Pediatr Pulmonol. 2002;34(1):37-41.

58. Lebecque P, Lapierre JG, Lamarre A, et al. Diffusion capacity and oxygen desaturation effects on exercise in patients with cystic fibrosis. Chest. 1987;91(5):693-97.

59. Levy WC, Arena R, Wagoner LE, et al. Prognostic impact of the addition of ventilatory efficiency to the Seattle Heart Failure Model in patients with heart failure. J Card Fail. 2012;18(8):614-19.

60. Macintyre N, Crapo RO, Viegi G, et al. Standardisation of the single-breath determination of carbon monoxide uptake in the lung. Eur Respir J. 2005;26(4):720-35.

61. Mahon AD, Vaccaro P. Ventilatory threshold and VO_{2max} changes in children following endurance training. Med Sci Sports Exerc. 1989;21(4):425-31.

62. Maldonado M, Portela LOC. Analysis of physiological variables during acute hypoxia and maximal stress test in adolescents clinically diagnosed with mild intermittent or mild persistent asthma. J Bras Pneumol. 2011;37(6):712-19.

63. Marinov B, Mandadzhieva S, Kostianev S. Oxygen-uptake efficiency slope in healthy 7- to 18-year-old children. Pediatr Exerc Sci. 2007;19(2):159-70.

64. Mengelkoch LJ, Martin D, Lawler J. A review of the principles of pulse oximetry and accuracy of pulse oximeter estimates during exercise. Phys Ther. 1994;74(1):40-49.

65. Miller A, Enright PL. PFT interpretive strategies: American Thoracic Society/European Respiratory Society 2005 guideline gaps. Respir Care. 2012;57(1):127-33; discussion 133-35.

66. Miller MR, Hankinson J, Brusasco V, et al. Standardisation of spirometry. Eur Respir J. 2005;26(2):319-38.

67. Moraes EZC de, Trevisan ME, Baldisserotto S de V, et al. Children and adolescents with mild intermittent or mild persistent asthma: aerobic capacity between attacks. J Bras Pneumol. 2012;38(4):438-44.

68. Moser C, Tirakitsoontorn P, Nussbaum E, et al. Muscle size and cardiorespiratory response to exercise in cystic fibrosis. Am J Respir Crit Care Med. 2000;162(5):1823-27.

69. Nagano Y, Baba R, Kuraishi K, et al. Ventilatory control during exercise in normal children. Pediatr Res. 1998;43(5):704-7.

70. Neder JA, Nery LE, Peres C, et al. Reference values for dynamic responses to incremental cycle ergometry in males and females aged 20 to 80. Am J Respir Crit Care Med. 2001;164(8 Pt 1):1481-86.

71. Nixon PA, Joswiak ML, Fricker FJ. A six-minute walk test for assessing exercise tolerance in severely ill children. J Pediatr. 1996;129(3):362-66.

72. Nixon PA, Orenstein DM, Kelsey SF, et al. The prognostic value of exercise testing in patients with cystic fibrosis. N Engl J Med. 1992;327(25):1785-88.

73. Nixon, PA. Pulmonary function. In: Armstrong N, Van Mechelen W, editors. Textbook of paediatric exercise science and medicine. Oxford: Oxford University Press; 2008. p. 67-76.

74. Nixon PA. Role of exercise in the evaluation and management of pulmonary disease in children and youth. Med Sci Sports Exerc. 1996;28(4):414-20.

75. Nourry C, Deruelle F, Guinhouya C, et al. High-intensity intermittent running training improves pulmonary function and alters exercise breathing pattern in children. Eur J Appl Physiol. 2005;94(4):415-23.

76. Nourry C, Fabre C, Bart F, et al. Evidence of exercise-induced arterial hypoxemia in prepubescent trained children. Pediatr Res. 2004;55(4):674-81.

77. Ohuchi H, Kato Y, Tasato H, et al. Ventilatory response and arterial blood gases during exercise in children. Pediatr Res. 1999;45(3):389-96.

78. Ondrak KS, McMurray RG. Exercise-induced breathing patterns of youth are related to age and intensity. Eur J Appl Physiol. 2006;98(1):88-96.

79. Orenstein DM, Curtis SE, Nixon PA, et al. Accuracy of three pulse oximeters during exercise and hypoxemia in patients with cystic fibrosis. Chest. 1993;104(4):1187-90.

80. Orenstein DM, Nixon PA. Exercise performance and breathing patterns in cystic fibrosis: male-female differences and influence of rest-

ing pulmonary function. Pediatr Pulmonol. 1991;10(2):101-5.

81. Paridon SM, Alpert BS, Boas SR, et al. Clinical stress testing in the pediatric age group: a statement from the American Heart Association Council on Cardiovascular Disease in the Young, Committee on Atherosclerosis, Hypertension, and Obesity in Youth. Circulation. 2006;113(15):1905-20.

82. Pellegrino R, Viegi G, Brusasco V, et al. Interpretative strategies for lung function tests. Eur Respir J. 2005;26(5):948-68.

83. Pianosi P, Wolstein R. Carbon dioxide chemosensitivity and exercise ventilation in healthy children and in children with cystic fibrosis. Pediatr Res. 1996;40(3):508-13.

84. Pianosi PT, Johnson JN, Turchetta A, et al. Pulmonary function and ventilatory limitation to exercise in congenital heart disease. Congenit Heart Dis. 2009;4(1):2-11.

85. Pinet C, Cassart M, Scillia P, et al. Function and bulk of respiratory and limb muscles in patients with cystic fibrosis. Am J Respir Crit Care Med. 2003;168(8):989-94.

86. Quanjer PH, Borsboom GJ, Brunekreef B, et al. Spirometric reference values for white European children and adolescents: Polgar revisited. Pediatr Pulmonol. 1995;19(2):135-42.

87. Quanjer PH, Stanojevic S, Cole TJ, et al. Multiethnic reference values for spirometry for the 3-95-yr age range: the global lung function 2012 equations. Eur Respir J. 2012;40(6):1324-43.

88. Ratel S, Duche P, Hennegrave A, et al. Acid-base balance during repeated cycling sprints in boys and men. J Appl Physiol. 2002;92(2):479-85.

89. Rausch CM, Taylor AL, Ross H, et al. Ventilatory efficiency slope correlates with functional capacity, outcomes, and disease severity in pediatric patients with pulmonary hypertension. Int J Cardiol. 2013;169(6):445-48.

90. Regnis JA, Alison JA, Henke KG, et al. Changes in end-expiratory lung volume during exercise in cystic fibrosis relate to severity of lung disease. Am Rev Respir Dis. 1991;144(3 Pt 1):507-12.

91. Reybrouck T, Mertens L, Kalis N, et al. Dynamics of respiratory gas exchange during exercise after correction of congenital heart disease. J Appl Physiol. 1996;80(2):458-63.

92. Reybrouck T, Weymans M, Stijns H, et al. Ventilatory anaerobic threshold in healthy children. Age and sex differences. Eur J Appl Physiol Occup Physiol. 1985;54(3):278-84.

93. Rhodes J, Ubeda Tikkanen A, Jenkins KJ. Exercise testing and training in children with congenital heart disease. Circulation. 2010;122(19):1957-67.

94. Robbins PA, Conway J, Cunningham DA, et al. A comparison of indirect methods for continuous estimation of arterial PCO_2 in men. J Appl Physiol. 1990;68(4):1727-31.

95. Rosenfeld M, Allen J, Arets BHGM, et al. An official American Thoracic Society workshop report: optimal lung function tests for monitoring cystic fibrosis, bronchopulmonary dysplasia, and recurrent wheezing in children less than 6 years of age. Ann Am Thorac Soc. 2013;10(2):S1-S11.

96. Rowe SM, Miller S, Sorscher EJ. Cystic fibrosis. N Engl J Med. 2005;352(19):1992-2001.

97. Rowland TW, Cunningham LN. Development of ventilatory responses to exercise in normal white children. A longitudinal study. Chest. 1997;111(2):327-32.

98. Santuz P, Baraldi E, Filippone M, et al. Exercise performance in children with asthma: is it different from that of healthy controls? Eur Respir J. 1997;10(6):1254-60.

99. Shephard RJ, Bar-Or O. Alveolar ventilation in near maximum exercise. Data on pre-adolescent children and young adults. Med Sci Sports. 1970;2(2):83-92.

100. Stanojevic S, Wade A, Cole TJ, et al. Spirometry centile charts for young Caucasian children: the Asthma UK Collaborative Initiative. Am J Respir Crit Care Med. 2009;180(6):547-52.

101. Stanojevic S, Wade A, Stocks J, et al. Reference ranges for spirometry across all ages: a new approach. Am J Respir Crit Care Med. 2008;177(3):253-60.

102. Stein R, Selvadurai H, Coates A, et al. Determination of maximal voluntary ventilation in children with cystic fibrosis. Pediatr Pulmonol. 2003;35(6):467-71.

103. Sue DY, Wasserman K, Moricca RB, et al. Metabolic acidosis during exercise in patients with chronic obstructive pulmonary disease. Use of the V-slope method for anaerobic threshold determination. Chest. 1988;94(5):931-38.

104. Ten Harkel ADJ, Takken T, Van Osch-Gevers M, et al. Normal values for cardiopulmonary exercise testing in children. Eur J Cardiovasc Prev Rehabil. 2011;18(1):48-54.

105. Thin AG, Dodd JD, Gallagher CG, et al. Effect of respiratory rate on airway deadspace ventilation during exercise in cystic fibrosis. Respir Med. 2004;98(11):1063-70.

106. Tirakitsoontorn P, Nussbaum E, Moser C, et al. Fitness, acute exercise, and anabolic and catabolic mediators in cystic fibrosis. Am J Respir Crit Care Med. 2001;164(8 Pt 1):1432-37.

107. Vilozni D, Szeinberg A, Barak A, et al. The relation between age and time to maximal

bronchoconstriction following exercise in children. Respir Med. 2009;103(10):1456-60.

108. Wanger J, Clausen JL, Coates A, et al. Standardisation of the measurement of lung volumes. Eur Respir J. 2005;26(3):511-22.

109. Washington RL, van Gundy JC, Cohen C, et al. Normal aerobic and anaerobic exercise data for North American school-age children. J Pediatr. 1988;112(2):223-33.

110. Wasserman K, McIlroy MB. Detecting the threshold of anaerobic metabolism in cardiac patients during exercise. Am J Cardiol. 1964;14:844-52.

111. Wasserman K, Whipp BJ. Excercise physiology in health and disease. Am Rev Respir Dis. 1975;112(2):219-49.

112. Weiler JM, Bonini S, Coifman R, et al. American Academy of Allergy, Asthma & Immunology Work Group report: exercise-induced asthma. J Allergy Clin Immunol. 2007;119(6):1349-58.

113. Wells GD, Wilkes DL, Schneiderman JE, et al. Skeletal muscle metabolism in cystic fibrosis and primary ciliary dyskinesia. Pediatr Res. 2011;69(1):40-45.

114. Welsh L, Roberts RGD, Kemp JG. Fitness and physical activity in children with asthma. Sports Med. 2004;34(13):861-70.

Chapter 12

1. Aboulhosn J, Child JS. Left ventricular outflow obstruction: subaortic stenosis, bicuspid aortic valve, supravalvar aortic stenosis, and coarctation of the aorta. Circulation. 2006;114:2412-22.

2. Babaee Bigi MA, Aslani A. Aortic root size and prevalence of aortic regurgitation in elite strength trained athletes. Am J Cardiol. 2007;100:528-30.

3. Basso C, Boschello M, Perrone C, et al. An echocardiographic survey of primary school children for bicuspid aortic valve. Am J Cardiol. 2004;93:661-63.

4. Beekman RH, Katz BP, Moorehead-Steffens C, et al. Altered baroreceptor function in children with systolic hypertension after coarctation repair. Am J Cardiol. 1983;52:112-17.

5. Bertovic DA, Waddell TK, Gatzka CD, et al. Muscular strength training is associated with low arterial compliance and high pulse pressure. Hypertension. 1999;33:1385-91.

6. Bonow RO, Carabello B, De Leon AC, et al. ACC/AHA guidelines for the management of patients with valvular heart disease: A report of the American College of Cardiology/American Heart Association Task Force on Practice Guidelines (Committee on Management of Patients with Valvular Heart Disease). J Am Coll Cardiol. 1998;32:1486–1582.

7. Bove T, De Meulder F, Vandenplas G, et al. Midterm assessment of the reconstructed arteries after the arterial switch operation. Ann Thorac Surg. 2008; 85(3):823-30.

8. Braith RW, Schofield RS, Hill JA, et al. Exercise training attenuates progressive decline in brachial artery reactivity in heart transplant recipients. J Heart Lung Transplant. 2008;27:52-59.

9. Brown DW, Dipilato AE, Chong EC, et al. Sudden unexpected death following balloon valvuloplasty for congenital aortic stenosis. J Am Coll Cardiol. 2010;56(23):1939-46.

10. Cameron JD, Dart AM. Exercise training increases total systemic arterial compliance in humans. Am J Physiol. 1994;266:H693-701.

11. Chen CR, Cheng TO, Huang T, et al. Percutaneous balloon valvuloplasty for pulmonic stenosis in adolescents and adults. N Engl J Med. 1996;335:21-25.

12. Chin TK, Perloff JK, Williams RG, et al. Isolated noncompaction of left ventricular myocardium: a study of eight cases. Circulation. 1990;82:507-13.

13. Cohen MS, Marino BS, McElhinney DB, et al. Neo-aortic root dilation and valve regurgitation up to 21 years after staged reconstruction for hypoplastic left heart syndrome. J Am Coll Cardiol. 2003;42(3):533-40.

14. Davis JA, McBride MG, Chrisant MR, et al. Longitudinal assessment of cardiovascular exercise performance after pediatric heart transplantation. J Heart Lung Transplant. 2006;25:626-33.

15. de Divitiis M, Pilla C, Kattenhorn M, et al. Vascular dysfunction after repair of coarctation of the aorta: impact of early surgery. Circulation. 2001;104:165-70.

16. de Koning WB, van Osch-Gevers M, Ten Harkel AD, et al. Follow-up outcomes 10 years after arterial switch operation for transposititon of the great arteries: Comparison of cardiological health status and health-related quality of life to those of the a normal reference population. Eur J Pediatr. 2008;167:995-1004.

17. Derrick GP, Narang I, White PA, et al. Failure of stroke volume augmentation during exercise and dobutamine stress is unrelated to load-independent indexes of right ventricular performance after the Mustard operation. Circulation. 2000;102(19 Suppl 3): III154-III159.

18. Diller GP, Dimopoulos K, Okonko D, et al. Exercise intolerance in adult congenital heart disease: Comparative severity, correlates, and prognostic implication. Circulation. 2005;112:828-35.

19. Dipchand AI, Manlhiot C, Russell JL, et al. Exercise capacity improves with time in pediatric heart transplant recipients. J Heart Lung Transplant. 2009;28:585-90.

20. Driscoll DJ, Staats B, Heise C, et al. Functional single ventricle: cardiorespiratory response to exercise. J Am Coll Cardiol. 1984;4:337-42.

21. Driscoll DJ, Durongpisitkul K. Exercise testing after the Fontan operation. Pediatr Cardiol. 1999;20(1):57-59.

22. Durongpisitkul K, Driscoll DJ, Mahoney DW, et al. Cardiorespiratory response to exercise after modified Fontan operation. J Am Coll Cardiol. 1997;29(4):785-90.

23. Fredriksen PM, Pettersen E, Thaulow E. Declining aerobic capacity of patients with arterial and atrial switch procedures. Pediatr Cardiol. 2009;30:166-71.

24. Frigiola A, Tsang V, Bull C, et al. Biventricular response after pulmonary valve replacement for right ventricular outflow tract dysfunction: is age a predictor of outcome? Circulation. 2008;118(14 Suppl):S182-90.

25. Galanti G, Stefani L, Toncelli L, et al. Effects of sports activity in athletes with bicuspid aortic valve and mild aortic regurgitation. Br J Sports Med. 2010;44:275-79.

26. Gatzoulis MA, Elliott, JT, Guru V, et al. Right and left ventricular systolic function late after repair of tetralogy of Fallot. Am J Cardiol. 2000; 86(12):1352-57.

27. Gatzoulis MA, Clark, AL, Cullen S, et al. Right ventricular diastolic function 15 to 35 years after repair of tetralogy of Fallot. Restrictive physiology predicts superior exercise performance. Circulation. 1995;91(6):1775-81.

28. Gatzoulis MA, Balaji S, Webber SA, et al. Risk factors for arrhythmia and sudden cardiac death late after repair of tetralogy of Fallot: a multicentre study. Lancet. 2000;356:975-81.

29. Ghai A, Silversides C, Harris L, et al. Left ventricular dysfunction is a risk factor for sudden cardiac death in adults late after repair of tetralogy of Fallot. J Am Coll Cardiol. 2002;40:1675-80.

30. Giardini A, Specchia S, Tacy TA, et al. Usefulness of cardiopulmonary exercise to predict long-term prognosis in adults with repaired tetralogy of Fallot. Am J Cardiol. 2007;99(10):1462-67.

31. Giardini A, Specchia S, Coutsoumbas, et al. Impact of pulmonary regurgitation and right ventricular dysfunction on oxygen uptake recovery kinetics in repaired tetralogy of Fallot. Eur J Heart Fail. 2006; 8(7):736-43.

32. Giardini A, Khambadkone S, Rizzo N, et al. Determinants of exercise capacity after arterial switch operation for transposition of the great arteries. Am J Cardiol. 2009;104:1007-12.

33. Gildein P, Mocellin R, Kaufmehl K. Oxygen uptake transient kinetics during constant-load exercise in children after operations of ventricular septal defect, tetralogy of Fallot, transposition of the great arteries, or tricuspid atresia. Am J Cardiol. 1994;74:166-69.

34. Goldberg B, Fripp RR, Lister G, et al. Effect of physical training on exercise performance of children following surgical repair of congenital heart disease. Pediatrics. 1981;68:691-99.

35. Goldberg DJ, Avitabile CM, McBride MG, et al. Exercise capacity in the Fontan circulation. Cardiol Young. 2013;23:823-29.

36. Graham J, Thomas P, Driscoll DJ, et al. 36th Bethesda conference: recommendations for determining eligibility for competition in athletes with cardiovascular abnormalities. Task force 2: congenital heart disease. J Am Coll Cardiol. 2005;45:1326-33.

37. Gratz A, Hess J, Hager A. Self-estimated physical functioning poorly predicts actual exercise capacity in adolescents and adults with congenital heart disease. Eur Heart J. 2009;30:497-504.

38. Hanson E, Eriksson BO, Sivertsson R. Baroreceptor reflexes after coarctectomy. Clin Physiol. 1981;1:503-9.

39. Harrison D, Liu P, Walters J. Cardiopulmonary function of adult patients late after Fontan repair. J Am Coll Cardiol. 1995;26:1016-21.

40. Hauser M, Bengel FM, Kuhn A, et al. Myocardial blood flow and flow reserve after coronary reimplantation in patients after arterial switch and Ross operation. Circulation. 2001;103(14):1875-80.

41. Haykowsky M, Eves N, Figgures L, et al. Effect of exercise training on $\dot{V}O_2$ peak and left ventricular systolic function in recent cardiac transplant recipients. Am J Cardiol. 2005;95:1002-4.

42. Hogarty AN, Leahy A, Zhao H, et al. Longitudinal evaluation of cardiopulmonary performance during exercise after bone marrow transplantation in children. J Pediatr. 2000;136(3):311-17.

43. Horneffer PJ, Zahka KG, Rowe SA, et al. Long-term results of total repair of tetralogy of Fallot during childhood. Ann Thorac Surg. 1990;50(2):179-83; discussion 183-85.

44. Hövels-Gürich HH, Krämer NA, Schulze A, et al. The cardiologic status of youngsters and young adults after neonatal arterial switch operation with a simple transposition of big vessels. Clin Res Cardiol. 2008;97:671.

45. Ichida F, Hamamichi Y, Miyawaki T, et al. Clinical features of isolated noncompaction of the ventricular myocardium: long-term clinical course, hemodynamic properties, and genetic background. J Am Coll Cardiol. 1999;34:233-40.

46. James FW. Exercise responses in aortic stenosis. Prog Pediatr Cardiol. 1993;2(3):1-7.

47. Kakiyama T, Sugawara J, Murakami H, et al. Effects of short-term endurance training on aortic distensibility in young males. Med Sci Sports Exerc. 2005;37:267-71.

48. Kakiyama T, Matsuda M, Koseki S. Effect of physical activity on the dispensability of the aortic wall in healthy males. Angiology. 1998;49:749-57.

49. Kato H, Sugimura T, Akagi T. Long-term consequences of Kawasaki disease: A 10 to 21 year follow-up study of 594 patients. Circulation. 1996;94:1379–85.

50. Lambert J, Ferguson RJ, Gervais A, et al. Exercise capacity, residual abnormalities and activity habits following total correction for tetralogy of Fallot. Cardiology. 1980;66(2):120-31.

51. MacLellan-Tobert SG, Driscoll DJ, Mottram CD, et al. Exercise tolerance in patients with Ebstein's anomaly. J Am Coll Cardiol. 1997;29:1615-22.

52. Mahle WT, McBride MG, Paridon SM. Exercise performance after the arterial switch operation for D-transposition of the great arteries. Am J Cardiol. 2001;87:753-58.

53. Mahle WT, McBride MG, Paridon SM. Exercise performance in tetralogy of Fallot: the impact of primary complete repair in infancy. Pediatr Cardiol. 2002;23:224-29.

54. Marelli, AJ, Alejos, JC. Exercise response in atrial septal defect. Prog Pediatr Cardiol. 1993;2(3):20-23.

55. Maron BJ, Doerer JJ, Haas TS, et al. Sudden deaths in young competitive athletes: analysis of 1866 deaths in the United States, 1980-2006. Circulation. 2009;119:1085-92.

56. Maron BJ. Hypertrophic cardiomyopathy. In: Allen HG, Shaddy RE, Driscoll DJ, Feltes TF, editors. Heart disease in infants, children, and adolescents. Philadelphia: Lippincott Williams & Wilkins; 2008. p. 1172-94.

57. Moons P, De Bleser L, Budts W, et al. Health status, functional abilities, and quality of life after the Mustard or Senning operation. Ann Thorac Surg. 2004;77(4): 1359-65; discussion 1365.

58. Mulla N, Paridon SM, Sullivan NM, et al. Cardiopulmonary performance during exercise following repair of tetralogy of Fallot with absent pulmonary valve. Pediatr Cardiol. 1995;16:120-26.

59. Mulla N, Simpson P, Sullivan NM, et al. Determinants of aerobic capacity during exercise following complete repair of tetralogy of Fallot with a transannular patch. Pediatr Cardiol. 1997;18(5):350-56.

60. Murakami T, Takei K, Ueno M, et al. Aortic reservoir function after arterial switch operation in elementary school-aged children. Circ J. 2008;72(8):1291-95.

61. Newberger JW, Takahashi M, Gerber MA. Diagnosis, treatment, and long-term management of Kawasaki disease: a statement for health professionals from the Committee on Rheumatic Fever, Endocarditis and Kawasaki Disease, Council on Cardiovascular Disease in the Young. Circulation. 2004;110:2747–71.

62. Nir A, Driscoll DJ, Mottram C. Cardiorespiratory response to exercise after the Fontan operation: a serial study. J Am Coll Cardiol. 1993;22:216-20.

63. Nishimura RA, Pieroni DR, Bierman FZ, et al. Second natural history study of congenital heart defects. Pulmonary stenosis: Echocardiography. Circulation. 1993;87:173-79.

64. Nistri S, Basso C, Marzari C, et al. Frequency of bicuspid aortic valve in young male conscripts by echocardiogram. Am J Cardiol. 2005;96:718-21.

65. Norozi K, Buchhorn R, Alpers V, et al. Relation of systemic ventricular function quantified by myocardial performance index (Tei) to cardiopulmonary exercise capacity in adults after Mustard procedure for transposition of the great arteries. Am J Cardiol. 2005;96(12):1721-25.

66. Ohuchi H, Ohashi H, Takasugi H, et al. Restrictive ventilatory impairment and arterial oxygenation characterize rest and exercise ventilation in patients after the Fontan operation. Pediatr Cardiol. 2004;25:513-21.

67. Oskarsson G, Pesonen E, Munkhammar P, et al. Normal coronary flow reserve after arterial switch operation for transposition of the great arteries: an intracoronary Doppler guidewire study. Circulation. 2002;106(13):1696-1702.

68. Paridon SM, Fisher DJ. Regulation of myocardial blood flow and oxygen consumption. In: Garson A, Bricker JT, Fisher DJ, Neish SR, editors. Science and practice of pediatric cardiology. Philadelphia: Lea & Febiger; 1990. p. 250-65.

69. Paridon SM. Congenital heart disease: cardiac performance and adaptations to exercise. Pediatr Exerc Sci. 1997;9:308-23.

70. Paridon SM. Exercise response in tetralogy of Fallot and pulmonary atresia with ventricular septal defect. Prog Pediatr Cardiol. 1993;2(3):35-43.

71. Paridon SM, Humes RA, Pinsky WW. The role of chronotropic impairment during exercise after the Mustard operation. J Am Coll Cardiol. 1991;17(3):729-32.

72. Paridon SM, Mitchell PD, Blaufox A, et al. A cross-sectional study of exercise performance during the first two decades of life following the Fontan procedure. J Am Coll Cardiol. 2008;8:52(2):99-107.

73. Paridon SM, Galioto FM, Vincent JA, et al. Exercise capacity and incidence of myocardial perfusion defects after Kawasaki disease in children and adolescents. J Am Coll Cardiol. 1995;25:1420-24.

74. Pasquali SK, Marino BS, McBride MG, et al. Coronary pattern and age impact exercise performance late after the arterial switch operation. J Thorac Cardiovasc Surg. 2007;134:1207-12.

75. Pasquali SK, Marino BS, Pudusseri A, et al. Risk factors and comorbidities associated with obesity in children and adolescents following the arterial switch operation and Ross procedure. Am Heart J. 2009;158(3):473-79.

76. Pasquali SK, Marino, BS, Pudusseri A, et al. Risk factors associated with obesity after congenital heart surgery. Congenit Heart Dis. 2008;3:376-94.

77. Patel JN, Kavey RE, Pophal SG, et al. Improved exercise performance in pediatric heart transplant recipients after home exercise training. Pediatr Transplant. 2008;12:336-40.

78. Perrault H, Drblik SP, Montigny M, et al. Comparison of cardiovascular adjustments to exercise in adolescents 8 to 15 years of age after correction of tetralogy of Fallot, ventricular septal defect, or atrial septal defect. Am J Cardiol. 1989;64:213-17.

79. Reybrouck T, Weymans M, Stijns H, et al. Ventilatory anaerobic threshold for evaluating exercise performance in children with congenital left to right intracardiac shunt. Pediatr Cardiol. 1986;7:19-24.

80. Reybrouck T, Rogers R, Weymans M, et al. Serial cardiorespiratory exercise testing in patients with congenital heart disease. Eur J Pediatr. 1995;154:801-6.

81. Reybrouck T, Eyskens B, Mertens L, et al. Cardiorespiratory exercise function after the arterial switch operation for transposition of the great arteries. Eur Heart J. 2001;22:1052-59.

82. Rhodes J, Hijazi ZM, Marx GR, et al. Aerobic exercise function in patients with persistent coronary artery aneurysms secondary to Kawasaki disease. Pediatr Cardiol. 1996;17(4):226-30.

83. Richard R, Verdier JC, Duvallet A, et al. Chronotropic competence in endurance trained heart transplant recipients: heart rate is not a limiting factor for exercise capacity. J Am Coll Cardiol. 1999;33:192-97.

84. Roos-Hesselink JW, Meijboom FJ, Spitaels SE, et al. Long-term outcome after surgery for pulmonary stenosis (a longitudinal study of 22-33 years). Eur Heart J. 2006;27(4):482-488.

85. Roos-Hesselink JW, Meijboom FJ, Spitaels SE, et al. Decline in ventricular function and clinical condition after Mustard repair for transposition of the great arteries (a prospective study of 22-29 years). Eur Heart J. 2004; 25(14):1264-70.

86. Rowe SA, Zahka KG, Manollo TA, et al. Lung function and pulmonary regurgitation limit exercise capacity in postoperative tetralogy of Fallot. J Am Coll Cardiol. 1991;17:461-66.

87. Rowland TW. Aerobic exercise testing protocols. In: Rowland TW, editor. Pediatric laboratory exercise testing. Champaign, IL: Human Kinetics; 1993. p. 19-42.

88. Russo CF, Cannata A, Lanfranconi M, et al. Is aortic wall degeneration related to bicuspid aortic valve anatomy in patients with valvular disease? J Thorac Cardiovasc Surg. 2008;136:937-42.

89. Schaefer BM, Lewin MB, Stout KK, et al. The bicuspid aortic valve: an integrated phenotypic classification of leaflet morphology and aortic root shape. Heart. 2008;94:1634-38.

90. Silber JH, Cnaan A, Clark BJ, et al. Enalapril to prevent cardiac function decline in long-term survivors of pediatric cancer exposed to anthracyclines. J Clin Oncol. 2004;22(5):820-28.

91. Silberbach M. Bicuspid aortic valve and thoracic aortic aneurysm: toward a unified theory. J Am Coll Cardiol. 2009;53:2296-97.

92. Singh TP, Curran TJ, Rhodes J. Cardiac rehabilitation improves heart rate recovery following peak exercise in children with repaired congenital heart disease. Pediatr Cardiol. 2007;28:276-79.

93. Steeds RP, Oakley D. Predicting late sudden death from ventricular arrhythmia in adults following surgical repair of tetralogy of Fallot. QJM. 2004;97:7-13.

94. Suda K, Iemura M, Nishiono H, et al. Long-term prognosis of patients with Kawasaki disease complicated by giant coronary aneurysms: a single-institution experience. Circulation. 2011;123:1836-42.

95. Sugawara J, Otsuki T, Tanabe T, et al. Physical activity duration, intensity, and arterial stiffening in postmenopausal women. Am J Hypertens. 2006;19:1032-36.

96. Sugawara J, Otsuki T, Tanabe T, et al. The effects of low-intensity single-leg exercise on regional arterial stiffness. Jpn J Physiol. 2003;53:239-41.

97. Sutton NJ, Peng L, Lock JE, et al. Effect of pulmonary artery angioplasty on exercise function after repair of tetralogy of Fallot. Am Heart J. 2008;155(1):182-86.

98. Tadros TM, Klein MD, Shapira OM. Ascending aortic dilatation associated with bicuspid aortic valve: pathophysiology, molecular biology, and clinical implications. Circulation. 2009;119:880-90.

99. Takkunen O, Mattila S, Nieminen MS, et al. Cardiorespiratory function after correction for tetralogy of Fallot: modifying effect of previous shunt operation. Scand J Thorac Cardiovasc Surg. 1987;21(1):21-26.

100. Toro-Salazar OH, Steinberger J, Thomas W, et al. Long-term follow-up of patients after coarctation of the aorta repair. Am J Cardiol. 2002;89:541-47.

101. Trojnarska O, Szyszka A, Gwizdala A, et al. Adults with Ebstein's anomaly—cardiopulmonary exercise testing and BNP levels exercise capacity and BNP in adults with Ebstein's anomaly. Int J Cardiol. 2006;111:92-97.

102. Wessel HU, Weiner MD, Paul MH, et al. Lung function in tetralogy of Fallot after intracardiac repair. J Thorac Cardiovasc Surg. 1981;82:616-28.

103. Wessel HU, Cunningham WJ, Paul MH, et al. Exercise performance in tetralogy of Fallot after intracardiac repair. J Thorac Cardiovasc Surg. 1980;80(4):582-93.

104. Wittlieb-Weber CA, Cohen MS, McBride MG, et al. Elevated left ventricular outflow tract velocities on exercise stress echocardiography may be a normal physiologic response in healthy youth. J Am Soc Echocardiogr. 2013;26(12):1372-78.

105. Wolfe RR, Bartle E, Daberkow E, et al. Exercise responses in ventricular septal defect. Prog Pediatr Cardiol. 1993;2(3):24-29.

Chapter 13

1. Abu-Hasan M, Tannous B, Weinberger M. Exercise-induced dyspnea in children and adolescents: if not asthma then what? Ann Allergy Asthma Immunol. 2005;94:366-71.

2. AlShati M, Cockcroft DW, Fenton ME. Exercise-induced hyperventilation: more common than appreciated. Ann Allergy Asthma Immunol. 2012;109:279-85.

3. Aleman M, Nieto JE, Benak J, et al. Tracheal collapse in American miniature horses:13 cases. J Am Vet Med Assoc. 2008;8:1302-6.

4. Anbar R. Self-hypnosis for management of chronic dyspnea in pediatric patients. Pediatrics. 2001;107:1-4.

5. Bever JC, Boas SR. Use of tidal loop analysis in suspected pediatric exercise induced vocal cord dysfunction. Med Sci Sports Exerc. 2006;38:S384.

6. Boogaard R, Huijsmans SH, Pijnenburg MWH, et al. Tracheomalacia and bronchomalacia in children: incidence and patient characteristics. Chest. 2005;128:3391-97.

7. Christensen PM, Heimdal JH, Christopher KL, et al. ERS/ELS/ACCP 2013 international consensus conference nomenclature on inducible laryngeal obstructions. Eur Respir Rev. 2015; 24:445-50.

8. Christopher KL, Wood RP, Eckert RC, et al. Vocal-cord dysfunction presenting as asthma. N Engl J Med. 1983;0308:1566-70.

9. Cropp GJ. Grading, time course, and incidence of exercise-induced airway obstruction and hyperinflation in asthmatic children. Pediatrics. 1975;56:868-79.

10. Echternach M, Delb W, Verse T, et al. Does isolated expiratory vocal cord dysfunction exist? Otolaryngol Head Neck Surg. 2008;138:805-6.

11. Finder JD. Primary bronchomalacia in infants and children. J Pediatr. 1997;130:59-66.

12. Gardner WN. The pathophysiology of hyperventilation disorders. Chest. 1996;109:516-34.

13. Gibson GJ. Obesity, respiratory function and breathlessness. Thorax. 200;55:41-44.

14. American Thoracic Society. Guidelines for methacholine and exercise challenge testing–1999. Am J Respir Crit Care Med. 2000;161:309-29.

15. Haldane JS, Poulton EP. The effects of want of oxygen on respiration. J Physiol. 1908;37:390-407.

16. Hammo AH, Weinberger MM. Exercise-induced hyperventilation: a pseudoasthma syndrome. Ann Allergy Asthma Immunol. 1999;82:574-78.

17. Kaufman C, Kelly AS, Kaiser DR, et al. Aerobic-exercise training improves ventilatory efficiency in overweight children. Pediatr Exerc Sci. 2007;19:89-92.

18. Kinnula VL, Sovijarvi ARA. Elevate ventilatory equivalents during exercise in patients with hyperventilation syndrome. Respiration. 1993;60:273-78.

19. Komarow HD, Young M, Neslon C, et al. Vocal cord dysfunction as demonstrated by impulse oscillometry. J Allergy Clin Immunol. 2013; 1:387-93.

20. Lindh M. Energy expenditure during walking in patient with scoliosis. Spine. 1978;3:122-34.

21. Maat RC, Roksubd OD, Halvorsen T, et al. Audiovisual assessment of exercise-induced laryngeal obstruction: reliability and validity of observations. Eur Arch Otorhinolaryngol. 2009;266:1929-36.

22. Mahut B, Fuchs-Clement D, Plantier L, et al. Cross-sectional assessment of exertional dyspnea in otherwise healthy children. Pediatr Pulmonol. In press, published early view on line.

23. Marinov B, Kostianev S, Turnovska T. Ventilatory efficiency and rate of perceived exertion in obese and non-obese children performing standardized exercise. Clin Physiol Funct Imaging. 2002;22:254-60.

24. Martinez-Llorens J, Ramirez M, Colomina MJ, et al. Muscle dysfunction and exercise limitation in adolescent idiopathic scoliosis. Eur Respir J. 2010; 36:393-400.

25. Mathers-Schmidt B. Paradoxical vocal fold motion: a tutorial on a complex disorder and the speech-language pathologist's role. Am J Speech Lang Pathol. 2001;10:111-25

26. Matheson GO, McKenzie DC. Breath holding during intense exercise:arterial blood gases, pH, and lactate. J Appl Physiol. 1988;64:1947-52.

27. Meek PM, Schwartzstein RM, Adams L, et al. Dyspnea—mechanisms, assessment, and management: a consensus document. Am J Respir Crit Care Med. 1999;159:321-40.

28. Meyer T, Faude O, Scharhag J, et al. Is lactic acidosis a cause of exercise induced hyperventilation at the respiratory compensation point? Br J Sports Med. 2004;38:622-25.

29. Milgrom H, Taussig LM. Keeping children with exercise-induced asthma active. Pediatrics. 1999;104:e38.

30. Morris MJ, Christopher KL. Diagnostic criteria for the classification of vocal cord dysfunction. Chest. 2010;138:1213-23.

31. National Heart, Lung, and Blood Institute. Expert panel report 3: guidelines for the diagnosis and management of asthma. NHLBI 2007, p. 363-364.

32. Norman AC, Drinkard B, McDuffie JR, et al. Influence of excess adiposity on exercise fitness and performance in overweight children and adolescents. Pediatrics. 2005;115:e690-96.

33. Parameswaran K, Todd DC, Soth M. Altered respiratory physiology in obesity. Can Respir J. 2006;13:203-10.

34. Paridon SM, Alpert BS, Boas SR, et al. Clinical stress testing in the pediatric age group. Circulation. 2006;113:1905-20.

35. Prado DM, Silva AG, Trombetta IC, et al. Weight loss associated with exercise training restores ventilatory efficiency in obese children. Int J Sports Med. 2009; 30:821-26.

36. Rafferty GF, Saisch SGN, Gardner WN. Relation of hypocapnic symptoms to rate of fall of end-tidal PCO_2 in normal subjects. Respir Med. 1992;86:335-40.

37. Reybrouck Y, Mertens L, Schepers D, et al. Assessment of cardiorespiratory exercise function in obese children and adolescents by body mass-independent parameters. Eur J Appl Physiol. 1997;75:478-83.

38. Seear M, Wensely D, West N. How accurate is the diagnosis of exercise-induced asthma amnong Vancouver school children? Arch Dis Child. 2005;902:898-902.

39. Shneerson J. Cardiac and respiratory responses to exercise in adolescent idiopathic scoliosis. Thorax. 1980;35:347-50.

40. Sterner JB, Morris MJ, Sill JM, et al. Inspiratory flow-volume curve evaluation for detecting upper airway disease. Respir Care. 2009;544:461-66.

41. Sullivan M, Heywood B, Beukelman D. A treatment for vocal cord dysfunction in female athletes: An outcome study. Laryngoscope. 2001; 111:1751-55.

42. Troosters T, Gosselink R, Decramer M. Deconditioning, and principles of training. In: Weisman I, Zeballos RJ, editors. Clinical exercise testing. New York: Karger; 2002. p. 60-61.

43. Van Dixhoorn J, Duivenvoorden HJ. Efficacy of Nijmegen questionnaire in recognition of the hyperventilation syndrome. J Psychosom Res. 1985;29:199-206.

44. Warnes E, Allen K. Biofeedback treatment of paradoxical vocal fold motion and respiratory distress in an adolescent girl. J Appl Behav Anal. 2005;38:529-32.

45. Wasserman K, Hansen JE, Sue DY, et al. Principles of exercise testing and interpretation. Philadelphia: Lippincott Williams & Wilkins; 2012. p. 125.

46. Weinstein DJ, Hull JE, Ritchie BL, et al. Exercise-associated excessive dynamic airway collapse in military personnel. Ann Am Thorac Soc. 2016; 13:1476-82.

47. Weiss P, Rundell KW. Imitators of exercise-induced bronchoconstriction. Allergy Asthma Clin Immunol. 2009;5:7.

48. Wilson J, Wilson E. Practical management: vocal cord dysfunction in athletes. Clin J Sport Med. 2006;16:357-60.

49. Zanconato S, Baraldi E, Santuz P, et al. Gas exchange during exercise in obese children. Eur J Pediatr. 1989;148:614-17.

Chapter 14

1. Bisset GS 3rd, Schwartz DC, Meyer RA, et al. Clinical spectrum and long-term follow-up of isolated mitral valve prolapse in 119 children. Circulation. 1980;62:423-29.

2. Bonhoeffer P, Bonnet D, Piechaud JF, et al. Coronary artery obstruction after the arterial switch operation for transposition of the great arteries in newborns. J Am Coll Cardiol. 1997;29:202-6.

3. Boon AW, Forton J. How to evaluate a child with chest pain. Curr Paediatr. 2004;14:64-70.

4. Brenner JI, Ringel RE, Berman MA. Cardiologic perspectives of chest pain in childhood: a referral problem? To whom? Pediatr Clin North Am. 1984;31:1241-58.

5. Danduran MJ, Earing MG, Sheridan DC, et al. Chest pain: characteristics of children/adolescents. Pediatr Cardiol. 2008;29:775-81.

6. Driscoll DJ, Glicklich LB, Gallen WJ. Chest pain in children: a prospective study. Pediatrics. 1976;57:648-51.

7. Ellestad MH. Predictive implications. In: Ellestad MH, editor. Stress testing. principles and practice. 5th ed. New York: Oxford University Press; 2003. p. 271-317.

8. Ellestad MH. Pediatric exercise testing. In: Ellestad MH, editor. Stress testing. principles and practice. 5th ed. New York: Oxford University Press; 2003. p. 404.

9. Evangelista JK, Parsons M, Renneburg AK. Chest pain in children: diagnosis through history and physical examination. J Pediatr Health Care. 2000;14:3-8.

10. Frescura C, Basso C, Thiene G, et al. Anomalous origin of coronary arteries and risk of sudden death: a study based on an autopsy population of congenital heart disease. Hum Pathol. 1998;29:689-95.

11. Friedman KG, Alexander ME. Chest pain and syncope in children: a practical approach to the diagnosis of cardiac disease. J Pediatr. Advance online publication. DOI: 10.1016/j.jpeds.2013.05.001.

12. Friedman KG, Kane DA, Rathod RH, et al. Management of pediatric chest pain using a standardized assessment and management plan. Pediatrics. 2011;128:239-45.

13. Fyfe DA. Chest pain in pediatric patients presenting to a cardiac clinic. Clin Pediatr. 1984;23:321-24.

14. Goodwin, JF. Exercise testing in hypertrophic cardiomyopathy. Prog Pediatr Cardiol 1993;2:61-64.

15. Hosenpud JD, Shipley GD, Wagner CR. Cardiac allograft vasculopathy: current concepts, recent developments and future directions. J Heart Lung Transplant. 1992;11:9-23.

16. Howell J. Xiphodynia: a report of three cases. *J Emerg Med.* 1992;10:435-38.

17. James FW. Exercise responses in aortic stenosis. Prog Pediatr Cardiol. 1993;2:1-7.

18. James, FW. Exercise testing in normal individuals and patients with cardiovascular disease. In: Engle MA, editor. Pediatric cardiovascular disease. Philadelphia: F.A. Davis; 1981. p. 227.

19. Kane DA, Fulton DR, Saleeb S, et al. Needles in hay: chest pain as the presenting symptom in children with serious underlying cardiac pathology. Congenit Heart Dis. 2010;5:366-73.

20. Kavey REW, Allada V, Daniels SR, et al. Cardiovascular risk reduction in high-risk pediatric patients. A scientific statement from the American Heart Association expert panel on Population and Prevention Science; the Councils on Cardiovascular Disease in the Young, Epidemiology and Prevention, Nutrition, Physical Activity and Metabolism, High Blood Pressure Research, Cardiovascular Nursing, and the Kidney in Heart Disease; and the Interdisciplinary Working Group on Quality of Care and Outcomes Research: endorsed by the American Academy of Pediatrics. Circulation. 2006;114:2710-38.

21. Kragel AH, Roberts WC. Anomalous origin of either the right or left main coronary artery from the aorta with subsequent coursing between aorta and pulmonary trunk: analysis of 32 necropsy cases. Am J Cardiol. 1988;62:771-77.

22. Kyle WB, Macicek SL, Lindle KA, et al. Limited utility of exercise stress tests in the evaluation of children with chest pain. Congenit Heart Dis. 2012;7:455-59.

23. Landwehr LP, Wood RP 2nd, Blager FB, et al. Vocal cord dysfunction mimicking exercise-induced bronchospasm in adolescents. Pediatrics. 1996;98:971-74.

24. Lipsitz JD, Masia C, Apfel H, et al. Noncardiac chest pain and psychopathology in children and adolescents. J Psychosom Res. 2005;59:185-88.

25. Loiselle KA, Lee JL, Gilleland J, et al. Factors associated with health care utilization among children with noncardiac chest pain and innocent heart murmurs. J Pediatr. Psychol. 2012;37:817-25.

26. Massin MM, Bourguignont A, Coremans C, et al. Chest pain in pediatric patients presenting to an emergency department or to a cardiac clinic. Clin Pediatr. 2004;43:231-38.

27. Nudel DB, Diamant S, Brady T, et al. Chest pain, dyspnea on exertion, and exercise induced asthma in children and adolescents. Clin Pediatr. 1987;26:388-92.

28. Pantell RH, Goodman Jr BW. Adolescent chest pain: a prospective study. Pediatrics. 1983;71:881-87.

29. Paridon SM, Albert BS, Boas SR, et al. Clinical stress testing in the pediatric age group: a statement from the American Heart Association Council on Cardiovascular Disease in the Young, Committee on Atherosclerosis, Hypertension, and Obesity in Youth. 2006;113:1905-20.

30. Pickering D. Precordial catch syndrome. Arch Dis Child. 1981;56:401-3.

31. Reddy SR, Singh HR. Chest pain in children and adolescents. Pediatr Rev. 2010;31:e1-e9.

32. Reddy SV, Forbes TJ, Chintala K. Cardiovascular involvement in Kawasaki disease. Images Paediatr Cardiol. 2005;7:1-9.

33. Saleeb SF, Li WY, Warren SZ, et al. Effectiveness of screening for life-threatening chest pain in children. Pediatrics. 2011;128:e1062-e1068.

34. Selbst SM. Consultation with the specialist: chest pain in children. Pediatr Rev. 1997;18:169-73.

35. Selbst SM, Ruddy RM, Clak BJ, et al. Pediatric chest pain: a prospective study. Pediatrics. 1988;82:319-23.

36. Singh TP, Di Carli MF, Sullivan NM, et al. Myocardial flow reserve in long-term survivors of repair of anomalous left coronary artery from pulmonary artery. J Am Coll Cardiol. 1998;31:437-43.

37. Verghese GR, Friedman KG, Rathod RH, et al. Resource utilization reduction for evaluation of chest pain in pediatrics using a novel standardized clinical assessment and management plan (SCAMP). J Am Heart Assoc. 2012;1:1-7.

38. Whitmer JT, James FW, Kaplan S, et al. Exercise testing in children before and after surgical treatment of aortic stenosis. Circulation. 1981;63:254-63.

39. Wiens L, Portnoy J, Sabath R, et al. Chest pain in otherwise healthy children and adolescents is frequently caused by exercise-induced asthma. Pediatrics.1992;90:350-53.

Chapter 15

1. Antzelevitch C, Brugada P, Borggrefe M, et al. Brugada syndrome: report of the second consensus conference. Circulation. 2005;111:659-70.

2. Bader YH, Link MS. Syncope in the athlete. Card Electrophysiol Clin. 2013;5:85-96.

3. Barth CW 3rd, Roberts WC. Left main coronary artery originating from the right sinus of Valsalva and coursing between the aorta and pulmonary trunk. J Am Coll Cardiol. 1986;7:366-73.

4. Basso C, Corrado D, Macus FL, et al. Arrhythmogenic right ventricular cardiomyopathy. Lancet. 2009;373:1289-1300.

5. Basso C, Maron BJ, Corrado D, et al. Clinical profile of congenital coronary artery anomalies with origin from the wrong aortic sinus leading to sudden death in young competitive athletes. J Am Coll Cardiol. 2000;35:1493-1501.

6. Bazett HC. An analysis of the time-relations of electrocardiograms. Heart. 1920;7:353-70.

7. Benditt DG, Nguyen JT. Syncope: therapeutic approaches. J Am Coll Cardiol. 2009;53:1741-51.

8. Calkins H, Seifert M, Morady F. Clinical presentation and long-term follow-up of athletes with exercise-induced vasodepressor syncope. Am Heart J. 1995;129:1159-64.

9. Casey DP, Joyner MJ. Local control of skeletal muscle blood flow during exercise: influence of available oxygen. J Appl Physiol. 2011;111:1527-38.

10. Childress MA, O'Connor FG, Levine BD. Exertional collapse in the runner: evaluation and management in fieldside and office-based settings. Clin Sports Med. 2010;29:459-76.

11. Colvicchi F, Ammirati F, Santini M. Epidemiology and prognostic implications of syncope in young competing athletes. Eur Heart J. 2004;25:1749-53.

12. Dalal D, Nasir K, Bomma C, et al. Arrhythmogenic right ventricular dysplasia: a United States experience. Circulation. 2005;112:3823-32.

13. Doyle EF, Arumugham P, Lara E, et al. Sudden death in young patients with congenital aortic stenosis. Pediatrics. 1974;53:481-89.

14. Driscoll DJ, Edwards WD. Sudden unexpected death in children and adolescents. J Am Coll Cardiol. 1985;5:118B-21B.

15. Dungan WT, Garson A Jr, Gillette PC. Arrhythmogenic right ventricular dysplasia: a cause of ventricular tachycardia in children with apparently normal hearts. Am Heart J. 1981;102:745-50.

16. Estes NA 3rd, Link MS, Cannom D, et al. Report of the NASPE policy conference on arrhythmias and the athlete. J Cardiovasc Electrophysiol. 2001;12:1208-19.

17. Freeman R, Wieling W, Axelrod RB, et al. Consensus statement on the definition of orthostatic hypotension, neutrally mediated syncope and the postural tachycardia syndrome. Clin Auton Res. 2011;21:69-72.

18. Gersh BJ, Maron BJ, Bonow RO, et al. 2011 ACCF/AHA guideline for the diagnosis and treatment of hypertrophic cardiomyopathy. Circulation. 2011;124:2761-96.

19. Goldenberg I, Moss AJ. Long QT syndrome. J Am Coll Cardiol. 2008;51:2291-2300.

20. Goodwin, JF. Exercise testing in hypertrophic cardiomyopathy. Prog Pediatr Cardiol 1993;2:61-64.

21. Grubb BP. Clinical practice. Neurocardiogenic syncope. N Engl J Med. 2005;352:1004-10.

22. Grubb BP, Temesy-Armos PN, Samoil D, et al. Tilt table testing in the evaluation and management of athletes with recurrent exercise-induced syncope. Med Sci Sports Exerc. 1993;25:24-28.

23. Hannon DW, Knilans TK. Syncope in children and adolescents. Curr Probl Pediatr. 1993;23:358-84.

24. Hastings JL, Levine BD. Syncope in the athletic patient. Progr Cardiovasc Dis. 2012;54:438-44.

25. Hobbs JB, Peterson DR, Moss AJ, et al. Risk of aborted cardiac arrest or sudden cardiac death during adolescence in the long-QT syndrome. JAMA. 2006;296:1249-54.

26. Holtzhausen LM, Noakes TD, Kroning B, et al. Clinical and biochemical characteristics of collapsed ultra-marathon runners. Med Sci Sports Exerc. 1994;26:1095-1101.

27. James FW. Exercise responses in aortic stenosis. Prog Pediatr Cardiol 1993;2:1-7.

28. James, FW. Exercise testing in normal individuals and patients with cardiovascular disease. In: Engle MA, editor. Pediatric cardiovascular disease. Philadelphia: F.A. Davis; 1981. p. 227.

29. Kapetanopoulos A, Kluger J, Maron BJ, et al. The congenital long QT syndrome and implications for young athletes. Med Sci Sports Exerc. 2006;38:816-25.

30. Kapoor WN. Evaluation and management of the patient with syncope. JAMA. 1992;268:2553-60.

31. Kienzele MG. Syncope: mechanisms and manifestations. Hosp Pract. 1990;25:77-85.

32. Kramer MR, Drori Y, Lev B. Sudden death in young soldiers: high incidence of syncope prior to death. Chest. 1988;93:345-47.

33. Kudenchuk PJ, McAnulty JH. Syncope: evaluation and treatment. Mod Concepts Cardiovasc Dis. 1985;54:25-29.

34. Leenhardt A, Lucet V, Denjoy I, et al. Catecholaminergic polymorphic ventricular tachycardia in children. A 7-year follow-up of 21 patients. Circulation. 1995;91:1512-19.

35. Lerman BB, Belardinelli L, West GA, et al. Adenosine-sensitive ventricular tachycardia: evidence suggesting cyclic AMP-mediated triggered activity. Circulation. 1986;74:270-80.

36. Link MS, Estes NA 3rd. How to manage athletes with syncope. Cardiol Clin. 2007;25:457-66.

37. Link MS, Wang PJ, Pandian NG, et al. An experimental model of sudden death due to low-energy chest-wall impact (commotio cordis). N Engl J Med. 1998;338:1805-11.

38. Marcus FL, McKenna WJ, Sherrill D, et al. Diagnosis of arrhythmogenic right ventricular cardiomyopathy/dysplasia: proposed modification of the task force criteria. Circulation. 2010;121:1533-41.

39. Maron BJ. Sudden death in young athletes. N Engl J Med. 2003;349:1064-75.

40. Maron BJ, Doerer JJ, Haas TS, et al. Sudden deaths in young competitive athletes: analysis of 1866 deaths in the United States, 1980-2006. Circulation. 2009;119:1085-92.

41. Maron BJ, Epstein SE, Roberts WC. Causes of sudden death in competitive athletes. J Am Coll Cardiol. 1986;7:204-14.

42. Maron BJ, Estes NA 3rd. Commotio cordis. N Engl J Med. 2010;362:917-27.

43. Maron BJ, Shirani J, Poliac LC, et al. Sudden death in young competitive athletes. Clinical, demographic, and pathological profiles. JAMA. 1996;276:199-204.

44. Moya A, Sutton R, Ammirati F, et al. Guidelines for the diagnosis and management of syncope (version 2009). Eur Heart J. 2009;30:2631-71.

45. Noakes TD. A modern classification of the exercise-related heat illnesses. J Sci Med Sport. 2008;11:33-39.

46. O'Connor FG, Levine BD, Childress MA, et al. Practical management: a systematic approach to the evaluation of exercise-related syncope in athletes. Clin J Sport Med 2009;19:429-34.

47. Pelliccia A, Di Paolo FM, Corrado D, et al. Evidence for efficacy of the Italian national pre-participation screening programme for identification of hypertrophic cardiomyopathy in competitive athletes. Eur Heart J. 2006;27: 2196-2200.

48. Roberts WO. A 12-yr profile of medical injury and illness for the Twin Cities Marathon. Med Sci Sports Exerc. 2000;32:1549-55.

49. Sneddon JF, Scalia G, Ward DE, et al. Exercise induced vasodepressor syncope. Br Heart J. 1994;71:554-57.

50. Speedy DB, Rogers IR, Noakes TD, et al. Exercise-induced hyponatremia in ultradistance triathletes is caused by inappropriate fluid retention. Clin J Sport Med. 2000;10:272-78.

51. Taylor AJ, Rogan KM, Virmani R. Sudden cardiac death associated with isolated congenital coronary artery anomalies. J Am Coll Cardiol. 1992;20:640-47.

52. Singh SK, Link MS, Wang PJ, et al. Syncope in the patient with nonischemic dilated cardiomyopathy. Pacing Clin Electrophysiol. 2004;27: 97-100.

53. Sivakumaran S, Krahn AD, Klein GJ, et al. A prospective randomized comparison of loop recorders versus Holter monitors in patients with syncope or presyncope. Am J Med. 2003;115:1-5.

54. Strickberger SA, Benson DW, Biaggioni I, et al. AHA/ACCF Scientific statement on the evaluation of syncope. Circulation. 2006;17;113:316-27.

55. Taylor AJ, Rogan KM, Virmani R. Sudden cardiac death associated with isolated congenital coronary artery anomalies. J Am Coll Cardiol. 1992;20:640-47.

56. Wilde AA, Antzelevitch C, Borggrefe M, et al. Proposed diagnostic criteria for the Brugada syndrome: consensus report. Circulation. 2002;106:2514-19.

57. Zipes DP, Ackerman MJ, Estes NA 3rd, et al. Task force 7: arrhythmias. J Am Coll Cardiol. 2005;45:1354-63.

Chapter 16

1. Katz BZ, Boas S, Shiraishi Y, et al. Exercise tolerance testing in a prospective cohort of adolescents with chronic fatigue syndrome and recovered controls following infectious mononucleosis. J Pediatr. 2010;157:468-72.

2. Katz BZ, Jason LA. Chronic fatigue syndrome following infections in adolescents. Curr Opin Pediatr. 2013;25:95-102.

3. Nudel DB, Gootman N, Nussbaum MP. Altered exercise performance and abnormal sympathetic responses to exercise in patients with anorexia nervosa. J Pediatr. 1984;105:34-37.

4. Rowland T. Iron deficiency in athletes: an update. Am J Lifestyle Med. 2012;6:319-27.

5. Takken T, Henneken T, van de Putte E, et al. Exercise testing in children and adolescents with chronic fatigue syndrome. Int J Sports Med. 2007;28:580-84.

6. Wilmore JH, Buskirk ER, DiGirolamo M, et al. Body composition (round table). Phys Sportsmed. 1986;14:144-62.

7. Winsley R, Matos N. Overtraining and elite young athletes. Med Sport Sci. 2011;56:97-105.

Chapter 17

1. Bawazir OA, Montgomery M, Harder J, et al. Midterm evaluation of cardiopulmonary effects of closed repair for pectus excavatum. J Pediatr Surg. 2005;40:863-67.

2. Beiser GD, Epstein SE, Stampfer M, et al. Impairment of cardiac function in patients with pecus excavatum, with improvement after operative correction. N Engl J Med. 1972;287:267-72.

3. Borowitz D, Cerny F, Zallen G, et al. Pulmonary function and exercise response in patients with pectus excavatum after Nuss repair. J Pediatr Surg. 2003;38:544-47.

4. Castellani C, Windhaber J, Schober PH, et al. Exercise performance testing patients with pectus excavatum before and after Nuss procedure. Pediatr Surg Int. 2010;26:659-63.

5. Chen Z, Amos EB, Luo H, et al. Comparative functional recovery after Nuss and Ravitch procedures for pectus excavatum repair: a meta-analysis. J Cardiothorac Surg. 2012;7:101.

6. Fabricius J, Davidsen HG, Hansen AT. Cardiac function in funnel chest: twenty-six patients investigated by cardiac catheterization. Dan Med Bull. 1957;4:251-57.

7. Haller JA, Loughlin GM. Cardiorespiratory function is significantly improved following corrective surgery for severe pectus excavatum. Proposed treatment guidelines. J Cardiovasc Surg. 2000;41:125-30.

8. Krueger T, Chassot PG, Christodoulou M, et al. Cardiac function assessed by transesophageal echocardiography during pectus excavatum repair. Ann Thorac Surg. 2010;89:243-44.

9. Jaroszewski D, Notrica D, McMahon L, et al. Current management of pectus excavatum: a review and update of therapy and treatment recommendations. J Am Board Fam Med. 2010;23:230-39.

10. Lesbo M, Tang M, Nielsen HH, et al. Compromised cardiac function in exercising teenagers with pectus excavatum. Interact Cardiovasc Thorac Surg. 2011;13:377-80.

11. Malek MH, Berger DE, Housh TJ, et al. Cardiovascular function following surgical repair of pectus excavatum: a meta-analysis. Chest. 2006;130:506-16.

12. Malek MH, Berger DE, Marelich WD, et al. Pulmonary function following surgical repair of pectus excavatum: a meta-analysis. Eur J Cardiothor Surg. 2006;30:637-43.

13. Malek MH, Coburn JW. Strategies for cardiopulmonary exercise testing of pectus excavatum patients. Clinics. 2008;63:245-54.

14. Malek MH, Fonkalsrud EW. Cardiorespiratory outcome after corrective surgery for pectus excavatum: a case study. Med Sci Sports Exerc. 2004;36:183-90.

15. Morshuis WJ, Folgering HT, Barentsz JO, et al. Exercise cardiorespiratory function before and one year after operation for pectus excavatum. J Thorac Cardiovasc Surg. 1994;107:1403-9.

16. Quigley PM, Haller JA, Jelus KL, et al. Cardiorespiratory function before and after corrective surgery in pectus excavatum. J Pediatr. 1996;128:638-43.

17. Rowland T, Moriarity K, Banever G. Effect of pectus excavatum deformity on cardiorespiratory fitness in adolescent boys. Arch Pediatr Adolesc Med. 2005;159:1069-73.

18. Sigalet DL, Montgomery M, Harder J. Cardiopulmonary effects of closed repair of pectus excavatum. J Pediatr Surg. 2003;38:380-85.

19. Swanson JW, Avansino JR, Phillips GS, et al. Correlating Haller index and cardiopulmonary disease in pectus excavatum. Am J Surg. 2012;203:660-64.

20. Tang M, Nielsen HH, Lesbo M, et al. Improved cardiopulmonary exercise function after modified Nuss operation for pectus excavatum. Eur J Cardiothor Surg. 2012;41:1063-67.

21. Zhao L, Feinberg MS, Gaides M, et al. Why is exercise capacity reduced in subjects with pectus excavatum? J Pediatr. 2000;136:163-67.

Chapter 18

1. Rowland T. Evolution of maximum oxygen uptake in children. In: Tomkinson GR, Olds TS, editors. Pediatric fitness: secular trends and geographic variability. Med Sport Sci. Basel: Karger; 2007. p. 200-209.

2. Armstrong N, Kirby BJ, McManus AM, et al. Aerobic fitness of prepubescent children. Ann Hum Biol. 1995;22:427-41.

3. Armstrong N, Welsman JR, Nevill AM, et al. Modeling growth and maturation changes in peak oxygen uptake in 11-13 yr olds. J Appl Physiol. 1999;87:2230-36.

4. Armstrong N, Williams J, Balding J, et al. The peak oxygen uptake of British children with reference to age, sex and sexual maturity. Eur J Appl Physiol Occup Physiol. 1991;62:369-75.

5. Åstrand P-O. Experimental studies of physical working capacity in relation to sex and age. Copenhagen: Munksgaard; 1952.

6. Barlow SE, Expert C. Expert committee recommendations regarding the prevention, assessment, and treatment of child and adolescent overweight and obesity: summary report. Pediatrics. 2007;120 Suppl 4:S164-92.

7. Beaver WL, Wasserman K, and Whipp BJ. A new method for detecting anaerobic threshold by gas exchange. J Appl Physiol. 1986;60:2020-27.

8. Bentzur KM, Kravitz L, Lockner DW. Evaluation of the BOD POD for estimating percent body fat in collegiate track and field female athletes: a comparison of four methods. J Strength Cond Res. 2008;22:1985-91.

9. Beunen G, Baxter-Jones AD, Mirwald RL, et al. Intraindividual allometric development of aerobic power in 8- to 16-year-old boys. Med Sci Sports Exerc. 2002;34:503-10.

10. Borghi E, de Onis M, Garza C, et al. Construction of the World Health Organization child growth standards: selection of methods for attained growth curves. Stat Med. 2006;25:247-65.

11. Brambilla P, Bedogni G, Moreno LA, et al. Cross-validation of anthropometry against magnetic resonance imaging for the assessment of visceral and subcutaneous adipose tissue in children. Int J Obes. 2006;30:23-30.

12. Breithaupt P, Adamo KB, Colley RC. The HALO submaximal treadmill protocol to measure cardiorespiratory fitness in obese children and youth: a proof of principle study. Appl Physiol Nutr Metab. 2012;37:308-14.

13. Broyles S, Katzmarzyk PT, Srinivasan SR, et al. The pediatric obesity epidemic continues unabated in Bogalusa, Louisiana. Pediatrics. 2010;125:900-5.

14. Colley RC, Garriguet D, Janssen I, et al. Physical activity of Canadian children and youth: accelerometer results from the 2007 to 2009 Canadian Health Measures Survey. Health Rep. 2011;22:15-23.

15. de Onis M, Onyango AW, Borghi E, et al. Comparison of the World Health Organization (WHO) child growth standards and the National Center for Health Statistics/WHO international growth reference: implications for child health programmes. Public Health Nutr. 2006;9:942-47.

16. Dencker M, Wollmer P, Karlsson MK, et al. Body fat, abdominal fat and body fat distribution related to VO(2PEAK) in young children. Int J Pediatr Obes. 2011;6:e597-e602.

17. Ekelund U, Franks PW, Wareham NJ, et al. Oxygen uptakes adjusted for body composition in normal-weight and obese adolescents. Obes Res. 2004;12:513-20.

18. Elberg J, McDuffie JR, Sebring NG, et al. Comparison of methods to assess change in children's body composition. Am J Clin Nutr. 2004;80:64-69.

19. Elliot DL, Goldberg L, Kuehl KS, et al. Metabolic evaluation of obese and nonobese siblings. J Pediatr. 1989;114:957-62.

20. Epstein Y, Rosenblum J, Burstein R, et al. External load can alter the energy cost of prolonged exercise. Eur J Appl Physiol Occup Physiol. 1988;57:243-47.

21. Fields DA, Goran MI, McCrory MA. Body-composition assessment via air-displacement plethysmography in adults and children: a review. Am J Clin Nutr. 2002;75:453-67.

22. Freedman DS, Wang J, Maynard LM, et al. Relation of BMI to fat and fat-free mass among children and adolescents. Int J Obes. 2005;29:1-8.

23. Freedman DS, Wang J, Ogden CL, et al. The prediction of body fatness by BMI and skinfold thicknesses among children and adolescents. Ann Hum Biol. 2007;34:183-94.

24. Galbraith RT, Gelberman RH, Hajek PC, et al. Obesity and decreased femoral anteversion in adolescence. J Orthop Res. 1987;5:523-28.

25. Henderson RC. Tibia vara: a complication of adolescent obesity. J Pediatr. 1992;121:482-86.

26. Khoury M, Manlhiot C, McCrindle BW. Role of the waist/height ratio in the cardiometabolic risk assessment of children classified by body mass index. J Am Coll Cardiol. 2013;62:742-51.

27. Kuczmarski RJ, Ogden CL, Grummer-Strawn LM, et al. CDC growth charts: United States. Adv Data. 2000;1-27.

28. Lazzer S, Boirie Y, Bitar A, et al. Relationship between percentage of $\dot{V}O_{2max}$ and type of physical activity in obese and non-obese adolescents. J Sports Med Phys Fitness. 2005;45:13-19.

29. Maffeis C, Schena F, Zaffanello M, et al. Maximal aerobic power during running and cycling in obese and non-obese children. Acta Paediatr. 1994;83:113-16.

30. Maffeis C, Schutz Y, Schena F, et al. Energy expenditure during walking and running in obese and nonobese prepubertal children. J Pediatr. 1993;123:193-99.

31. Marinov B, Kostianev S. Exercise performance and oxygen uptake efficiency slope in obese children performing standardized exercise. Acta Physiol Pharmacol Bulg. 2003;27:59-64.

32. McCrindle BW, Manlhiot C, Millar K, et al. Population trends toward increasing cardiovascular risk factors in Canadian adolescents. J Pediatr. 2010;157:837-43.

33. McCrory MA, Gomez TD, Bernauer EM, et al. Evaluation of a new air displacement plethysmo-

graph for measuring human body composition. Med Sci Sports Exerc. 1995;27:1686-91.

34. Mei Z, Grummer-Strawn LM, Wang J, et al. Do skinfold measurements provide additional information to body mass index in the assessment of body fatness among children and adolescents? Pediatrics. 2007;119:e1306-13.

35. Milano GE, Rodacki A, Radominski RB, et al. Scale of VO(2peak) in obese and non-obese adolescents by different methods. Arq Bras Cardiol. 2009;93:554-57, 598-602.

36. Mokha JS, Srinivasan SR, Dasmahapatra P, et al. Utility of waist-to-height ratio in assessing the status of central obesity and related cardiometabolic risk profile among normal weight and overweight/obese children: the Bogalusa Heart Study. BMC Pediatr. 2010;10:73.

37. Norman AC, Drinkard B, McDuffie JR, et al. Influence of excess adiposity on exercise fitness and performance in overweight children and adolescents. Pediatrics. 2005;115:e690-96.

38. Pritchett JW, and Perdue KD. Mechanical factors in slipped capital femoral epiphysis. J Pediatr Orthop. 1988;8:385-88.

39. Resaland GK, Mamen A, Anderssen SA, et al. Cardiorespiratory fitness and body mass index values in 9-year-old rural Norwegian children. Acta Paediatr. 2009;98:687-92.

40. Rowland TW. Aerobic exercise testing protocols. In: Rowland TW, editor. Pediatric laboratory exercise testing—clinical guidelines. Champaign, IL: Human Kinetics; 1993, p. 19-42.

41. Rowland TW. Pediatric laboratory exercise testing—clinical guidelines. Champaign, IL: Human Kinetics; 1993.

42. Singh GK, Kogan MD, van Dyck PC. Changes in state-specific childhood obesity and overweight prevalence in the United States from 2003 to 2007. Arch Pediatr Adolesc Med. 2010;164:598-607.

43. Skelton JA, Cook SR, Auinger P, et al. Prevalence and trends of severe obesity among US children and adolescents. Acad Pediatr. 2009;9:322-29.

44. Volpe Ayub B, Bar-Or O. Energy cost of walking in boys who differ in adiposity but are matched for body mass. Med Sci Sports Exerc. 2003;35:669-74.

45. Wearing SC, Hennig EM, Byrne NM, et al. The impact of childhood obesity on musculoskeletal form. Obes Rev. 2006;7:209-18.

46. Zanconato S, Baraldi E, Santuz P, et al. Gas exchange during exercise in obese children. Eur J Pediatr. 1989;148:614-17.

Chapter 19

1. Agiovlasitis S, Collier SR, Baynard T, et al. Autonomic response to upright tilt in people with and without Down syndrome. Res Dev Disabil. 2010;31:857-63.

2. Agiovlasitis S, Motl RW, Fahs CA, et al. Metabolic rate and accelerometer output during walking in people with Down syndrome. Med Sci Sports Exerc. 2011;43:1322-27.

3. Agiovlasitis S, Motl RW, Foley JT, et al. Prediction of energy expenditure from wrist accelerometry in people with and without Down syndrome. Adapt Phys Activ Q. 2012;29:179-90.

4. Agiovlasitis S, Motl RW, Ranadive SM, et al. Prediction of oxygen uptake during over-ground walking in people with and without Down syndrome. Eur J Appl Physiol. 2011;111:1739-45.

5. Agiovlasitis S, Pitetti KH, Guerra M, et al. Prediction of $\dot{V}O_{2peak}$ from the 20-m shuttle-run test in youth with Down syndrome. Adapt Phys Activ Q. 2011;28:146-56.

6. AAIDD. Definition of intellectual disability [Internet]. Washington, DC: American Association on Intellectual and Developmental Disabilities [cited April 2, 2014]. Available from: http://aaidd.org/intellectual-disability/definition#.UzzcYBC2zSg

7. Auxter D, Pfyfer J, Huettig C. Principles and methods of adapted physical education and recreation. New York: McGraw-Hill; 2001.

8. Bar-Or O, Skinner J, Bergsteinova V, et al. Maximal aerobic capacity of 6-15 year-old girls and boys with subnormal intelligence quotients. Acta Paediatr Scand Suppl. 1971;217:108-13.

9. Baynard T, Pitetti KH, Guerra M, et al. Age-related changes in aerobic capacity in individuals with mental retardation: A 20-yr review. Med Sci Sports Exerc. 2008;40:1984-89.

10. Cappelli-Bigazzi M, Santoro G, Battaglia C, et al. Endothelial cell function in patients with Down's syndrome. Am J Cardiol. 2004;94:392-95.

11. Costa V, Sommese L, Casamassimi A, et al. Impairment of circulating endothelial progenitors in Down syndrome. BMC Med Genomics. 2010;3:40.

12. Cowley PM, Ploutz-Snyder LL, Baynard T, et al. Physical fitness predicts functional tasks in individuals with Down syndrome. Med Sci Sports Exerc. 2010;42:388-93.

13. Cureton JK, Sloniger MA, O'Bannon JP, et al. A generalized equation for prediction of $\dot{V}O_2$ peak from 1-mile run/walk performance. Med Sci Sports Exerc. 1995;27:445-51.

14. Cureton K. Aerobic capacity. Dallas: Cooper Institute of Aerobics; 1994.

15. De S, Small J, Baur LA. Overweight and obesity among children with developmental disabilities. J Intellect Dev Disabil. 2008;33:43-47.

16. Eyman RK, Call TL. Life expectancy of persons with Down syndrome. Am J Ment Retard. 1991;95:603-12.

17. Fernhall B. Physical fitness and exercise training of individuals with mental retardation. Med Sci Sports Exerc. 1993;25:442-50.

18. Fernhall B. Mental retardation. In: Durstine JL, Moore G, editors. ACSM's exercise management for persons with chronic diseases and disabilities. 2nd ed. Champaign, IL: Human Kinetics; 1997. p. 221-26.

19. Fernhall B. Mental retardation. In: Durstine JL, Moore G, Painter P, et al., editors. ACSM's exercise management for persons with chronic diseases and disabilities. 3rd ed. Champaign, IL: Human Kinetics; 2003. p. 304-10.

20. Fernhall B. Mental retardation. In: LeMura L, von Duvillard S, editors. Clinical exercise physiology: application and physiological principles. Philadelphia, PA: Lippincott Williams & Wilkins; 2004. p. 617-27.

21. Fernhall B, Baynard T, Collier SR, et al. Catecholamine response to maximal exercise in persons with Down syndrome. Am J Cardiol. 2009;103:724-26.

22. Fernhall B, Figueroa A, Collier S, et al. Blunted heart rate response to upright tilt in people with Down syndrome. Arch Phys Med Rehabil. 2005;86:813-18.

23. Fernhall B, McCubbin J, Pitetti K, et al. Prediction of maximal heart rate in individuals with mental retardation. Med Sci Sports Exerc. 2001;33:1655-60.

24. Fernhall B, Mendonca GV, Baynard T. Reduced work capacity in individuals with Down syndrome: a consequence of autonomic dysfunction? Exerc Sport Sci Rev. 2013;41:138-47.

25. Fernhall B, Millar AL, Tymeson G, et al. Maximal exercise testing of mentally retarded adolescents and adults: reliability study. Arch Phys Med Rehabil. 1990;71:1065-68.

26. Fernhall B, Otterstetter M. Attenuated responses to sympathoexication in individuals with Down syndrome. J Appl Physiol. 2003;94:2158-65.

27. Fernhall B, Pitetti K. Leg strength is related to endurance run performance in children and adolescents with mental retardation. Pediatr Exerc Sci. 2000;12:324-33.

28. Fernhall B, Pitetti K. Limitations to work capacity in individuals with intellectual disabilities. Clin Exerc Physiol. 2001;3:176-85.

29. Fernhall B, Pitetti K, Millar AL, et al. Cross validation of the 20 m shuttle run in children with mental retardation. Adapt Phys Activ Q. 2000;17:402-12.

30. Fernhall B, Pitetti K, Rimmer JH, et al. Cardiorespiratory capacity of individuals with mental retardation including Down syndrome. Med Sci Sports Exerc. 1996;28:366-71.

31. Fernhall B, Pitetti K, Stubbs N, et al. Validity and reliability of the 1/2 mile run-walk as an indicator of aerobic fitness in children with mental retardation. Pediatr Exerc Sci. 1996;8:130-42.

32. Fernhall B, Pitetti KH, Vukovich MD, et al. Validation of cardiovascular fitness field tests in children with mental retardation. Am J Ment Retard. 1998;102:602-12.

33. Fernhall B, Tymeson G. Graded exercise testing of mentally retarded adults; a study of feasibility. Arch Phys Med Rehabil. 1987;63:363-65.

34. Fernhall B, Tymeson G. Validation of cardiovascular fitness field tests for persons with mental retardation. Adapt Phys Activ Q. 1988;5:49-59.

35. Fernhall B, Tymeson G, Millar AL, et al. Cardiovascular fitness testing and fitness levels of adolescents and adults with mental retardation including Down syndrome. Educ Train Ment Retard. 1989;68:363-65.

36. Fernhall B, Unnithan V. Physical activity, metabolic issues, and assessment. Phys Med Rehabil Clin N Am. 2002;13:925-47.

37. Figueroa A, Collier S, Baynard T, et al. Impaired vagal modulation of heart rate in individuals with Down syndrome. Clin Auton Res. 2005;15:45-50.

38. Guerra M, Llorens N, Fernhall B. Chronotropic incompetence in persons with Down syndrome. Arch Phys Med Rehabil. 2003;84:1604-8.

39. Guerra M, Roman B, Geronimo C, et al. Physical fitness levels of sedentary and active individuals with Down syndrome. Adapt Phys Activ Q. 2000;17:310-21.

40. Hauck JL. Strategies for adapting accelerometer wear for youth with disabilities. Res Q Exerc Sport. 2011;82(1):supplement.

41. Hauck JL, Ulrich DA. Acute effects of a therapeutic mobility device on physical activity and heart rate in children with Down syndrome. Res Q Exerc Sport. In press.

42. Hayden MF. Mortality among people with mental retardation living in the United States: research review and policy application. Ment Retard. 1998;36:345-59.

43. Heffernan KS, Baynard T, Goulopoulou S, et al. Baroreflex sensitivity during static exercise in individuals with Down syndrome. Med Sci Sports Exerc. 2005;37:2026-31.

44. Heffernan KS, Karas RH, Patvardhan EA, et al. Endothelium-dependent vasodilation is associated with exercise capacity in smokers and nonsmokers. Vasc Med. 2010;15:119-25.

45. Hu M, Yan H, Ranadive SM, et al. Arterial stiffness response to exercise in persons with and without Down syndrome. Res Dev Disabil. 2013;34:3139-47.

46. Iellamo F, Galante A, Legramante JM, et al. Altered autonomic cardiac regulation in individuals with Down syndrome. Am J Physiol Heart Circ Physiol. 2005;289:H2387-91.

47. MacDonald M., Esposito, P., Hauck, J., et al. Bicycle training for youth with Down syndrome and autism spectrum disorders. Focus Autism Other Dev Disabl. 2012;27(1):12-21.

48. Maksud M, Hamilton L. Physiological responses of EMR children to strenuous exercise. Am J Ment Defic. 1974;79:32-38.

49. Menear, K. Parents' perceptions of health and physical activity needs of children with Down syndrome. Downs Syndr Res Pract. 2007;12(1):60-68.

50. Millar AL, Fernhall B, Burkett LN, et al. Effect of aerobic training in adolescents with Down syndrome. Med Sci Sports Exerc. 1993;25:260-64.

51. Perkins EA, Moran JA. Aging adults with intellectual disabilities. JAMA. 2010;304:91-92.

52. Pitetti K, Fernhall B. Aerobic capacity as related to leg strength in youths with mental retardation. Pediatr Exerc Sci. 1997;9:223-36.

53. Pitetti K, Tan DM. Cardiorespiratory responses of mentally retarded adults to air brake ergometry and treadmill exercise. Arch Phys Med Rehabil. 1990;17:318-21.

54. Pitetti KH, Campbell KD. Mentally retarded individuals—a population at risk? Med Sci Sports Exerc. 1991;23:586-93.

55. Pitetti KH, Climstein M, Campbell KD, et al. The cardiovascular capacities of adults with Down syndrome: a comparative study. Med Sci Sports Exerc. 1992;24:13-19.

56. Pitetti KH, Climstein M, Mays MJ, et al. Isokinetic arm and leg strength of adults with Down syndrome: a comparative study. Arch Phys Med Rehabil. 1992;73:847-50.

57. Pitetti KH, Millar AL, Fernhall B. Reliability of peak performance treadmill test for children and adolescents with and without mental retardation. Adapt Phys Activ Q. 2000;17:322-32.

58. Pitetti KH, Yarmer DA, Fernhall B. Cardiovascular fitness and body composition of youth with and without mental retardation. Adapt Phys Activ Q. 2001;18:127-41.

59. Roizen NJ, Patterson D. Down's syndrome. Lancet. 2003;361:1281-89.

60. Teo-Koh SM, McCubbin JA. Relationship between peak $\dot{V}O_2$ and 1-mile walk test performance of adolescents with mental retardation. Ped Exerc Sci. 1999;11:144-57.

61. Yoshizawa S, Tadatoshi T, Honda H. Aerobic work capacity of mentally retarded boys and girls in junior high school. J Human Ergol. 1975;4:15-26.

Chapter 20

1. Alpert BS, Verrill DE, Flood NL, et al. Complications of ergometer exercise in children. Pediatr Cardiol. 1983;4(2):91-96.

2. Andersen SP, Sveen ML, Hansen RS, et al. Creatine kinase response to high-intensity aerobic exercise in adult-onset muscular dystrophy. Muscle Nerve. 2013;48(6):897-901.

3. Bax M, Goldstein M, Rosenbaum P, et al. Proposed definition and classification of cerebral palsy, April 2005. Dev Med Child Neurol. 2005;47(8):571-76.

4. Beenakker EA, Maurits NM, Fock JM, et al. Functional ability and muscle force in healthy children and ambulant Duchenne muscular dystrophy patients. Eur J Paediatr Neurol. 2005;9(6):387-93.

5. van den Berg-Emons RJ, van Baak MA, de Barbanson DC, et al. Reliability of tests to determine peak aerobic power, anaerobic power and isokinetic muscle strength in children with cerebral palsy. Dev Med Child Neurol. 1996;38:1117-25.

6. Birrer RB, Levine R. Performance parameters in children and adolescent athletes. Sports Med. 1987;4(3):211-27.

7. Bosma L. Functioneel meten van spierkracht en inspanningsvermogen bij kinderen met een Cerebrale Parese en GMFCS III. Utrecht: University Medical Centre Utrecht; 2007.

8. Bushby K, Muntoni F, Bourke JP. 107th ENMC international workshop: the management of cardiac involvement in muscular dystrophy and myotonic dystrophy. Neuromuscul Disord. 2003;13(2):166-72.

9. Bushby K, Finkel RS, Birnkrant DJ, et al. Diagnosis and management of Duchenne muscular dystrophy, part 2: implementation of multidisciplinary care. Lancet Neurol. 2010;9:177-89.

10. Finsterer J, Stollberger C. The heart in human dystrophinopathies. Cardiology. 2003;99(1):1-19.

11. Geiger R, Strasak A, Treml B, et al. Six-minute walk test in children and adolescents. J Pediatr. 2007;150(4):395-99, 399.e1-2.

12. Hoofwijk M, Unnithan VB, Bar-Or O. Maximal treadmill performance of children with cerebral palsy. Pediatr Exerc Sci. 1995;7:305-13.

13. Jansen M, De Jong M, Coes HM, et al. The assisted 6-minute cycling test to assess endurance in children with a neuromuscular disorder. Muscle Nerve. 2012;46(4):520-30.

14. Kilmer DD. Response to aerobic exercise training in humans with neuromuscular disease. Am J Phys Med Rehabil. 2002;81(11 Suppl):S148-50.

15. Maher CA, Williams MT, Olds TS. The six-minute walk test for children with cerebral palsy. Int J Rehabil Res. 2008;31(2):185-88.

16. Maltais D, Pierrynowski M, Galea V, et al. Physical activity level is associated with the O_2 cost of walking in cerebral palsy. Med Sci Sports Exerc. 2005;37(3):347-53.

17. Markert CD, Case LE, Carter GT, et al. Exercise and Duchenne muscular dystrophy: where we have been and where we need to go. Muscle Nerve. 2012;45(5):746-51.

18. McDonald CM, Henricson EK, Han JJ, et al. The 6-minute walk test as a new outcome measure in Duchenne muscular dystrophy. Muscle Nerve. 2010;41(4):500-10.

19. McDonald CM, Henricson EK, Abresch RT, et al. The 6-minute walk test and other clinical endpoints in Duchenne muscular dystrophy: reliability, concurrent validity, and minimal clinically important differences from a multicenter study. Muscle Nerve. 2013;48(3):357-68.

20. McDougall JD, Wenger HA, Green HJ. Physiological testing of the elite athlete. Ithaca, NY: Mouvement Pubns; 1982.

21. Noonan V, Dean E. Submaximal exercise testing: clinical application and interpretation. Phys Ther. 2000;80(8):782-807.

22. Palisano RJ, Rosenbaum P, Bartlett D, et al. Content validity of the expanded and revised Gross Motor Function Classification System. Dev Med Child Neurol. 2008;50(10):744-50.

23. Paridon SM, Alpert BS, Boas SR, et al. Clinical stress testing in the pediatric age group. Circulation. 2006;113:1905-20.

24. Rieckert H, Bruhm U, Schwalm U. Endurance training within a program of physical education in children predominantly with cerebral palsy. Med Welt. 1977;28:1694-1701.

25. Sockolov R, Irwin B, Dressendorfer RH, et al. Exercise performance in 6- to-11-year-old boys with Duchenne muscular dystrophy. Arch Phys Med Rehabil. 1977;58(5):195-201.

26. Sveen ML, Jeppesen TD, Hauerslev S, et al. Endurance training improves fitness and strength in patients with Becker muscular dystrophy. Brain. 2008;131(Pt 11):2824-31.

27. Takken T, Groen WG, Hulzebos EH, Ernsting CG, et al. Exercise stress testing in children with metabolic or neuromuscular disorders. International journal of pediatrics. 2010;2010:pii: 254829.

28. Thompson P, Beath T, Bell J, et al. Test-retest reliability of the 10-metre fast walk test and 6-minute walk test in ambulatory school-aged children with cerebral palsy. Dev Med Child Neurol. 2008;50(5):370-76.

29. Tirosh E, Bar-Or O, Rosenbaum P. New muscle power test in neuromuscular disease. Feasibility and reliability. Am J Dis Child. 1990;144(10):1083-87.

30. Unnithan VB, Katsimanis G, Evangelinou C, et al. Effect of strength and aerobic training in children with cerebral palsy. Med Sci Sports Exerc. 2007;39(11):1902-9.

31. van Brussel M, van der Net J, Hulzebos E, et al. The Utrecht approach to exercise in chronic childhood conditions: the decade in review. Pediatr Phys Ther. 2011;23(1):2-14.

32. Verschuren O, Takken T, Ketelaar M, et al. Reliability and validity of data for 2 newly developed shuttle run tests in children with cerebral palsy. Phys Ther. 2006;86(8):1107-17.

33. Verschuren O, Takken T, Ketelaar M, et al. Reliability for running tests for measuring agility and anaerobic muscle power in children and adolescents with cerebral palsy. Pediatr Phys Ther. 2007;19(2):108-15.

34. Verschuren O, Bloemen M, Kruitwagen C, et al. Reference values for anaerobic performance and agility in ambulatory children and adolescents with cerebral palsy. Dev Med Child Neurol. 2010;52(10):e222-28.

35. Verschuren O, Bloemen M, Kruitwagen C, et al. Reference values for aerobic fitness in children, adolescents, and young adults who have cerebral palsy and are ambulatory. Phys Ther. 2010;90(8):1148-56.

36. Verschuren O, Ketelaar M, Keefer D, et al. Identification of a core set of exercise tests for children and adolescents with cerebral palsy: a Delphi survey of researchers and clinicians. Dev Med Child Neurol. 2011;53(5):449-56.

37. Verschuren O, Ketelaar M, De Groot J, et al. Reproducibility of two functional field exercise tests for children with cerebral palsy who self-propel a manual wheelchair. Dev Med Child Neurol. 2013;55(2):185-90.

38. Verschuren O, Zwinkels M, Ketelaar M, et al. Reproducibility and validity of the 10-meter shuttle ride test in wheelchair-using children and adolescents with cerebral palsy. Phys Ther. 2013;93(7):967-74.

39. Verschuren O, Zwinkels M, Obeid J, et al. Reliability and validity of short-term performance tests for wheelchair-using children and adolescents with cerebral palsy. Dev Med Child Neurol. 2013;55(12):1129-35.

Index

Note: The italicized *f* and *t* following page numbers refer to figures and tables, respectively.

About the Editors

Thomas W. Rowland, MD, is a pediatric cardiologist at Baystate Medical Center in Springfield, Massachusetts and a professor of pediatrics at Tufts University School of Medicine. A graduate of the University of Michigan Medical School, Rowland is board certified in pediatrics and pediatric cardiology by the American Board of Pediatrics.

Rowland, who has had more than 150 journal articles published, is the author of four books: *Biologic Regulation of Physical Activity*; *Children's Exercise Physiology, Second Edition*; *Tennisology: Inside the Science of Serves, Nerves, and On-Court Dominance*; and *The Athlete's Clock*. He has served as editor of the journal *Pediatric Exercise Science* and as president of the North American Society for Pediatric Exercise Medicine (NASPEM) and was on the board of trustees of the American College of Sports Medicine (ACSM). He is past president of the New England chapter of the ACSM and received the Honor Award from that organization in 1993.

Rowland is a competitive tennis player and distance runner. He and his wife, Margot, reside in Longmeadow, Massachusetts.

The American College of Sports Medicine (ACSM), founded in 1954, is the largest sports medicine and exercise science organization in the world. With more than 50,000 members and certified professionals worldwide, ACSM is dedicated to improving health through science, education, and medicine. ACSM members work in a wide range of medical specialties, allied health professions, and scientific disciplines. Members are committed to the diagnosis, treatment, and prevention of sport-related injuries and the advancement of the science of exercise.

The ACSM promotes and integrates scientific research, education, and practical applications of sports medicine and exercise science to maintain and enhance physical performance, fitness, health, and quality of life.

The North American Society for Pediatric Exercise Medicine (NASPEM), founded in 1985, is a professional organization whose membership is composed of medical doctors, researchers, educators, and students interested in pediatric exercise. NASPEM is dedicated to the mission of promoting exercise science, physical activity, and fitness in the health and medical care of children and adolescents. That mission is accomplished in part through scientific meetings, a scholarly journal (*Pediatric Exercise Science*), collaborative research, student aid in the form of grants and awards, and a training program database.

About the Contributors

Bruce Alpert, MD
University of Tennessee Health Science Center, retired
Le Bonheur Children's Hospital
Memphis, TN, USA

Neil Armstrong, PhD, DSc
Children's Health and Exercise Research Centre
University of Exeter
Exeter, UK

Laura Banks, PhD
Faculty of Kinesiology and Physical Education
University of Toronto
Toronto, ON, Canada

Tracy Baynard, PhD
Department of Kinesiology and Nutrition
University of Illinois at Chicago
Chicago, IL, USA

Steven R. Boas, MD
The Sports Physiology Center of Chicago
Glenview, IL, USA
Feinberg School of Medicine
Northwestern University
Chicago, IL, USA

Shannon S.D. Bredin, PhD
Physical Activity Promotion and Chronic Disease Prevention Unit, School of Kinesiology
University of British Columbia
Vancouver, BC, Canada

Julie Brothers, MD
Assistant Professor, Division of Cardiology
Perelman School of Medicine
University of Pennsylvania
Medical Director, Lipid Heart Clinic
The Children's Hospital of Philadelphia
Philadelphia, PA, USA

Bo Fernhall, PhD
Department of Kinesiology and Nutrition
University of Illinois at Chicago
Chicago, IL, USA

Robert P. Garofano, EdD
Director, Pediatric Cardiopulmonary Exercise Laboratory
Morgan Stanley Children's Hospital
New York Presbyterian
Columbia University Medical Center
New York, NY, USA

Janke de Groot, PT, PhD
University of Applied Science Utrecht
Shared Utrecht Pediatric Exercise Research Laboratory
Utrecht, The Netherlands

Michael G. McBride, PhD
Director, Exercise Physiology and Cardiac Rehabilitation
The Children's Hospital of Philadelphia
Philadelphia, PA, USA

Brian W. McCrindle, MD, MPH
Preventive Cardiology, Labatt Family Heart Centre
The Hospital for Sick Children
University of Toronto
Toronto, ON, Canada

Ali M. McManus, PhD
Centre for Heart, Lung and Vascular Health, School of Health and Exercise Sciences
University of British Columbia
Kelowna, BC, Canada

Patricia A. Nixon, PhD
Professor, Department of Health and Exercise Science
Wake Forest University
Winston-Salem, NC, USA

Stephen M. Paridon, MD
Professor, Pediatric Cardiology
Perelman School of Medicine
University of Pennsylvania
Medical Director, Exercise Physiology Laboratory
The Children's Hospital of Philadelphia
Philadelphia, PA, USA

Ranjit Philip, MD
Pediatric Cardiology, Department of Pediatrics
University of Tennessee Health Science Center
Le Bonheur Children's Hospital
Memphis, TN, USA

Thomas W. Rowland, MD
Baystate Medical Center
Springfield, MA, USA

Richard J. Sabath III, EdD
The Ward Family Heart Center
The Children's Mercy Hospital, retired
Kansas City, MO, USA

Tim Takken, PhD
Child Development & Exercise Center
Wilhelmina Children's Hospital
Shared Utrecht Pediatric Exercise Research Laboratory
University Medical Centre Utrecht
Utrecht, The Netherlands

Amy Lynne Taylor, PhD
Children's Clinical Research Organization
Children's Hospital Colorado
Aurora, CO, USA

Kelli M. Teson, PhD
The Ward Family Heart Center
The Children's Mercy Hospital
Kansas City, MO, USA

Olaf Verschuren, PT, PhD
Brain Center Rudolf Magnus and Center of Excellence for Rehabilitation Medicine
University Medical Center Utrecht and De Hoogstraat Rehabilitation
Centre of Excellence, Rehabilitation Centre 'De Hoogstraat'
Shared Utrecht Pediatric Exercise Research Laboratory
Utrecht, The Netherlands

Darren E.R. Warburton, PhD
Physical Activity Promotion and Chronic Disease Prevention Unit, School of Kinesiology, Faculty of Education, Experimental Medicine Program, Faculty of Medicine
University of British Columbia
Vancouver, BC, Canada

David A. White, PhD
The Ward Family Heart Center
The Children's Mercy Hospital
Kansas City, MO, USA